THE ANALY

VOLU

Probability

Guy Lebanon

First Edition

2012

Probability. The Analysis of Data, Volume 1.

First Edition, First Printing, 2013

http://theanalysisofdata.com

In memory of Alex Lebanon

Contents

Preface

The Analysis of Data Project

The Analysis of Data (TAOD) project provides educational material in the area of data analysis.

- The project features comprehensive coverage of all relevant disciplines including probability, statistics, computing, and machine learning.

- The content is almost self-contained and includes mathematical prerequisites and basic computing concepts.

- The R programming language is used to demonstrate the contents. Full code is available, facilitating reproducibility of experiments and letting readers experiment with variations of the code.

- The presentation is mathematically rigorous, and includes derivations and proofs in most cases.

- HTML versions are freely available on the website http://theanalysisofdata.com. Hardcopies are available at affordable prices.

Volume 1: Probability

This volume focuses on probability theory. There are many excellent textbooks on probability, and yet this book differs from others in several ways.

- Probability theory is a wide field. This book focuses on the parts of probability that are most relevant for statistics and machine learning.

- The book contains almost all of the mathematical prerequisites, including set theory, metric spaces, linear algebra, differentiation, integration, and measure theory.

- Almost all results in the book appear with a proof.

- Probability textbooks are typically either elementary or advanced. This book strikes a balance by attempting to avoid measure theory where possible, but resorting to measure theory and other advanced material in a few places where they are essential.

- The book uses R to illustrate concepts. Full code is available in the book, facilitating reproducibility of experiments and letting readers experiment with variations of the code.

I am not aware of a single textbook that covers the material from probability theory that is necessary and sufficient for an in-depth understanding of statistics and machine learning. This book represents my best effort in that direction.

Since this book is part of a series of books on data analysis, it does not include any statistics or machine learning. Such content is postponed to future volumes.

Website

A companion website (http://theanalysisofdata.com) contains an HTML version of this book, errata, and additional multimedia material. The website will also link to additional TAOD volumes as they become available.

Mathematical Appendices

A large part of the book contains six appendices on mathematical prerequisites. Probability requires knowledge of many branches of mathematics, including calculus, linear algebra, set theory, metric spaces, measure, and Lebesgue integration. Instead of referring the reader to a large collection of math textbooks we include here all of the necessary prerequisites. References are provided in the notes sections at the end of each chapter for additional resources.

Dependencies

The diagram below indicates the dependencies between the different chapters of the book. Appendix chapters are shaded and dependencies between appendix chapters and regular chapters are marked by dashed arrows. It is not essential to strictly adhere to this dependency as many chapters require only a brief familiarity with some issues in the chapters that they depend on.

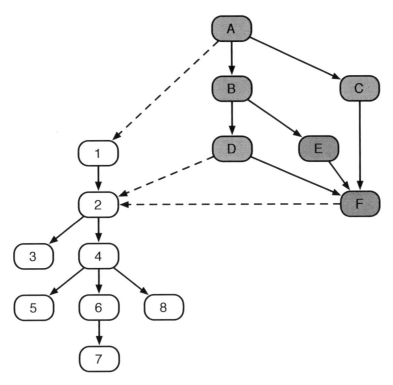

Starred sections correspond to material that requires measure theory, or that can be better appreciated with knowledge of measure theory.

R Code

The book contains many fragments of R code, aimed to illustrate probability theory and its applications. The code is included so that the reader can reproduce the results as well as modify the code and run variations of it. In order to appreciate the code and modify it, the reader will need a basic understanding of the R programming language and R graphics. A good introduction to R is available from the CRAN website at http://cran.r-project.org/doc/manuals/R-intro.pdf. Alternatively, two chapters from volume 2 of the analysis of data series (R Programming and R Graphics) are freely available online at http://theanalysisofdata.com.

The book uses a variety of R graphics packages, including base, lattice, and ggplot2. The most frequently used package is ggplot2, which is described in detail in [49] (see also http://had.co.nz/ggplot2/ and the R Graphics chapter available at http://theanalysisofdata.com.)

To ensure that the code fragments run locally as they appear in the book, install and bring into scope the required packages using the install.packages and library R functions. For example, to install and load the ggplot2 package type the following commands in the R prompt.

```
# this is a comment
install.packages("ggplot2")   # installs the package ggplot2
library("ggplot2")   # brings ggplot2 into scope
```

The code fragments throughout the book are annotated with output displayed by the R interpreter. This output is displayed following two hash symbols which are interpreted as comments by R (see below).

```
a = pi
print(a)   # note the ## symbols preceding the output below

## [1] 3.142

print(a + 1)

## [1] 4.142
```

This format makes it easy to copy a code fragment from an HTML page and paste it directly into R (the output following the hash symbols will be interpreted as comments and not produce a syntax error).

Acknowledgements

The following people made technical suggestions that helped improve the contents: Krishnakumar Balasubramanian, Rohit Banga, Joshua Dillon, Sanjeet Hajarnis, Oded Green, Yi Mao, Seungyeon Kim, Joonseok Lee, Fuxin Li, Nishant Mehta, Yaron Rachlin, Parikshit Ram, Kaushik Rangadurai, Neil Slagle, Mingxuan Sun, Brian Steber, Gena Tang, Long Tran. In addition, many useful comments were received through a discussion board during my fall 2011 class computational data analysis at Georgia Tech. These comments were mostly anonymized, but some commentators who identified themselves appear above.

Katharina Probst, Neil Slagle and Laura Usselman edited portions of this book, and made many useful suggestions. The book features a combination of text, equations, graphs, and R code, made possible by the knitr package. I thank Yihui Xie for implementing knitr, and for his help through the knitr Google discussion group. Katharina Probst helped with web development and design.

Mathematical Notations

Logic

$\forall$	for all
$\exists$	there exists
$\Rightarrow$	implies
$\Leftrightarrow$	if and only if
$\overset{\text{def}}{=}$	defined as

For example, $\forall a > 0, 1/a > 0$ reads "for every $a > 0$ we have $1/a > 0$" and $\forall a > 0, \exists b > 0, \ a/b = 1 \Rightarrow b/a = 1$ reads "for all $a > 0$ there exists $b > 0$ such that, whenever $a/b = 1$, we also have $b/a = 1$".

Sets and Functions (Chapter A)

Ω	sample space
ω	an element of the sample space Ω
I_A	indicator function $I_A(x) = 1$ if $x \in A$ and 0 otherwise
δ_{ij}	Kronercker's delta: $\delta_{ij} = 1$ if $i = j$ and 0 otherwise
$\mathbb{N}$	natural numbers $\{1, 2, 3, \ldots\}$
$\mathbb{Z}$	integers $\{\ldots, -2, -1, 0, 1, 2, \ldots\}$
$\mathbb{Q}$	rational numbers
$A \times B$	Cartesian product of two sets
A^k	repeated Cartesian product (k times) of a set: $A \times \cdots \times A$
A^c	complement of a set: $\Omega \setminus A$
2^A	power set of the set A
$|A|$	number of elements in a finite set
$f^{-1}(A)$	pre-image of the function f: $\{x : f(x) \in A\}$
$f \equiv g$	$f(x) = g(x)$ for all x
$x^{(n)}$	sequence of elements in a set
$A_n \nearrow A$	convergence of a sequence of increasing sets
$A_n \searrow A$	convergence of a sequence of decreasing sets

Combinatorics (Section 1.6)

$n!$	factorial function $n(n-1) \cdots 2 \cdot 1$
$(n)_r$	$n!/(n-r)!$
$\binom{n}{r}$	$n!/(r!(n-r)!)$

Metric spaces (Chapter B)

$\mathbb{R}$	real numbers
$\mathbb{R}^d$	set of d-dimensional vectors of real numbers
$B_a(\boldsymbol{x})$	open ball of radius a centered at $\boldsymbol{x}$
$\boldsymbol{x}$	vector in $\mathbb{R}^d$
x_i	the i-component of the vector $\boldsymbol{x}$
$\boldsymbol{x}^{(n)}$	sequence of vectors in $\mathbb{R}^d$
$x_i^{(j)}$	the i-component of the vector $\boldsymbol{x}^{(j)}$
$\langle \boldsymbol{x}, \boldsymbol{y} \rangle$	inner product $\sum_i x_i y_i$
$\|\boldsymbol{x}\|$	Euclidean norm $\sqrt{\sum_i x_i^2}$
$d(\boldsymbol{x}, \boldsymbol{y})$	Euclidean distance $\sqrt{\sum_i (x_i - y_i)^2}$
$\boldsymbol{x}^{(n)} \to \boldsymbol{x}$	convergence of a sequence
$f_n \to f$	pointwise convergence of a sequence of functions
$f_n \nearrow f$	pointwise convergence of an increasing sequence of functions
$f_n \searrow f$	pointwise convergence of a decreasing sequence of functions
$O(f)$	big O growth notation
$o(f)$	little o growth notation

Note in particular that we denote vectors in bold face, for example $\boldsymbol{x}$, and the scalar components of such vectors using subscripts (non-bold face) $\boldsymbol{x} = (x_1, \ldots, x_d)$. We refer to sequence of vectors using superscripts, for example $\boldsymbol{x}^{(n)}$ and their scalar components as $\boldsymbol{x}^{(n)} = (x_1^{(n)}, \ldots, x_d^{(n)})$.

Probability (Chapters 1, 2, 4, 8)

P	probability function
P_E	probability conditioned on the event E
F_X	cumulative distribution function (cdf) corresponding to the RV X
f_X	probability density function (pdf) corresponding to the RV X
p_X	probability mass function (pmf) corresponding to the RV X
E	expectation
Var	variance
Cov	covariance
std	standard deviation
$\sim$	distributed according to
$\overset{\text{iid}}{\sim}$	independent identically distributed (iid) sampling
$\mathbb{P}_k$	simplex of probability functions over $\Omega = \{1, \ldots, k\}$
i.o.	infinitely often
$\overset{\text{as}}{\to}$	convergence with probability 1
$\overset{\text{p}}{\to}$	convergence in probability
$\rightsquigarrow$	convergence in distribution

Matrices (Chapter C)

$A^\top$	matrix transpose
A^n	the matrix A raised to the n-power
A_{ij}^n	the i, j element of the matrix A^n
A^{-1}	inverse of matrix A
I	identity matrix
tr A	trace of the matrix A
det A	determinant of the matrix A
row A	row space of the matrix A
col A	column space of the matrix A
dim A	dimension of a linear space
rank A	rank of the matrix A
diag($\boldsymbol{v}$)	diagonal matrix whose diagonal is given by the vector $\boldsymbol{v}$
$A \otimes B$	Kronecker product of two matrices A, B

All vectors are assumed to be column vectors unless stated explicitly otherwise. For example, if $\boldsymbol{x}$ is an arbitrary vector and A a matrix the expression $\boldsymbol{x}^\top A\boldsymbol{x}$ represents a scalar.

Differentiation (Chapter D)

df/dx	derivative
d^2f/dx^2	second order derivative
$\partial f/\partial x_i$	partial derivative
∇f	gradient vector of partial derivatives of $f : \mathbb{R}^k \to \mathbb{R}$
$\nabla \boldsymbol{f}$	Jacobian matrix of partial derivatives of $\boldsymbol{f} : \mathbb{R}^k \to \mathbb{R}^m$
$\partial^2 f/\partial x_i \partial x_j$	second order partial derivative
$\nabla^2 f$	Hessian matrix of second order partial derivatives of $f : \mathbb{R}^k \to \mathbb{R}$

We consider the gradient vector ∇f as a row vector. This ensures that the notations ∇f and $\nabla \boldsymbol{f}$ are consistent.

Measure and Integration (Chapters E, F)

$\sigma(\mathcal{C})$	σ algebra generated by the set of sets $\mathcal{C}$
μ	measure
$\int d\mu$	Lebesgue integral with respect to measure μ
$\int d\mathsf{P}$	Lebesgue integral with respect to probability measure P
$\int dF_X$	Lebesgue integral with respect to probability measure corresponding to cdf F_X
$\int dx$	Lebesgue integral with respect to Lebesgue measure or Riemann integral
$\mathcal{B}(\mathbb{R}^d)$	Borel σ-algebra over metric space $\mathbb{R}^d$
$\mathcal{F}_1 \otimes \mathcal{F}_1$	product σ-algebra
$\mu_1 \times \mu_2$	product measure
a.e.	almost everywhere (except on a set of measure zero)

Chapter 1

Basic Definitions

This chapter covers the most basic definitions of probability theory and explores some fundamental properties of the probability function.

1.1   Sample Space and Events

Our starting point is the concept of an abstract random experiment. This is an experiment whose outcome is not necessarily determined before it is conducted. Examples include flipping a coin, the outcome of a soccer match, and the weather. The set of all possible outcomes associated with the random experiment is called the sample space. Events are subsets of the sample space, or in other words sets of possible outcomes. The probability function assigns real values to events in a way that is consistent with our intuitive understanding of probability. Formal definitions appear below.

Definition 1.1.1. A sample space Ω associated with a random experiment is the set of all possible outcomes of the experiment.

A sample space can be finite, for example

$$\Omega = \{1, \ldots, 10\}$$

in the experiment of observing a number from 1 to 10. Or Ω can be countably infinite, for example

$$\Omega = \{0, 1, 2, 3, \ldots\}$$

in the experiment of counting the number of phone calls made on a specific day. A sample space may also be uncountably infinite, for example

$$\Omega = \{x \; : \; x \in \mathbb{R}, \, x \geq 0\}$$

in the experiment of measuring the height of a passer-by.

The notation $\mathbb{N}$ corresponds to the natural numbers $\{1, 2, 3, \ldots\}$, and the notation $\mathbb{N} \cup \{0\}$ corresponds to the set $\{0, 1, 2, 3, \ldots\}$. The notation $\mathbb{R}$ corresponds to the real numbers and the notation $\{x : x \in \mathbb{R}, x \geq 0\}$ corresponds to the non-negative real numbers. See Chapter A in the appendix for an overview of set theory, including the notions of a power set and countably infinite and uncountably infinite sets.

In the examples above, the sample space contained unachievable values (number of people and height are bounded numbers). A more careful definition could have been used, taking into account bounds on the number of potential phone calls or height. For the sake of simplicity, we often use simpler sample spaces containing some unachievable outcomes.

Definition 1.1.2. An event E is a subset of the sample space Ω, or in other words a set of possible outcomes.

In particular, the empty set $\emptyset$ and the sample space Ω are events. Figure 1.1 shows an example of a sample space Ω and two events $A, B \subset \Omega$ that are neither $\emptyset$ nor Ω. The R code below shows all possible events of an experiment with $\Omega = \{a, b, c\}$. There are $2^{|\Omega|}$ such sets, assuming Ω is finite (see Chapter A for more information on the power set).

```
# bring the sets package into scope (install
# it first using install.packages('sets') if
# needed)
library(sets)
Omega = set("a", "b", "c")
# display a set containing all possible
# events of an experiment with a sample
# space Omega
2^Omega

## {{}, {"a"}, {"b"}, {"c"}, {"a", "b"},
##   {"a", "c"}, {"b", "c"}, {"a", "b",
##   "c"}}
```

Example 1.1.1. *In the random experiment of tossing a coin three times and observing the results (heads or tails), with ordering, the sample space is the set*

$$\Omega = \{HHH, HHT, HTH, HTT, THH, THT, TTH, TTT\}.$$

The event

$$E = \{HHH, HHT, HTT, HTH\} \subset \Omega$$

describes "a head was obtained in the first coin toss." In this case both the sample space Ω and the event E are finite sets.

Example 1.1.2. *Consider a random experiment of throwing a dart at a round board without missing the board. Assuming the radius of the board is 1, the sample space is the set of all two dimensional vectors inside the unit circle*

$$\Omega = \left\{ (x, y) \, : \, x, y \in \mathbb{R}, \, \sqrt{x^2 + y^2} < 1 \right\}.$$

An event describing a bullseye hit may be

$$E = \left\{ (x, y) \, : \, x, y \in \mathbb{R}, \, \sqrt{x^2 + y^2} < 0.1 \right\} \subset \Omega.$$

In this case both the sample space Ω and the event E are uncountably infinite.

For an event E, the outcome of the random experiment $\omega \in \Omega$ is either in E ($\omega \in E$) or not in E ($\omega \notin E$). In the first case, we say that the event E occurred, and in the second case we say that the event E did not occur. $A \cup B$ is the event of either A or B occurring and $A \cap B$ is the event of both A and B occurring. A^c (in the complement, the universal set is taken to be Ω: $A^c = \Omega \setminus A$) is the event that A did not occur. If the events A, B are disjoint ($A \cap B = \emptyset$), the two events cannot happen at the same time, since no outcome of the random experiment belongs to both A and B. If $A \subset B$, then B occurring implies that A occurs as well.

1.2 The Probability Function

Definition 1.2.1. Let Ω be a sample space associated with a random experiment. A probability function P is a function that assigns real numbers to events $E \subset \Omega$ satisfying the following three axioms.

1.
$$\mathsf{P}(E) \geq 0 \qquad \text{for all } E.$$

2.
$$\mathsf{P}(\Omega) = 1$$

3. If $E_n, n \in \mathbb{N}$, is a sequence of pairwise disjoint events ($E_i \cap E_j = \emptyset$ whenever $i \neq j$), then

$$\mathsf{P}\left(\bigcup_{i=1}^{\infty} E_i \right) = \sum_{i=1}^{\infty} \mathsf{P}(E_i).$$

Some basic properties of the probability function appear below.

Proposition 1.2.1.
$$\mathsf{P}(\emptyset) = 0.$$

Proof. Using the second and third axioms of probability,

$$1 = P(\Omega) = P(\Omega \cup \emptyset \cup \emptyset \cup \cdots) = P(\Omega) + P(\emptyset) + P(\emptyset) + \cdots$$
$$= 1 + P(\emptyset) + P(\emptyset) + \cdots,$$

implying that $P(\emptyset) = 0$ (since $P(E) \geq 0$ for all E). ∎

Proposition 1.2.2 (Finite Additivity of Probability). *For every finite sequence* $E_1, \ldots, E_N$ *of pairwise disjoint events* $(E_i \cap E_j = \emptyset$ *whenever* $i \neq j)$,

$$P(E_1 \cup \cdots \cup E_N) = P(E_1) + \cdots + P(E_N).$$

Proof. Setting $E_k = \emptyset$ for $k > N$ in the third axiom of probability, we have

$$P(E_1 \cup \cdots \cup E_N) = P\left(\bigcup_{i=1}^{\infty} E_i\right) = \sum_{i=1}^{\infty} P(E_i) = P(E_1) + \cdots + P(E_N) + 0.$$

The last equality above follows from the previous proposition. ∎

Proposition 1.2.3.
$$P(A^c) = 1 - P(A).$$

Proof. By finite additivity,

$$1 = P(\Omega) = P(A \cup A^c) = P(A) + P(A^c).$$

∎

Proposition 1.2.4.
$$P(A) \leq 1.$$

Proof. The previous proposition implies that $P(A^c) = 1 - P(A)$. Since all probabilities are non-negative $P(A^c) = 1 - P(A) \geq 0$, proving that $P(A) \leq 1$. ∎

Proposition 1.2.5. *If* $A \subset B$ *then*

$$P(B) = P(A) + P(B \setminus A)$$
$$P(B) \geq P(A).$$

Proof. The first statement follows from finite additivity:

$$P(B) = P(A \cup (B \setminus A)) = P(A) + P(B \setminus A).$$

The second statement follows from the first statement and the non-negativity of the probability function. ∎

Proposition 1.2.6 (Principle of Inclusion-Exclusion).

$$P(A \cup B) = P(A) + P(B) - P(A \cap B).$$

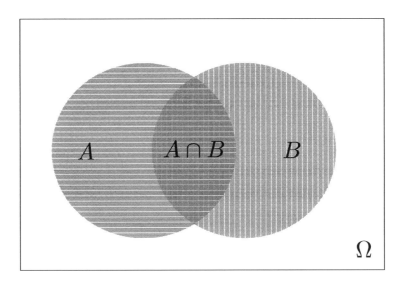

Figure 1.1: Two circular sets A, B, their intersection $A \cap B$ (gray area with horizontal and vertical lines), and their union $A \cup B$ (gray area with either horizontal or vertical lines or both). The set $\Omega \setminus (A \cup B) = (A \cup B)^c = A^c \cap B^c$ is represented by white color.

Proof. Using the previous proposition, we have

$$
\begin{aligned}
\mathsf{P}(A \cup B) &= \mathsf{P}((A \setminus (A \cap B)) \cup (B \setminus (A \cap B)) \cup (A \cap B)) \\
&= \mathsf{P}((A \setminus (A \cap B)) + \mathsf{P}(B \setminus (A \cap B))) + \mathsf{P}(A \cap B) \\
&= \mathsf{P}(A) - \mathsf{P}(A \cap B) + \mathsf{P}(B) - \mathsf{P}(A \cap B) + \mathsf{P}(A \cap B) \\
&= \mathsf{P}(A) + \mathsf{P}(B) - \mathsf{P}(A \cap B).
\end{aligned}
$$

∎

Figure 1.1 illustrates the Principle of Inclusion-Exclusion. Intuitively, the probability function $\mathsf{P}(A)$ measures the size of the set A (assuming a suitable definition of size). The size of the set A plus the size of the set B equals the size of the union $A \cup B$ plus the size of the intersection $A \cap B$: $\mathsf{P}(A) + \mathsf{P}(B) = \mathsf{P}(A \cup B) + \mathsf{P}(A \cap B)$ (since the intersection $A \cap B$ is counted twice in $\mathsf{P}(A) + \mathsf{P}(B)$).

Definition 1.2.2. For a finite sample space Ω, an event containing a single element $E = \{\omega\}, \omega \in \Omega$ is called an elementary event.

If the sample space is finite $\Omega = \{\omega_1, \ldots, \omega_n\}$, it is relatively straightforward to define probability functions by defining the n probabilities of the elementary events. More specifically, for a sample space with n elements, suppose that we are given a set of n non-negative numbers $\{p_\omega : \omega \in \Omega\}$ that sum to one. There exists then a unique probability function P over events such that $P(\{\omega\}) = p_\omega$. This probability is defined for arbitrary events through the finite additivity property

$$P(E) = \sum_{\omega \in E} P(\{\omega\}) = \sum_{\omega \in E} p_\omega.$$

A similar argument holds for sample spaces that are countably infinite.

The R code below demonstrates such a probability function, defined on $\Omega = \{1, 2, 3, 4\}$ using $p_1 = 1/2$, $p_2 = 1/4$, $p_3 = p_4 = 1/8$.

```
# sample space
Omega = c(1, 2, 3, 4)
# probabilities of 4 elementary events
p = c(1/2, 1/4, 1/8, 1/8)
# make sure they sum to 1
sum(p)
```

```
## [1] 1
```

```
# define an event 1,4 using a binary
# representation
A = c(1, 0, 0, 1)
# compute probability of A using
# probabilities of elementary events
sum(p[A == 1])
```

```
## [1] 0.625
```

1.3   The Classical Probability Model on Finite Spaces

In the classical interpretation of probability on finite sample spaces, the probabilities of all elementary events $\{\omega\}, \omega \in \Omega$, are equal. Since the probability function must satisfy $P(\Omega) = 1$ we have

$$P(\{\omega\}) = |\Omega|^{-1}, \qquad \text{for all} \quad \omega \in \Omega.$$

This implies that under the classical model on a finite Ω, we have

$$P(E) = \frac{|E|}{|\Omega|}.$$

Example 1.3.1. *Consider the experiment of throwing two distinct dice and observing the two faces with order. The sample space is*

$$\Omega = \{1, \ldots, 6\} \times \{1, \ldots, 6\} = \{(x, y) : x, y \in \{1, 2, \ldots, 6\}\}$$

(see Chapter A in the appendix for the notation of a Cartesian product of two sets). Since Ω has 36 elements, the probability of the elementary event $E = \{(4, 4)\}$ is $P(E) = 1/|\Omega| = 1/36$. The probability of getting a sum of 9 in both dice is

$$P(sum = 9) = P(\{(6, 3), (3, 6), (4, 5), (5, 4)\}) = \frac{|\{(6, 3), (3, 6), (4, 5), (5, 4)\}|}{36}$$

$$= \frac{4}{36}.$$

The classical model in this case is reasonable, assuming the dice are thrown independently and are fair.

The R code below demonstrates the classical model and the resulting probabilities on a small Ω.

```
Omega = set(1, 2, 3)
# all possible events
2^Omega

## {{}, {1}, {2}, {3}, {1, 2}, {1, 3}, {2,
##  3}, {1, 2, 3}}

# size of all possible events
sapply(2^Omega, length)

## [1] 0 1 1 1 2 2 2 3

# probabilities of all possible events under
# the classical model
sapply(2^Omega, length)/length(Omega)

## [1] 0.0000 0.3333 0.3333 0.3333 0.6667 0.6667
## [7] 0.6667 1.0000
```

Note that the sequence of probabilities above does not sum to one since it contains probabilities of non-disjoint events. The R code below demonstrates this below for a larger set using by graphing the histogram of sizes and probabilities.

```
library(ggplot2)
Omega = set(1, 2, 3, 4, 5, 6, 7, 8, 9, 10)
# histogram of sizes of all possible events
qplot(sapply(2^Omega, length), xlab = "event sizes")
```

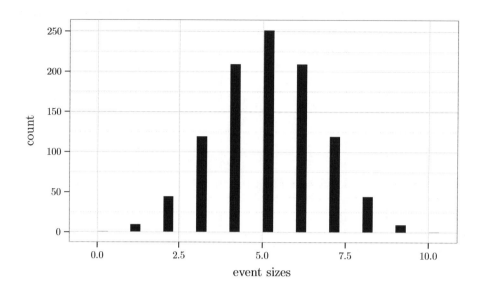

```
# histogram of probabilities of all possible
# events under classical model
probs = sapply(2^Omega, length)/length(Omega)
qplot(probs, xlab = "event probabilities")
```

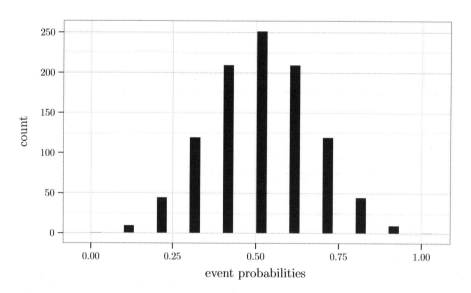

The left-most and right-most bars represent two sets with probabilities 0 and 1, respectively. These sets are obviously $\emptyset$ and Ω.

1.4 The Classical Probability Model on Continuous Spaces

For a continuous sample space of dimension n (for example $\Omega = \mathbb{R}^n$), we define the classical probability function as

$$P(A) = \frac{\text{vol}_n(A)}{\text{vol}_n(\Omega)},$$

where $\text{vol}_n(S)$ is the n-dimensional volume[1] of the set S.

Example 1.4.1. *In an experiment measuring the weight of residents in a particular geographical region, the sample space could be $\Omega = (0, 1000) \subset \mathbb{R}^1$ (assuming our measurement units are pounds and people weigh less than 1000 pounds). The probability of getting a measurement between 150 and 250 (in the classical model) is the ratio of the 1-dimensional volumes or lengths:*

$$P((150, 250)) = \frac{|250 - 150|}{|1000 - 0|} = 0.1.$$

The classical model in this case is highly inaccurate and not likely to be useful.

Example 1.4.2. *Assuming the classical model on the sample space of Example 1.1.2, the probability of hitting the bullseye is*

$$P\left(\left\{(x, y) : \sqrt{x^2 + y^2} < 0.1\right\}\right) = \frac{\pi \, 0.1^2}{\pi \, 1^2} = 0.01$$

(since the area of a circle of radius r is $\pi \cdot r^2$). The classical model in this case assumes that the person throwing the darts does not make any attempt to hit the center. For most dart throwers this model is inaccurate.

1. For the classical model to apply, the sample space Ω must by finite or be continuous with a finite non-zero volume.

2. The classical model (on both finite and continuous spaces) satisfies the three axioms defining a probability function.

3. A consequence of the classical model on continuous spaces is that the probability of an elementary event is zero (the volume of a single element is 0).

4. In the next two chapters we will explore a number of alternative probability models that may be more accurate than the classical model.

---

[1]The 1-dimensional volume of a set $S \subset \mathbb{R}$ is its length. The 2-dimensional volume of a set $S \subset \mathbb{R}^2$ is its area. The 3-dimensional volume of a set $S \subset \mathbb{R}^3$ is its volume. In general, the n-dimensional volume of A is the n-dimensional integral of the constant function 1 over the set A.

1.5   Conditional Probability and Independence

Definition 1.5.1. The conditional probability of an event A given an event B with $P(B) > 0$ is

$$P(A \mid B) = \frac{P(A \cap B)}{P(B)}.$$

If $P(A) > 0$ and $P(B) > 0$ we have

$$P(A \cap B) = P(A \mid B) P(B) = P(B \mid A) P(A).$$

Intuitively, $P(A \mid B)$ is the probability of A occurring assuming that the event B occurred. In accordance with that intuition, the conditional probability has the following properties.

1. If $B \subset A$, then $P(A \mid B) = P(B)/P(B) = 1$.

2. If $A \cap B = \emptyset$, then $P(A \mid B) = 0/P(B) = 0$.

3. If $A \subset B$ then $P(A \mid B) = P(A)/P(B)$.

4. The conditional probability may be viewed as a probability functions

$$P_A(E) \stackrel{\text{def}}{=} P(E \mid A)$$

   satisfying Definition 1.2.1 (Exercise 1.7). In addition, all the properties and intuitions that apply to probability functions apply to P_A as well.

5. Assuming the event A occurred, P_A generally has better forecasting abilities than P.

As mentioned above, conditional probabilities are usually intuitive. The following example from [16], however, shows a counter-intuitive situation involving conditional probabilities. This demonstrates that intuition should not be a substitute for rigorous computation.

Example 1.5.1. *Consider families with two children where the gender probability of each child is symmetric (1/2). We select a family at random and consider the sample space describing the gender of the children $\Omega = \{MM, MF, FM, FF\}$. We assume a classical model, implying that the probabilities of all 4 elementary events are 1/4.*

We define the event that both children in the family are boys as $A = \{MM\}$, the event that a family has a boy as $B = \{MF, FM, MM\}$, and the event that the first child is a boy as $C = \{MF, MM\}$.

Given that the first child is a boy, the probability that both children are boys is

$$P(A \mid C) = P(A \cap C)/P(C) = P(A)/P(C) = (1/4)/(1/2) = 1/2.$$

This matches our intuition. Given that the family has a boy, the probability that both children are boys is the counterintuitive

$$P(A \mid B) = \frac{P(A \cap B)}{P(B)} = \frac{P(A)}{P(B)} = (1/4)/(3/4) = 1/3.$$

Definition 1.5.2. Two events A, B are independent if $P(A \cap B) = P(A) P(B)$. A finite number of events $A_1, \ldots, A_n$ are independent if

$$P(A_1 \cap \cdots A_n) = P(A_1) \cdots P(A_n)$$

and are pairwise independent if every pair $A_i, A_j, i \neq j$ are independent.

The following definition generalizes independence to an arbitrary collection of events, indexed by a (potentially infinite) set Θ.

Definition 1.5.3. Multiple events $A_\theta, \theta \in \Theta$ are pairwise independent if every pair of events is independent. Multiple events $A_\theta, \theta \in \Theta$ are independent if for every $k > 0$ and for every size k-subset of distinct events $A_{\theta_1}, \ldots, A_{\theta_k}$, we have

$$P(A_{\theta_1} \cap \ldots \cap A_{\theta_k}) = P(A_{\theta_1}) \cdots P(A_{\theta_k}).$$

Note that pairwise independence is a strictly weaker condition than independence.

In agreement with our intuition, conditioning on an event that is independent of A does not modify the probability of A:

$$P(A \mid B) = P(A) P(B) / P(B) = P(A).$$

On the other hand, two disjoint events cannot occur simultaneously and should therefore be dependent. Indeed, in this case $P(A \mid B) = 0 \neq P(A)$ (assuming that $P(A)$ and $P(B)$ are non-zero).

Example 1.5.2. *We consider a random experiment of throwing two dice independently and denote by A the event that the first throw resulted in 1, by B the event that the sum in both throws is 3, and by C the event that the second throw was even. Assuming the classical model, the events A, B are dependent*

$$P(A \cap B) = P(B \mid A) P(A) = (1/6)(1/6) \neq (1/6)(2/36) = P(A) P(B),$$

while A and C are independent

$$P(A \cap C) = P(C \mid A) P(A) = (1/2)(1/6) = P(A) P(C).$$

Proposition 1.5.1. *If A, B are independent, then so are the events A^c, B, the events A, B^c, and the events A^c, B^c.*

Proof. For example,

$$P(A^c \cap B) = P(B \setminus A) = P(B) - P(A \cap B) = P(B) - P(A) P(B)$$
$$= (1 - P(A)) P(B) = P(A^c) P(B).$$

The other parts of the proof are similar. ∎

Proposition 1.5.2 (Bayes Theorem). *If* $P(B) \neq 0$ *and* $P(A) \neq 0$, *then*

$$P(A \mid B) = \frac{P(B \mid A) P(A)}{P(B)}.$$

Proof.

$$P(A \mid B) P(B) = P(A \cap B) = P(B \cap A) = P(B \mid A) P(A).$$

■

Example 1.5.3. *We consider the following imaginary voting pattern of a group of 100 Americans, classified according to their party and whether they live in a city or a small town. The last row and last column capture the sum of the columns and the sum of the rows, respectively.*

	City	Small Town	Total
Democrats	*30*	*15*	*45*
Republicans	*20*	*35*	*55*
Total	*50*	*50*	*100*

We consider the experiment of drawing a person at random and observing the vote. The sample space contains 100 elementary events and we assume a classical model, implying that each person may be selected with equal $1/100$ probability.

Defining A as the event that a person selected at random lives in the city, and B as the event that a person selected at random is a democrat, we have

$$P(A \cap B) = 30/100$$
$$P(A^c \cap B) = 15/100$$
$$P(A \cap B^c) = 20/100$$
$$P(A^c \cap B^c) = 35/100$$
$$P(A) = 50/100$$
$$P(B) = 45/100$$
$$P(A \mid B) = 0.3/0.45$$
$$P(A \mid B^c) = 0.2/0.55$$
$$P(B \mid A) = 0.3/0.5$$
$$P(B \mid A^c) = 0.15/0.5.$$

Since A, B are dependent, conditioning on city dwelling raises the probability that a randomly drawn person is democrat from $P(B) = 0.45$ to $P(B \mid A) = 0.6$.

Proposition 1.5.3 (General Multiplication Rule).

$$P(A_1 \cap \cdots \cap A_n) = P(A_1) P(A_2 \mid A_1) P(A_3 \mid A_2 \cap A_1) \cdots P(A_n \mid A_1 \cap \cdots \cap A_{n-1}).$$

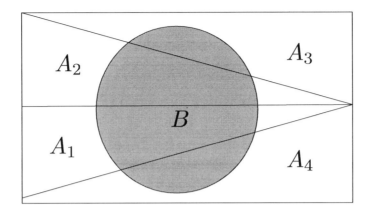

Figure 1.2: The partition $A_1, \ldots, A_4$ of Ω induces a partition $B \cap A_i$, $i = 1, \ldots, 4$ of B (see Proposition 1.5.4).

Proof. Using induction and $P(A \cap B) = P(A \mid B) P(B)$, we get

$$P(A_1 \cap \cdots \cap A_n) = P(A_n \mid A_1 \cap \cdots \cap A_{n-1}) P(A_1 \cap \cdots \cap A_{n-1})$$
$$= \cdots$$
$$= P(A_1) P(A_2 \mid A_1) P(A_3 \mid A_2 \cap A_1) \cdots P(A_n \mid A_1 \cap \cdots \cap A_{n-1}).$$

∎

Proposition 1.5.4 (The Law of Total Probability)**.** *If $A_i, i \in S$, form a finite or countably infinite partition of Ω (see Definition A.1.12)*

$$P(B) = \sum_{i \in S} P(A_i) P(B \mid A_i).$$

Proof. The partition $A_i, i \in S$, of Ω induces a partition $B \cap A_i$, $i \in S$, of B. The result follows from countable additivity (third probability axiom) or finite additivity applied to that partition

$$P(B) = P\left(\bigcup_{i \in S} (B \cap A_i) \right) = \sum_{i \in S} P(A_i \cap B) = \sum_{i \in S} P(A_i) P(B \mid A_i).$$

∎

Figure 1.2 illustrates the above proposition and its proof.

The definition below extends the notion of independence to multiple experiments.

Definition 1.5.4. Consider n random experiments with sample spaces $\Omega_1, \ldots, \Omega_n$. The set $\Omega = \Omega_1 \times \cdots \times \Omega_n$ (see Chapter A for a definition of the cartesian product

×) is the sample space expressing all possible results of the experiments. The experiments are independent if for all sets $A_1 \times \cdots \times A_n$ with $A_i \subset \Omega_i$,

$$P(A_1 \times \cdots \times A_n) = P(A_1) \cdots P(A_n).$$

In the equation above, the probability function on the left hand side is defined on $\Omega_1 \times \cdots \times \Omega_n$ and the probability functions on the right hand side are defined on Ω_i, $i = 1, \ldots, n$.

Example 1.5.4. *In two independent die throwing experiments* $\Omega = \{1, \ldots, 6\} \times \{1, \ldots, 6\}$ *and*

$$P(\textit{first die is 3, second die is 4}) = P(\textit{first die is 3}) \, P(\textit{second die is 4}) = \frac{1}{6} \cdot \frac{1}{6} = \frac{1}{36}.$$

Chapter 4 contains an extended discussion of probabilities associated with multiple experiments.

1.6   Basic Combinatorics for Probability

Some knowledge of combinatorics is essential for probability. For example, computing the probability $P(E)$ in the classical model on finite sample spaces $P(E) = |E|/|\Omega|$ is equivalent to the combinatorial problem of enumerating the elements in E and Ω.

In the case of a two-stage experiment where stage 1 has k outcomes and stage 2 has l outcomes and every combination of results in the two stages is possible, the total number of combinations of results of stage 1 and 2 is $k \cdot l$. A formal generalization appears below.

Definition 1.6.1. A k-tuple over the sets $S_1, \ldots, S_k$ is a finite ordered sequence $(s_1, \ldots, s_k)$ such that $s_i \in S_i$.

Proposition 1.6.1. *There are* $\prod_{j=1}^{k} |S_j|$ *ways to form k-tuples over the finite sets* $S_1, \ldots, S_k$. *In particular if* $S_1 = \cdots = S_k = S$ *there are* $|S|^k$ *possible k-tuples.*

Proof. A k-tuple is characterized by picking one element from each group. There are n_1 elements in group 1, n_2 in group 2, and so on. Since choices in one group do not constrain the choices in other groups, the number of possible choices is $\prod_{j=1}^{k} |S_j|$.   ∎

Example 1.6.1. *The R code below generates all possible 3-tuples over* $S_1 = \{1, 2\}$, $S_2 = \{1, 2, 3\}$, *and* $S_3 = \{1, 2\}$. *There are* $2 \cdot 3 \cdot 2 = 12$ *such possibilities.*

```
expand.grid(S1 = 1:2,  S2 = 1:3,  S3 = 1:2)

##      S1 S2 S3
## 1    1  1  1
```

```
## 2     2   1   1
## 3     1   2   1
## 4     2   2   1
## 5     1   3   1
## 6     2   3   1
## 7     1   1   2
## 8     2   1   2
## 9     1   2   2
## 10    2   2   2
## 11    1   3   2
## 12    2   3   2
```

Definition 1.6.2. Assuming that n is a positive integer and $r \leq n$ is another positive integer, we use the following notation:

$$n! \overset{\text{def}}{=} n \cdot (n-1) \cdot (n-2) \cdots 2 \cdot 1$$

$$(n)_r \overset{\text{def}}{=} \frac{n!}{(n-r)!}$$

$$\binom{n}{r} \overset{\text{def}}{=} \frac{(n)_r}{r!} = \frac{n!}{r!(n-r)!}.$$

We refer to the function $f(n) = n!$ as the factorial function and to $\binom{n}{r}$ as n-choose-r.

The factorial function grows very rapidly as $n \to \infty$. The R code below shows the considerable magnitude of $n!$ even for small n.

```
factorial(1:8)
```

```
## [1]       1       2       6      24     120     720    5040
## [8] 40320
```

The following proposition shows that the growth rate of $n!$ is similar to the growth rate of $(n/e)^n$ (see Definition B.2.1 for a definition of the limit notation below).

Proposition 1.6.2 (Stirling's Formula).

$$\lim_{n \to \infty} \frac{n!}{n^{n+1/2}e^{-n}} = \sqrt{2\pi}.$$

Proofs are available in [16] and [37].

Proposition 1.6.3. *The factorial function grows faster than any exponential:*

$$\lim_{n \to \infty} \frac{a^n}{n!} = 0 \qquad \textit{for all } a > 0.$$

Proof. Using

$$\lim_{n\to\infty} \frac{a^n}{n^{n+1/2}e^{-n}} = \lim_{n\to\infty} \frac{1}{\sqrt{n}}\left(\frac{ae}{n}\right)^n = 0, \qquad a > 0,$$

and Proposition 1.6.2, we get

$$\lim_{n\to\infty} \frac{a^n}{n!} = 0, \qquad a > 0. \tag{1.1}$$

■

The proposition above is illustrated by the following graph comparing the growth rate on a log scale of the factorial with Stirling's approximation, an exponential function, and a linear function x. Stirling's approximation overlaps the factorial line indicating extremely good approximation.

```
x = 1:100
R = stack(list(`$x!$` = lfactorial(x), Stirling = log(2 *
      pi)/2 + (x + 1/2) * log(x) - x, `exp($x$)` = x,
      `$x$` = log(x)))
names(R) = c("lf", "Function")
R$x = x
qplot(x, lf, color = Function, lty = Function,
      geom = "line", xlab = "$x$", ylab = "$\\log(f(x))$",
      data = R, size = I(2), main = "Growth Rate on Log Scale")
```

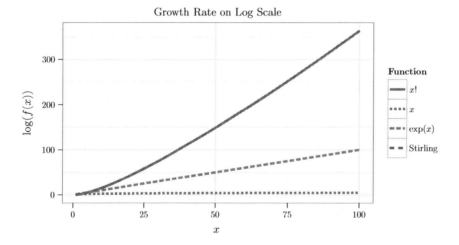

Proposition 1.6.4. *The number of r-tuples over a finite set S in which no element appears twice is $(|S|)_r$ and the number of different orderings of n elements is $n!$.*

Proof. The first statement is a direct corollary of Proposition 1.6.1, where each value in the r tuple is selected from the population of remaining or unselected items. The second statement follows from the first $(n = r = |S|)$. ■

The following code generates all possible orderings of the letters a, b, and c. There are $3! = 6$ such orderings.

```
# generate all 6 permutations over three
# letters
library(gtools)
permutations(3, 3, letters[1:3])

##       [,1] [,2] [,3]
## [1,] "a"  "b"  "c"
## [2,] "a"  "c"  "b"
## [3,] "b"  "a"  "c"
## [4,] "b"  "c"  "a"
## [5,] "c"  "a"  "b"
## [6,] "c"  "b"  "a"
```

Example 1.6.2 (The Birthday Paradox). *There are 365^r possible assignments of birthdays to r people. Using the previous proposition, the number of assignments of birthdays to r people, assuming that all birthdays are different, is $(365)_r$. Under the classical probability model, the probability $P(A_r)$ that a group of r people will have all different birthdays is*

$$P(A_r) = \frac{|A_r|}{|\Omega|} = \frac{(365)_r}{365^r}.$$

For example, $P(A_{30}) \approx 0.294$, implying that it is likely to find recurring birthdays in a group of 30 people. The graph below shows how the probability of having different birthdays decays to zero as r increases. The median (the value at which the probability is approximately 1/2) is $r = 23$. The name "The Birthday Paradox" is sometimes associated with this example, since it is intuitively likely that 23 people will all have different birthdays with high probability.

The following R code graphs the probability of r people having all different birthdays as a function of r.

```
# perform calculation on log-scale to avoid
# overflow
r = 1:50
p = exp(lfactorial(365) - lfactorial(365 - r) -
    r * log(365))
qplot(x = r, y = p, size = I(2), xlab = "$r$",
    ylab = "$\\P(A_r)$")
```

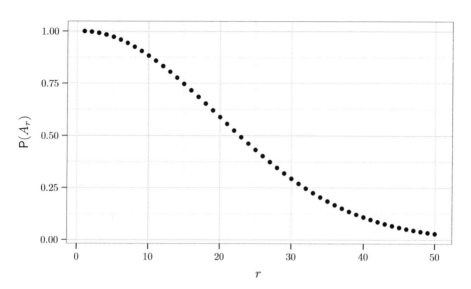

Proposition 1.6.5. *A population of n elements has $\binom{n}{r}$ different subsets of size r. Equivalently there are $\binom{n}{r}$ ways to select r elements out of n distinct elements with no element appearing twice (selection without replacement) if order is neglected.*

Proof. There are $(n)_r$ ways to select r elements out of n elements if ordering matters (number of r-tuples over n elements). Since there are $r!$ possible orderings of the selected values, the number we are interested in times $r!$ equals $(n)_r$. Dividing $(n)_r$ by $r!$ completes the proof. ∎

Example 1.6.3. *We use R below to enumerate all possible subsets of size 3 out of a set of size 4. There are ten columns listing these subsets in accordance with $\binom{5}{3} = 20/2$.*

```
# list all possible combinations of 3 out of
# 5 letters
combn(letters[1:5], 3)
```

```
##      [,1] [,2] [,3] [,4] [,5] [,6] [,7] [,8]
## [1,] "a"  "a"  "a"  "a"  "a"  "a"  "b"  "b"
## [2,] "b"  "b"  "b"  "c"  "c"  "d"  "c"  "c"
## [3,] "c"  "d"  "e"  "d"  "e"  "e"  "d"  "e"
##      [,9] [,10]
## [1,] "b"  "c"
## [2,] "d"  "d"
## [3,] "e"  "e"
```

Example 1.6.4. *In poker, a hand is a subset of 5 cards (order does not matter) out of 52 distinct cards. The cards have face values (1-13) and suits (clubs, spades, hearts, diamonds). There are* $|\Omega| = \binom{52}{5}$ *different hands at poker since this is the number of subsets of size 5 from the 52 distinct cards. The probability that a random hand has five different face values under the classical model is*

$$4^5 \binom{13}{5} / \binom{52}{5} \approx 0.507$$

as face values are chosen in $\binom{13}{5}$ *ways and there are four suits possible for each of the five face values.*

Example 1.6.5. *The number of sequences of length* $p+q$ *containing* p *zeros and* q *ones is* $\binom{p+q}{p}$ *(choosing* p *among* $p+q$ *sequence positions and assigning them to zero values causes the remaining positions to be automatically assigned to one values).*

Example 1.6.6. *Assuming that the U.S. Senate has 60 male senators and 40 female senators, the probability under the classical probability model of selecting an all-male committee of 3 senators is*

$$P(E) = \frac{|E|}{|\Omega|} = \frac{\text{number of samples of 3 out of 60 without order and replacement}}{\text{number of samples of 3 out of 100 without order and replacement}}$$

$$= \frac{\binom{60}{3}}{\binom{100}{3}} = \frac{60 \cdot 59 \cdot 58}{3 \cdot 2} \frac{3 \cdot 2}{100 \cdot 99 \cdot 98} \approx 0.211.$$

Intuitively, if the frequency of all male committees is significantly larger than 21%, we may conclude that the classical model is inappropriate.

Proposition 1.6.6 (Binomial Theorem).

$$(x+y)^n = \sum_{k=0}^{n} \binom{n}{k} x^{n-k} y^k.$$

Proof. Expanding the expression

$$(x+y)^n = (x+y)(x+y)\cdots(x+y)$$

we see that it contains many additive terms, each corresponding to a pick of x or y from each of the product terms above. Collecting equal additive terms $x^{n-k}y^k$ for $k = 0, \ldots, n$ (the sum of the two exponents must be n) we have $\binom{n}{k}$ possible selections of k choices of y out of n leading to the term $\binom{n}{k} x^{n-k} y^k$. Repeating this argument for all possible $k = 0, \ldots, n$ completes the proof.   ∎

The description above corresponds to choosing r elements out of n distinct elements, or alternatively placing n distinct elements into two bins — one with r elements and one with $n-r$ elements. A useful generalization is placing n distinct elements in k bins with r_i elements being placed at bin i, for $i = 1, \ldots, k$.

Proposition 1.6.7. *The number of ways to deposit n distinct objects into k bins with r_i objects in bin i, $i = 1, \ldots, k$, is the multinomial coefficient $n!/(r_1! \cdots r_k!)$.*

Proof. Repeated use of Proposition 1.6.5 shows that the number is

$$\binom{n}{r_1}\binom{n-r_1}{r_2}\binom{n-r_1-r_2}{r_3}\cdots\binom{n-r_1-\cdots-r_{k-2}}{r_{k-1}}.$$

Canceling common factors in the numerators and denominators completes the proof. ∎

Example 1.6.7. *A throw of twelve dice can result in 6^{12} different outcomes. The event A that each face appears twice can occur in as many ways as twelve dice can be arranged in six groups of two each. Assuming the classical probability model, the above proposition implies that*

$$P(A) = \frac{|A|}{|\Omega|} = \frac{12!/(2^6)}{6^{12}} \approx 0.0034.$$

The following inequality is useful in bounding the probability of complex events in terms of the probability of multiple simpler events.

Proposition 1.6.8 (Boole's Inequality). *For a finite or countably infinite set of events A_i, $i \in C$*

$$P\left(\bigcup_{i \in C} A_i\right) \leq \sum_{i \in C} P(A_i).$$

Proof. For two events the proposition holds since $P(A \cup B) = P(A) + P(B) - P(A \cap B)$ (Principle of Inclusion-Exclusion). The case of a finite number of sets holds by induction. The case of a countably infinite number of sets follows from Proposition E.2.1 in the appendix. ∎

1.7 Probability and Measure Theory*

Definition 1.2.1 appears to be formal, and yet is not completely rigorous. It states that a probability function P assigns real values to events $E \subset \Omega$ in a manner consistent with the three axioms. The problem is that the domain of the probability function P is not clearly specified. In other words, if P is a function $P : \mathcal{F} \to \mathbb{R}$ from a set $\mathcal{F}$ of subsets of Ω to $\mathbb{R}$, the set $\mathcal{F}$ is not specified. The importance of this issue stems from the fact that the three axioms need to hold for all sets in $\mathcal{F}$.

At first glance this appears to be a minor issue that can be solved by choosing $\mathcal{F}$ to be the power set of Ω: 2^Ω. This works nicely whenever Ω is finite or countably infinite. But selecting $\mathcal{F} = 2^\Omega$ does not work well for uncountably infinite Ω such as continuous spaces. It is hard to come up with useful functions $\mathsf{P} : 2^\Omega \to \mathbb{R}$ that satisfy the three axioms for all subsets of Ω.

A satisfactory solution that works for uncountably infinite Ω is to define $\mathcal{F}$ to be a σ-algebra of subsets of Ω (see Section E.1) that is smaller than 2^Ω. In particular, when $\Omega \subset \mathbb{R}^d$, the Borel σ-algebra (Definition E.5.1) is sufficiently large to include the "interesting" subsets of Ω and yet is small enough to not restrict P too much.

We also note that a probability function P is nothing but a measure μ on a measurable space $(\Omega, \mathcal{F})$ (Definition E.2.1) satisfying $\mu(\Omega) = 1$. In other words, the triplet $(\Omega, \mathcal{F}, \mathsf{P})$ is a measure space (see Definition E.2.1) where $\mathcal{F}$ is the σ-algebra of measurable sets and P is a measure satisfying $\mathsf{P}(\Omega) = 1$. Thus, the wide array of mathematical results from measure theory (Chapter E) and Lebesgue integration (Chapter F.3) are directly applicable to probability theory.

1.8 Notes

Our exposition follows the axiomatic view of probability, promoted by A. Kolmogorov. Alternative viewpoints are available, including the frequency viewpoint ($\mathsf{P}(A)$ is the frequency of A occurring in a long sequence of repetitive experiments) and the subjective viewpoint ($\mathsf{P}(A)$ measures the belief that A will occur).

More information on the basic concepts of probability is available in nearly any probability textbook. One example is Feller's first volume [16], which inspired a generation of probabilists as well as several of the examples in this chapter. Examples of books with rigorous coverage of probability theory are [17, 10, 5, 1, 33, 25]. Elementary exposition that avoids measure theory is available in most undergraduate probability textbooks, such as [48, 36, 14]. More information on combinatorics is available in combinatorics textbooks, for example [35] (undergraduate level) and [42] (graduate level).

1.9 Exercises

1. Extend the argument at the end of Section 1.2 and characterize probability functions on a countably infinite Ω using a sequence of non-negative numbers that sum to one. What is the problem with extending this argument further to uncountably infinite Ω?

2. Can there be a classical probability model on sample spaces that are countably infinite? Provide an example or prove that it is impossible.

3. Complete the proof of Proposition 1.5.1.

4. Describe a sample space consistent with the experiment of drawing a hand in poker. Write the events E corresponding to drawing three aces and drawing a full house (and their sizes $|E|$). What is the event corresponding to the intersection of the two events above, and what is its size and probability under the classical model?

5. Formulate a theory of probability that mirrors the standard theory, with the only difference that the second axiom would be $P(\Omega) = 2$. How would the propositions throughout the chapter change (if at all)?

6. Show a situation where we have three events that are independent but not mutually independent. Hint: Look for a probability function satisfying

$$P(A) = P(B) = P(C) = 1/3$$
$$P(A \cap B) = P(A \cap C) = P(B \cap C) = 1/9 = P(A \cap B \cap C).$$

7. Prove that $P_E(A) = P(A \mid E)$ is a probability function if $P(E) \neq 0$.

8. Consider the experiment of throwing three fair six-sided dice independently and observing the results without order. Identify the sample space, and the most and least probable elements of it.

9. Repeat the previous exercise, if the results are observed with order.

10. Generalize Proposition 1.2.6 (Principle of Inclusion-Exclusion) to a union of three sets. Can you further generalize it to a union of an arbitrary number of sets?

Chapter 2

Random Variables

In this chapter we continue studying probability theory, covering random variables and their associated functions: cumulative distribution functions, probability mass functions, and probability density functions.

2.1 Basic Definitions

Definition 2.1.1. A random variable (RV) X is a function from the sample space Ω to the real numbers $X : \Omega \to \mathbb{R}$. Assuming $E \subset \mathbb{R}$ we denote the event $\{\omega \in \Omega : X(\omega) \in E\} \subset \Omega$ by $\{X \in E\}$ or just $X \in E$ (see Figure 2.1).

- RVs are typically denoted using upper case letters, such as X, Y, Z.

- We sometimes use the notation $\{X \in E\}$ or $\{X \in E\}$ loosely. For example, $X = x$ corresponds to $\{X \in \{x\}\}$ and $a < X < b$ corresponds to $\{X \in (a, b)\}$.

- While X denotes a function, the notations $X \in A$ and $X = x$ correspond to subsets of Ω.

An RV $X : \Omega \to \mathbb{R}$ defines a new probability function P' on a new sample space $\Omega' = \mathbb{R}$:

$$\mathsf{P}'(E) \stackrel{\text{def}}{=} \mathsf{P}(X \in E), \qquad E \subset \mathbb{R} = \Omega'. \tag{2.1}$$

Verifying that P' satisfies the three probability axioms is straightforward. We can often leverage the conceptually simpler $(\mathbb{R}, \mathsf{P}')$ in computing probabilities $\mathsf{P}(A)$ of events $A = \{X \in E\}$, ignoring (Ω, P).

Example 2.1.1. *In Example 1.3.1 (throwing two fair dice, $\Omega = \{\omega = (a, b) : a \in \{1, \ldots, 6\}, b \in \{1, \ldots, 6\}\}$, and $\mathsf{P}(A) = |A|/36$), the RV $X(a, b) = a + b$ (sum*

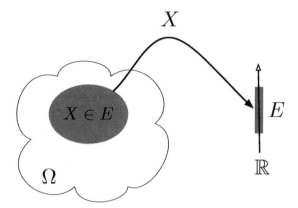

Figure 2.1: A random variable X is a mapping from Ω to $\mathbb{R}$. The set $X \in E$ or $\{X \in E\}$, $E \subset \mathbb{R}$, corresponds to $\{\omega \in \Omega : X(\omega) \in E\}$ and $P(X \in E) = P(\{\omega \in \Omega : X(\omega) \in E\})$.

of the faces of the two dice) exhibits the following events and their associated probabilities.

$$\{X = 3\} = \{(1,2),(2,1)\}, \qquad\qquad P(\{X = 3\}) = 2/36$$
$$\{X = 4\} = \{(2,2),(3,1),(1,3)\}, \qquad P(\{X = 4\}) = 3/36$$
$$\{X < 3\} = \{(1,1)\}, \qquad\qquad\quad P(\{X < 3\}) = 1/36$$
$$\{X < 2\} = \emptyset, \qquad\qquad\qquad\quad P(\{X < 2\}) = 0.$$

Example 2.1.2. *In Example 1.1.1 (tossing three fair coins), the RV X counting the number of heads gives the following events and their associated probabilities.*

$$P(X = 0) = P(\{\omega \in \Omega : X(\omega) = 0\}) = P(\{TTT\}) = 0.5^3 = 1/8$$
$$P(X = 1) = P(\{\omega \in \Omega : X(\omega) = 1\}) = P(\{HTT, THT, TTH\}) = 3 \cdot 0.5^3 = 3/8$$
$$P(X = 2) = P(\{\omega \in \Omega : X(\omega) = 2\}) = P(\{HHT, HHH, HTH\}) = 3 \cdot 0.5^3 = 3/8$$
$$P(X = 3) = P(\{\omega \in \Omega : X(\omega) = 3\}) = P(\{HHH\}) = 0.5^3 = 1/8.$$

Definition 2.1.2. A random variable X is discrete if $P(X \in K) = 1$ for some finite or countably infinite set $K \subset \mathbb{R}$. A random variable X is continuous if $P(X = x) = 0$ for all $x \in \mathbb{R}$.

An RV can be discrete, continuous, or neither. If $\{X(\omega) : \omega \in \Omega\}$ is finite or countably infinite, X is discrete. This happens in particular if Ω is finite or countably infinite.

Example 2.1.3. *In the dart throwing experiment of Example 1.1.2, the sample space is $\Omega = \{(x,y) : \sqrt{x^2 + y^2} < 1\}$. The RV Z, which is 1 for a bullseye* *and 0 otherwise, is discrete since* $P(Z \in \{0,1\}) = 1$. *The RV R, defined as* $R(x,y) = \sqrt{x^2 + y^2}$, *represents the distance of between the dart position and the origin. Since the area of a circle* $\{R = r\} = \{(x,y) : x, y \in \mathbb{R}, \sqrt{x^2 + y^2} = r\}$ *is zero, we have for all r, $P(R = r) = 2D\text{-}area(\{R = r\})/(\pi 1^2) = 0$ (under the classical model), implying that R is a continuous RV. We also have*

$$P(R < r) = P(\{(x,y) : x, y \in \mathbb{R}, \sqrt{x^2 + y^2} < r\}) = \frac{\pi r^2}{\pi 1^2} = r^2$$

$$P(0.2 < R < 0.5) = P(\{(x,y) : x, y \in \mathbb{R}, 0.2 < \sqrt{x^2 + y^2} < 0.5\}) = \frac{\pi(0.5^2 - 0.2^2)}{\pi 1^2}.$$

We also have (recall that an area of a circle of radius r is πr^2)

$$P(X < r) = P(\{(x,y) : x, y \in \mathbb{R}, \sqrt{x^2 + y^2} < r\}) = \frac{\pi r^2}{\pi 1^2} = r^2$$

$$P(0.2 < X < 0.5) = P(\{(x,y) : x, y \in \mathbb{R}, 0.2 < \sqrt{x^2 + y^2} < 0.5\}) = \frac{\pi(0.5^2 - 0.2^2)}{\pi 1^2}.$$

Definition 2.1.3. We define the cumulative distribution function (cdf) with respect to an RV X to be $F_X : \mathbb{R} \to \mathbb{R}$ given by $F_X(x) = P(X \leq x)$.

Definition 2.1.4. Given a discrete RV X, we define the associated probability mass function (pmf) $p_X : \mathbb{R} \to \mathbb{R}$ to be $p_X(x) = P(X = x)$.

Definition 2.1.5. Given a continuous RV X, we define the associated probability density function (pdf) $f_X : \mathbb{R} \to \mathbb{R}$ as the derivative of the cdf $f_X = F'_X$ where it exists, and 0 elsewhere.

The cdf is useful for both discrete and continuous RVs, the pmf is useful for discrete RVs only, and the pdf is useful for continuous RVs only. Notice that in each of the three functions above, we denote the applicable RV with an uppercase subscript and the function argument with a lowercase letter, and the two usually correspond; for example, consider $p_X(x) = P(X = x)$.

We have the following relationship between the cdf and the pdf

$$f_X(r) = \begin{cases} F'_X(r) & F'_X(r) \text{ exists} \\ 0 & \text{otherwise} \end{cases}$$

$$F_X(x) = \int_{-\infty}^{x} f_X(r)\, dr.$$

where the first equation follows from the definition of a pdf and the second line follows from Proposition F.2.2.

Example 2.1.4. *In the coin-tossing experiment (Example 2.1.2), we have*

$$p_X(x) = P(X = x) = \begin{cases} 1/8 & x = 0 \\ 3/8 & x = 1 \\ 3/8 & x = 2 \\ 1/8 & x = 3 \\ 0 & else \end{cases} \qquad F_X(x) = P(X \le x) = \begin{cases} 0 & x < 0 \\ 1/8 & 0 \le x < 1 \\ 1/2 & 1 \le x < 2 \\ 7/8 & 2 \le x < 3 \\ 1 & 3 \le x \end{cases}.$$

The following R code illustrates these pmf and cdf functions.

```
D = data.frame(x = c(-1, 0, 1, 2, 3, 4), y = c(0,
    1/8, 3/8, 3/8, 1/8, 0))
qplot(x, y, data = D, xlab = "$x$", ylab = "$p_X(x)$") +
    geom_linerange(aes(x = x, ymin = 0, ymax = y))
```

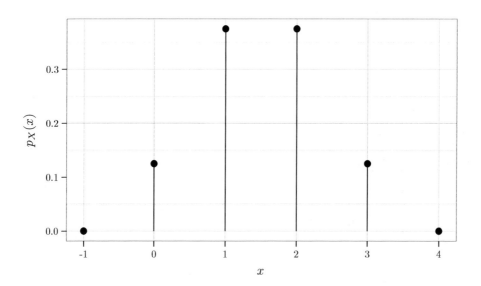

```
plot(ecdf(c(0, 1, 1, 1, 2, 2, 2, 3)), verticals = FALSE,
    lwd = 3, main = "", xlab = "$x$", ylab = "$F_X(x)$")
grid()
```

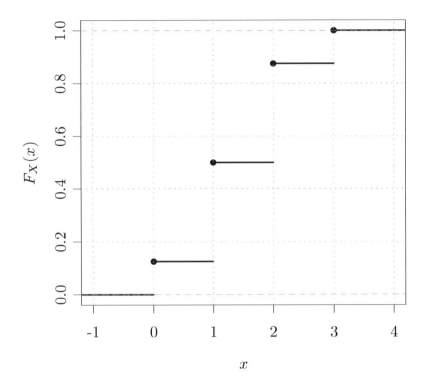

In the previous figure, the filled circles above clarify the value of F_X at the discontinuity points above, for example $F_X(3) = 1$. Note how the F_X above is monotonic increasing from 0 to 1, and is continuous from the right, but not continuous from the left.

Example 2.1.5. *In the dart-throwing experiment (Example 2.1.3), we have*

$$F_R(r) = P(R \le r) = \begin{cases} 0 & r < 0 \\ \pi r^2/(\pi 1^2) = r^2 & 0 \le r < 1 \\ 1 & 1 \le r \end{cases},$$

$$f_R(r) = \frac{dF_R(r)}{dr} = \begin{cases} 0 & r < 0 \\ 2r & 0 \le r < 1 \\ 0 & 1 \le r \end{cases}.$$

The following R code illustrates these pdf and cdf functions.

```
x = seq(-0.5, 1.5, length = 100)
y = x^2
y[x < 0] = 0
y[x > 1] = 1
z = 2 * x
z[x < 0] = 0
z[x > 1] = 0
D = stack(list(`$f_R(r)$` = z, `$F_R(r)$` = y))
names(D) = c("probability", "Function")
D$x = x
qplot(x, y = probability, geom = "line", xlab = "$r$",
     ylab = "", lty = Function, color = Function,
     data = D, size = I(1.5))
```

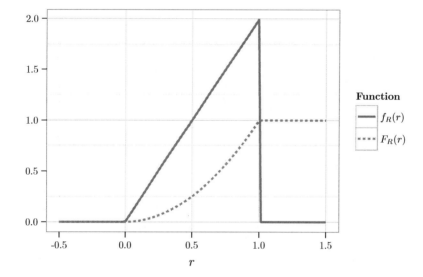

Proposition 2.1.1. *The pmf p_X of a discrete RV X satisfies*

1. $p_X(x) \geq 0$

2. The set $\{x \in \mathbb{R} : p_X(x) \neq 0\}$ is finite or countably infinite

3. $1 = \sum_x p_X(x)$

4. $\mathrm{P}(A) = \sum_{x \in A} p_X(x)$,

where the summations above contain only the finite or countably infinite number of terms for which p_X is nonzero. Conversely, a function p_X that satisfies the above properties is a valid pmf for some discrete RV.

Proof. Statement (1) follows from the nonnegativity of P and (2) follows from the definition of a discrete RV. Statements (3) and (4) follow from (2) and the third axiom of probability functions. The converse statement holds since the probability P defined in (4) satisfies the probability axioms. ∎

Lemma 2.1.1 (Continuity of the Probability Function). *Let $A_n, n \in \mathbb{N}$ be a sequence of monotonic sets with a limit A. That is, either $A_1 \subset A_2 \subset A_3 \subset \cdots$ with $A = \cup_n A_n$ or $\cdots \subset A_3 \subset A_2 \subset A_1$ with $A = \cap_n A_n$. Then $\lim_{n\to\infty} P(A_n) = P(A)$.*

Proof*. The result follows directly from Proposition E.2.1 in the appendix. ∎

Proposition 2.1.2. *Denoting the left and right hand limits of F_X by $F_X(a^-) = \lim_{x\to a^-} F_X(x)$ and $F_X(a^+) = \lim_{x\to a^+} F_X(x)$, respectively (see Definition B.2.11), we have*

1. $F_X(a^-) = P(X < a)$,

2. F_X *is monotonically increasing: if $a < b$, $F_X(a) \leq F_X(b)$,*

3. $0 = \lim_{a\to-\infty} F_X(a) < \lim_{a\to\infty} F_X(a) = 1$, *and*

4. F_X *is continuous from the right; that is for all $a \in \mathbb{R}$, $F_X(a) = F_X(a^+)$.*

A function satisfying (2), (3), and (4) above is a cdf for some RV X.

Proof. Statement (2) follows since $A \subset B$ implies that $P(A) \leq P(B)$. Statements (1), (3), (4) follow Lemma 2.1.1:

$$F_X(a^-) = \lim_{n\to\infty} P'((-\infty, a - 1/n]) = P'(\cup_{n\in\mathbb{N}}(-\infty, a - 1/n]) = P'((-\infty, a))$$

$$\lim_{n\to\infty} P'((-\infty, -n]) = P'(\cap_{n\in\mathbb{N}}(-\infty, -n]) = P'(\emptyset) = 0$$

$$\lim_{n\to\infty} P'((-\infty, n]) = P'(\cup_{n\in\mathbb{N}}(-\infty, n]) = P'(\mathbb{R}) = 1$$

$$P'((-\infty, a]) = P'(\cap_n(-\infty, a + 1/n]) = \lim_{n\to\infty} P'((-\infty, a + 1/n]) = F_X(a^+).$$

We can infer the converse since a function satisfying (2), (3), and (4) defines a unique probability measure on $\mathbb{R}$, described and proved in the appendix chapter on measure theory (Chapter E) in Definition E.5.2 and Proposition E.5.5. ∎

Corollary 2.1.1. *For all RV X and $a < b$,*

$$P(a < X \leq b) = F_X(b) - F_X(a)$$
$$P(a \leq X \leq b) = F_X(b) - F_X(a^-)$$
$$P(a < X < b) = F_X(b^-) - F_X(a)$$
$$P(a \leq X < b) = F_X(b^-) - F_X(a^-)$$
$$P(X = x) = F_X(x) - F_X(x^-).$$

Proof. These statements follow directly from the previous proposition. ∎

Corollary 2.1.2. *The function F_X is continuous if and only if X is a continuous RV. In this case,*
$$F_X(a^-) = F_X(a) = F_X(a^+).$$

Proof. An RV X is continuous if and only if $0 = P(X = x) = F_X(x) - F_X(x^-)$ for all $x \in \mathbb{R}$, or in other words, the cdf is continuous from the left. Since the cdf is always continuous from the right, the continuity of X is equivalent to the continuity of F_X. ∎

Definition 2.1.6. The quantile function associated with an RV X is
$$Q_X(r) = \inf\{x : F_X(x) = r\}.$$

The reason for the infimum in the definition above is that F_X may not be a one-to-one function, and there may be multiple values x for which $F_X(x) = r$. Recall that for a continuous X, the cdf F_X is continuous and monotonic increasing from 0 at $-\infty$ to 1 at ∞ (see Definition B.2.5). If the cdf is strictly monotonically increasing (see Definition B.2.5), F_X is one-to-one and hence invertible and $Q_X(r) = F_X^{-1}(r)$. The value, $Q_X(1/2)$, at which 50% of the probability mass lies above and 50% of the probability mass lie below is called the median.

Proposition 2.1.3. *Let X be a continuous RV. Then*

1. *f_X is a nonnegative function, and*

2. *$\int_{\mathbb{R}} f_X(x)dx = 1$.*

A piecewise continuous function f_X satisfying the above properties is a valid pdf for some continuous RV.

Proof. Statement (1) follows from the fact that the pdf is a derivative of a nondecreasing function (the cdf). Statement (2) follows from
$$\int_{-\infty}^{\infty} f_X(x)dx = \lim_{n\to\infty} \int_{-n}^{n} f_X(x)dx = \lim_{n\to\infty} P(-n \le X \le n)$$
$$= \lim_{n\to\infty} (F_X(n) - F_X(-n)) = 1 - 0.$$

The converse follows from the existence of a cdf via
$$F_X(x) = \int_{-\infty}^{x} f_X(r)\,dr$$

(see Proposition F.2.2) and the converse part of Proposition 2.1.2. ∎

Since the pdf is the derivative of the cdf we have
$$f_X(x) = \lim_{\Delta\to 0} \frac{F_X(x+\Delta) - F_X(x)}{\Delta}.$$

This implies that for a small $\Delta > 0$

$$f_X(x) \approx \frac{\mathsf{P}(x < X < x + \Delta)}{\Delta}, \quad \text{or} \tag{2.2}$$

$$f_X(x)\Delta \approx \mathsf{P}(x < X < x + \Delta). \tag{2.3}$$

The above derivations shows that $f_X(x)\Delta$ is approximately the probability that X lies within an interval of width Δ around x.

Proposition 2.1.4.

$$\mathsf{P}(X \in A) = \begin{cases} \sum_{x \in A} p_X(x) & X \text{ is discrete} \\ \int_A f_X(x) dx & X \text{ is continuous} \end{cases}. \tag{2.4}$$

Proof. In the discrete case,

$$\mathsf{P}(X \in A) = \mathsf{P}(\cup_{x \in A}\{X = x\}) = \sum_{x \in A} \mathsf{P}(X = x) = \sum_{x \in A} p_X(x).$$

The continuous case follows from the fact that the pdf is the derivative of the cdf and the relationship between a derivative and the definite integral (Proposition F.2.2). ∎

Example 2.1.6. *In the coin-toss experiment (Example 2.1.2), we computed the pmf of X to be $p_X(0) = 1/8, p_X(1) = 3/8, p_X(2) = 3/8, p_X(3) = 1/8$, so indeed $\sum_{x \in X} p_X(x) = 1$. Using the proposition above, we have*

$$\mathsf{P}(X > 0) = \sum_{x \in \{1,2,...\}} p_X(x) = p_X(1) + p_X(2) + p_X(3) = 7/8.$$

Example 2.1.7. *In the dart-throwing experiment (Example 2.1.3), we computed $F_R(r) = r^2$ and $f_R(r) = 2r$ for $r \in (0,1)$. Using the proposition above, the probability of hitting the bullseye is*

$$\mathsf{P}(R < 0.1) = \mathsf{P}(R \in (-\infty, 0.1)) = \int_{-\infty}^{0.1} f_R(r)\, dr = \int_0^{0.1} 2r\, dr = r^2 \Big|_0^{0.1} = 0.01$$

in accordance with Example 2.1.3.

2.2 Functions of a Random Variable

Definition 2.2.1. Composing an RV $X : \Omega \to \mathbb{R}$ with $g : \mathbb{R} \to \mathbb{R}$ defines a new RV $g \circ X : \Omega \to \mathbb{R}$, which we denote by $g(X)$. In other words, $g(X)(\omega) = g(X(\omega))$ (see Figure 2.2).

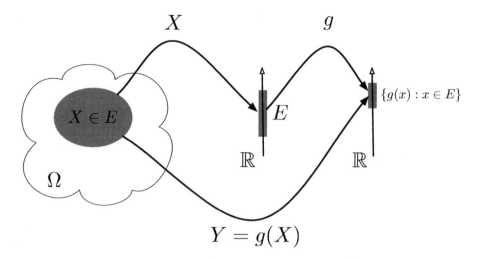

Figure 2.2: Given an RV X, the RV $Y = g(X)$ is a mapping from Ω to $\mathbb{R}$ realized by $\omega \mapsto g(X(\omega))$. Contrast this with Figure 2.1.

We can compute probabilities of events relating to $Y = g(X)$ through probabilities of events concerning X:

$$
\begin{aligned}
\mathsf{P}(Y \in B) &= \mathsf{P}(\{\omega \in \Omega : Y(\omega) \in B\}) \qquad\qquad (2.5)\\
&= \mathsf{P}(\{\omega \in \Omega : g(X(\omega)) \in B\})\\
&= \mathsf{P}(\{\omega \in \Omega : X(\omega) \in g^{-1}(B)\})\\
&= \mathsf{P}(X \in g^{-1}(B))
\end{aligned}
$$

where $g^{-1}(B) = \{r \in \mathbb{R} : g(r) \in B\}$. In other words, if we can compute quantities such as $\mathsf{P}(X \in A)$, we can also compute quantities such as $\mathsf{P}(g(X) \in B)$. This implies a relationship between the cdfs, pdfs, and pmfs of X and $Y = g(X)$, on which we elaborate below.

2.2.1 Discrete $g(X)$

For a discrete RV X and $Y = g(X)$, equation (2.5) implies

$$
p_Y(y) = \mathsf{P}(Y = y) = \mathsf{P}(g(X) = y) = \mathsf{P}(X \in g^{-1}(\{y\})) = \sum_{x:g(x)=y} p_X(x).
$$

In particular, if g is one-to-one, then $p_Y(y) = p_X(x)$ for the x satisfying $g(x) = y$.

Example 2.2.1. *In the coin-toss experiment (Example 2.1.2), we computed that* $p_X(0) = p_X(3) = 1/8, p_X(1) = p_X(2) = 3/8$ *and* $p_X(x) = 0$ *otherwise. The function* $g(r) = r^2 - 2$ *is one-to-one on the set A for which* $\mathsf{P}(X \in A) = 1$

$(A = \{0, 1, 2, 3\})$. *We therefore have the following pmf of* $Y = g(X) = X^2 - 2$

$$p_Y(y) = \begin{cases} p_X(0) = 1/8 & y = -2 \\ p_X(1) = 3/8 & y = -1 \\ p_X(2) = 3/8 & y = 2 \\ p_X(3) = 1/8 & y = 7 \\ 0 & otherwise \end{cases}.$$

By contrast, the function

$$h(r) = \begin{cases} 0 & r < 3/2 \\ 1 & r \geq 3/2 \end{cases}$$

is not one-to-one on the set $A = \{0, 1, 2, 3\}$, *suggesting that more than one value of* x *contributes to each value of* y. *Specifically,* $h(0) = h(1) = 0$ *and* $h(2) = h(3) = 1$, *implying that the pmf of the RV* $Z = h(X)$ *is*

$$p_Z(r) = \begin{cases} p_X(0) + p_X(1) = 1/8 + 3/8 = 1/2 & r = 0 \\ p_X(2) + p_X(3) = 1/8 + 3/8 = 1/2 & r = 1 \\ 0 & otherwise \end{cases}.$$

2.2.2   Continuous $g(X)$

If X is a continuous RV, $Y = g(X)$ can be discrete, continuous, or neither. If Y is a discrete RV, its pmf can be determined by the formula

$$p_Y(y) = \int_{\{r:g(r)=y\}} f_X(r) dr.$$

Example 2.2.2. *In the dart-throwing experiment (Example 2.1.3), the RV* $Y = g(R)$, *where* $g(r) = 0$ *if* $r < 0.1$ *and* $g(r) = 1$ *otherwise, is a discrete RV with*

$$p_Y(y) = \begin{cases} P(g(R) = 0) = \int_0^{0.1} 2r\, dr = 0.01 & y = 0 \\ \int_{0.1}^1 2r\, dr = 0.99 & y = 1 \\ 0 & otherwise \end{cases}.$$

If $g(X)$ is continuous, then we can compute the density $f_{g(X)}$ in three ways.

1. Compute the cdf $F_Y(y)$ for all $y \in \mathbb{R}$ using (2.5)

$$F_Y(y) = P(Y \leq y) = P(g(X) \leq y)$$

and proceed to compute the pdf $f_{g(X)}$ by differentiating $F_{g(X)}$.

2. Compute the density of $Y = g(X)$ directly from f_X when $g : \mathbb{R} \to \mathbb{R}$ is differentiable and one-to-one on the domain of f_X, using the change of variables technique (Proposition F.2.3)

$$f_{g(X)}(y) = \frac{1}{|g'(g^{-1}(y))|} f_X(g^{-1}(y)). \tag{2.6}$$

The requirement that g be one-to-one is essential since otherwise the inverse function g^{-1} fails to exist.

3. Compare the moment generating function (defined in Section 2.4) of Y to that of a known RV.

Example 2.2.3. *We can apply the first of the three techniques to the dart-throwing experiment (Example 2.1.3). The RV $Y = g(R) = R^2$ ($g(r) = r^2$) is a continuous RV with*

$$F_Y(y) = \begin{cases} P(R^2 \leq y) = P(\{\omega \in \Omega : R(\omega) \leq \sqrt{y}\}) = F_R(\sqrt{y}) = \frac{\pi(\sqrt{y})^2}{\pi \cdot 1} & y \in (0,1) \\ 0 & y < 0 \\ 1 & y \geq 1 \end{cases}$$

$$f_Y(y) = F'_Y(y) = \begin{cases} 1 & y \in (0,1) \\ 0 & otherwise \end{cases}.$$

Since the function $g(r) = r^2$ is increasing and differentiable in $(0,1)$ and $P(R \in (0,1)) = 1$, we can also use the second alternative:

$$f_Y(y) = \frac{1}{|g'(\sqrt{y})|} f_R(\sqrt{y}) = \frac{1}{2\sqrt{y}} 2\sqrt{y} = 1$$

for $0 < y < 1$ and 0 otherwise.

Example 2.2.4. *Let X be a continuous RV with a cdf F_X strictly increasing on the range of X. Using (2.6), the RV $Y = F_X(X)$ has the pdf*

$$f_Y(y) = \frac{1}{f_X(F_X^{-1}(y))} f_X(F_X^{-1}(y)) = \begin{cases} 1 & y \in [0,1] \\ 0 & otherwise \end{cases}.$$

The second branch above ($y \notin [0,1]$) follows since F_X is not strictly monotonic increasing outside of $[0,1]$ and therefore (2.6) does not apply outside of $[0,1]$. In that region we identify $P(Y \leq a)$ for $a \notin [0,1]$ is constant, implying that its derivative, the pdf, is zero.

The transformation $X \mapsto Y = F_X(X)$ maps a continuous RV X into an RV $F_X(X)$ with a a pdf that is constant over $[0,1]$ and 0 elsewhere. This RV corresponds to the classical probability model on $(0,1)$ (see Section 3.7). We will encounter this RV again in the next chapter.

Note that the same reasoning does not hold when X is discrete since in that case F_X is not one-to-one and F_X^{-1} does not exist.

Example 2.2.5. *Consider an RV X with $f_X(x) = \exp(-x)$ for $x \geq 0$ and 0 otherwise. Using (2.6), the pdf of $Y = g(X) = \log(X)$ is*

$$f_Y(y) = \frac{1}{|1/e^y|} \exp(-\exp(y)) = \exp(y - \exp(y)).$$

(Note that $g(x) = \log(x)$, $g^{-1}(y) = \exp(y)$, $g'(x) = 1/x$.) The following R code and graph contrasts the pdfs of X and $Y = \log(X)$.

```
x = seq(-5, 5, length = 100)
Y1 = exp(-x)
Y1[x < 0] = 0
Y2 = exp(x - exp(x))
D = stack(list(`$f_X(r)$` = Y1, `$f_{\\log(X)}(r)$` = Y2))
D$x = x
qplot(x, values, xlab = "$r$", ylab = "", geom = "line",
    color = ind, lty = ind, data = D, size = I(1.5))
```

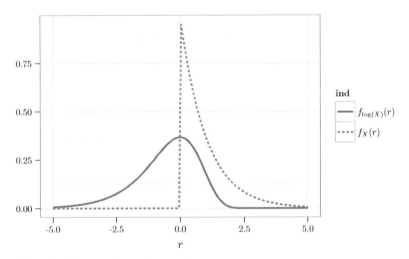

The calculation above shows that

$$f_{g(X)}(r) \neq g \circ f_X(r) = g(f_X(r)).$$

The right hand side above is often mistakenly taken to be the density of $g(X)$, but is in most cases not even a density function (for example in this example $g(f_X(r))$ would assume negative values.)

2.3 Expectation and Variance

Given a random variable, we often compute the expectation and variance, two important summary statistics. The expectation describes the average value and the variance describes the spread (amount of variability) around the expectation.

Definition 2.3.1. The expectation of an RV X is the real number

$$E(X) = \begin{cases} \sum_x x\, p_X(x) & X \text{ is a discrete RV} \\ \int_{-\infty}^{\infty} x\, f_X(x)\, dx & X \text{ is a continuous RV} \end{cases},$$

provided the sum or integral exists.

Example 2.3.1. *The expectation of an RV expressing the result of throwing a fair die is*

$$E(X) = 1 p_X(1) + \cdots 6 p_X(6) = \frac{1}{6}(1 + \cdots + 6) = 21/6 = 3.5.$$

Intuitively, the expectation gives the longterm average of many throws of the dice

$$average = \frac{1 \cdot (\#\text{-of times we get } 1) + \cdots + 6 \cdot (\#\text{-of-times-we-get } 6)}{k}$$

$$= 1 \cdot \text{frequency-of-getting } 1 + \cdots + 6 \cdot \text{frequency-of-getting } 6 \to E(X),$$

as $n \to \infty$, *assuming that the frequency of getting* i *in the long term converges to* $p(X = i)$. *We elaborate on this topic in Section 8.6.*

Proposition 2.3.1.

$$E(g(X)) = \begin{cases} \sum_y y p_{g(X)}(y) & g(X) \text{ is discrete} \\ \int y f_{g(X)}(y) dy & g(X) \text{ is continuous} \end{cases} \tag{2.7}$$

$$= \begin{cases} \sum_x g(x) p_X(x) & X \text{ is a discrete RV} \\ \int_{-\infty}^{\infty} g(x) f_X(x) dx & X \text{ is a continuous RV} \end{cases}. \tag{2.8}$$

Proof. The first equality follows from definition and the second equality follows since the value $g(x)$ occurs when x occurs, which happens with probability $p_X(x)$ in the discrete case or density $f_X(x)$ in the continuous case. ∎

Proposition 2.3.2. *For any constants* $a, b \in \mathbb{R}$, *the expectation of* $Y = aX + b$ *is*

$$E(aX + b) = a\, E(X) + b.$$

Proof. Using Equation (2.8)

$$E(aX + b) = \sum_x (ax + b) p_X(x) = a \sum_x x p_X(x) + b \sum_x p_X(x) = a\, E(x) + b \cdot 1$$

for a discrete RV. In the case of a continuous RV, we replace the summation with an integral. ∎

The proposition above specifically states that

$$\mathsf{E}(X + b) = \mathsf{E}(X) + b, \text{ and}$$
$$\mathsf{E}(aX) = a\,\mathsf{E}(X).$$

The variance measures the amount of variability of the RV X around $\mathsf{E}(X)$.

Definition 2.3.2. The variance of an RV X is the expectation of the RV $Y = (X - \mathsf{E}(X))^2$:

$$\mathsf{Var}(X) = \mathsf{E}((X - \mathsf{E}(X))^2).$$

The standard deviation of an RV X is $\mathsf{std}(X) = \sqrt{\mathsf{Var}(X)}$.

Since variance measures squared deviation about the expectation, an RV with a narrow pmf or pdf should exhibit low variance and an RV with a wide pmf or pdf should exhibit large variance.

Example 2.3.2. *Consider the classical probability model on (a, b) and its associated random variable X. We have $\mathsf{P}(X \in A) = |A|/(b - a)$ and*

$$F_X(x) = \begin{cases} (x - a)/(b - a) & x \in (a, b) \\ 0 & x \le a \\ 1 & x \ge b \end{cases}$$

$$f_X(x) = \begin{cases} 1/(b - a) & x \in (a, b) \\ 0 & otherwise \end{cases}$$

The expectation of X

$$\mathsf{E}(X) = \int_a^b \frac{x}{(b - a)}\,dx = \frac{1}{b - a}\frac{1}{2}x^2\Big|_{x=a}^{x=b} = \frac{b^2 - a^2}{2(b - a)} = \frac{(b - a)(b + a)}{2(b - a)} = \frac{a + b}{2}$$

is the middle point of the interval (a, b). The variance

$$\mathsf{Var}(X) = \mathsf{E}((X - \mathsf{E}\,X)^2) = \int_a^b \frac{1}{(b - a)}(x - (a + b)/2)^2\,dx$$

$$= \frac{1}{b - a}\int_a^b (x^2 - x(a + b) + (a + b)^2/4)dx$$

$$= \frac{1}{b - a}\left(\frac{b^3 - a^3}{3} - (a + b)\frac{b^2 - a^2}{2} + \frac{(a + b)^2}{4}(b - a)\right)$$

$$= \frac{b^3 - a^3}{3(b - a)} - (a + b)\frac{b + a}{2} + \frac{(a + b)^2}{4}$$

$$= \frac{(a - b)(a^2 + ab + b^2)}{3(b - a)} - \frac{(a + b)^2}{4}$$

$$= \frac{a^2 + b^2 - 2ab}{12}$$

$$= \frac{(a - b)^2}{12}$$

grows with the interval length $|b - a|$, *confirming our intuition that narrow pdfs and wide pdfs exhibit low and wide variance, respectively.*

Proposition 2.3.3.

$$\text{Var}(X) = E(X^2) - (E(X))^2.$$

Proof. Using the linearity of expectations (Proposition 2.3.2),

$$\text{Var}(X) = E((X - E(X))^2) = E(X^2 - 2X\,E(X) + (E(X))^2)$$
$$= E(X^2) - 2\,E(X)\,E(X) + (E(X))^2$$
$$= E(X^2) - (E(X))^2.$$

∎

Proposition 2.3.4. *For any constants* $a, b \in \mathbb{R}$,

$$\text{Var}(aX + b) = a^2\,\text{Var}(X). \tag{2.9}$$

Proof. We decompose the proof to the following two cases:

$$\text{Var}(aX) = E(a^2 X^2) - (E\,aX)^2 = a^2\,E(X^2) - (a\,E(X))^2 = a^2(E(X^2) - (E\,X)^2)$$
$$= a^2\,\text{Var}(X)$$
$$\text{Var}(X + b) = E((X + b - E(X + b))^2) = E((X + b - E(X) - b))^2) = \text{Var}(X).$$

It then follows that $\text{Var}(aX + b) = \text{Var}(aX) = a^2\,\text{Var}(X)$. ∎

We can consider a constant $b \in \mathbb{R}$ to be a deterministic RV such that $b : \Omega \to \mathbb{R}$ and $b(\omega) = b$. The outcome of such a random variable is pre-determined, or "deterministic". The corresponding expectation and variance are

$$E(b) = \sum_x b p_X(x) = b$$
$$\text{Var}(b) = E(b - E(b)))^2 = E(0) = 0.$$

2.4   Moments and the Moment Generating Function

Definition 2.4.1. The k-moment of an RV X is $E(X^k)$.

Definition 2.4.2. The moment generating function (mgf) of the RV X is the function

$$m : \mathbb{R} \to \mathbb{R}, \qquad m(t) = E(\exp(tX)).$$

The name mgf stems from the fact that its derivatives, evaluated at 0 produce the moments.

Proposition 2.4.1.
$$m^{(k)}(0) = \mathsf{E}(X^k)$$

Proof. Using the Taylor series expansion (Proposition D.2.1) of e^z around 0, we have

$$m'(0) = \frac{d}{dt} \mathsf{E}\left(1 + tX + \frac{t^2 X^2}{2!} + \cdots\right)\Big|_{t=0} = \mathsf{E}\left(\frac{d}{dt}\left(1 + tX + \frac{t^2 X^2}{2!} + \cdots\right)\right)\Big|_{t=0}$$

$$= \mathsf{E}(X).$$

Similarly, the second derivative produces the second moment

$$m''(0) = \frac{d^2}{dt^2} \mathsf{E}\left(1 + tX + \frac{t^2 X^2}{2!} + \cdots\right)\Big|_{t=0}$$

$$= \mathsf{E}\left(\frac{d^2}{dt^2}\left(1 + tX + \frac{t^2 X^2}{2!} + \cdots\right)\right)\Big|_{t=0}$$

$$= \mathsf{E}(X^2),$$

and so on for the higher-order derivatives. ∎

In addition to producing the moments of X, the mgf is useful in identifying the distribution of X.

Proposition 2.4.2. *Let X_1, X_2 be two RVs with mgfs m_1, m_2. If $m_1(t) = m_2(t)$ for all $t \in (-\epsilon, +\epsilon)$, for some $\epsilon > 0$, then the two RVs have identical cdfs (and therefore identical pdfs or pmfs).*

Sketch of Proof*. According to a result from complex analysis, an analytic function (a function determined locally by its Taylor series) has a unique extension, called the analytics continuation, from $(-\epsilon, \epsilon)$ to the right half of the complex plane. Since the moment generating function $m(t)$ agrees with the characteristic function $\phi(-it)$ for $t \in \mathbb{R}$, and the characteristic function uniquely determines the distribution (see Chapter 8 for a description of the characteristic function and a proof that it uniquely determines the distribution), the moment generating function uniquely determines the distribution as well. ∎

The moments of an RV X, provided they exist, completely characterize the Taylor series of the mgf and, therefore, characterize the mgf. Together with the above proposition, the moments $\mathsf{E}(X^k), k \in \mathbb{N}$ (assuming the corresponding integrals or sums converge) completely characterize the distribution of X and the corresponding cdf, pdf, and pmf.

2.5   Random Variables and Measurable Functions*

As described in Section 1.7, a rigorous definition of P requires it to be defined on a σ-algebra $\mathcal{F}$ of events in Ω. Similarly, a rigorous definition of a random variable is a measurable function from the measurable space $(\Omega, \mathcal{F})$ to the measurable space $(\mathbb{R}, \mathcal{B}(\mathbb{R}))$ (see Definitions E.2.1 and E.5.1). This definition ensures that we can compute $P(X \in A)$ for all Borel sets $A \in \mathcal{B}(\mathbb{R})$.

Random variables may be discrete, continuous, or neither. The description in this chapter is simplified in that we consider only discrete or continuous random variables. This simplification enables us to develop the theory of random variables almost without reference to measure theory.

As mentioned in the chapter, a random variable $X : \Omega \to \mathbb{R}$ defines a new probability function P' on $\mathbb{R}$ such that $P'(A) = P(X \in A)$. The more precise statement is that X defines a measure space $(\mathbb{R}, \mathcal{B}(\mathbb{R}), P')$ where $P' = P\, X^{-1}$ is the transformed measure (see Definition F.3.7).

We have the following correspondence between RV notations and Lebesgue integrals.

RV Notation	Integral Notation
$P(A)$	$\int_A d\, P$
$E(X)$	$\int x\, P(dx)$
$P(X \in A)$	$\int_A d\, P\, X^{-1}$
$E(g(X))$	$\int g\, d\, P$

For example, the Lebesgue integral $\int_A P(dx)$ becomes $\int_A f_X(x)\, dx$ if X is a continuous RV and $\sum_{x \in A} p_X(x)$ if X is discrete. Thus, we can leverage a single notation to address discrete RVs, continuous RVs, and RVs that are neither discrete nor continuous, a significant convenience. Doing so, however, requires the significantly more complex mathematics of measure theory and Lebesgue integration.

2.6   Notes

Our exposition focuses on random variables that are either discrete or continuous. Random variables that are neither discrete nor continuous (for example the sum of a discrete RV and a continuous RV) require more careful treatment. They do not have a pdf function and they do not have a pmf function (all RVs have a cdf function though). A unified exposition is possible through the use of measure theory and Lebesgue integration (see Chapters E and F).

Note that we derive the expectation from the probability function. The indicator function $I_A(x)$ (equals 1 if $x \in A$ and 0 otherwise) may be used to reverse this since $E(I_A) = 1 \cdot P(X \in A) + 0 \cdot P(X \notin A) = P(X \in A)$. Specifically, given an expectation operator E, we can define $P(X \in A)$ to be equal to $E(I_A)$.

More information on random variables is available in nearly any probability textbook. An elementary exposition that avoids measure theory is available in most undergraduate probability textbooks, for example [48, 14]. Examples of rigorous measure-theoretic textbooks are [17, 10, 5, 1, 33, 25].

2.7 Exercises

1. Prove that P' from (2.1) is a probability function.

2. Give a concrete example of a random variable that is neither discrete nor continuous.

3. Prove Corollary 2.1.1.

4. Consider a continuous RV X whose density is $f_X(x) = x$ for $x \in (0, 1)$ and 0 otherwise. Compute its expectation and variance.

5. Consider an RV $Y = X^2$ where X is defined in (3). Compute the pdf of Y and its expectation using both formulas in Proposition 2.3.1.

6. We define the support size of a discrete RV X as the number of values it can achieve with nonzero probability (that number may be infinity). What is the relationship between the support sizes of a discrete RV X and of $g(X)$?

Chapter 3

Important Random Variables

In this chapter we describe several discrete and continuous random variables. These random variables are important for developing probability theory, and for applying it to solve real world problems.

We denote the distributions corresponding to the different random variables using abbreviations such as Ber or Bin. In cases where the distributions are parameterized, we attach the parameter or parameters to the abbreviation, for example $\text{Bin}(n, \theta)$. We use the notation $\sim$ to denote "distributed according to", for example $X \sim \text{Ber}(\theta)$ implies that the RV X follows the $\text{Ber}(\theta)$ distribution.

3.1 The Bernoulli Trial Random Variable

The Bernoulli trial RV, $X \sim \text{Ber}(\theta)$, where $\theta \in [0, 1]$, is characterized by the following pmf:

$$p_X(x) = \begin{cases} \theta & x = 1 \\ 1 - \theta & x = 0 \\ 0 & \text{otherwise} \end{cases}, \qquad \theta \in [0, 1].$$

The Bernoulli trial RV may be used to characterize the probability that an experiment (or trial) may either succeed, $X = 1$, or fail, $X = 0$, with probabilities θ, $1 - \theta$ respectively. A popular example of such an experiment is flipping a potentially biased coin, with success corresponding to heads and failure corresponding to tails.

The expectation and variance of $X \sim \text{Ber}(\theta)$ are:

$$\mathsf{E}(X) = 1\theta + 0(1 - \theta) = \theta$$
$$\mathsf{Var}(X) = \mathsf{E}(X^2) - E^2(X) = 1^2\theta + 0^2(1 - \theta) - \theta^2 = \theta(1 - \theta).$$

The R code below graphs the mass functions corresponding to three different θ parameters.

```
x = c(0, 1)
D = stack(list(`$\\theta=0.3$` = dbinom(x, 1,
        0.3), `$\\theta=0.5$` = dbinom(x, 1, 0.5),
      `$\\theta=0.9$` = dbinom(x, 1, 0.9)))
names(D) = c("mass", "theta")
D$x = x
qplot(x, mass, data = D, main = "Bernoulli pmf functions",
      geom = "point", stat = "identity", facets = . ~
          theta, xlab = "$x$", ylab = "$p_X(x)$") +
      geom_linerange(aes(x = x, ymin = 0, ymax = mass))
```

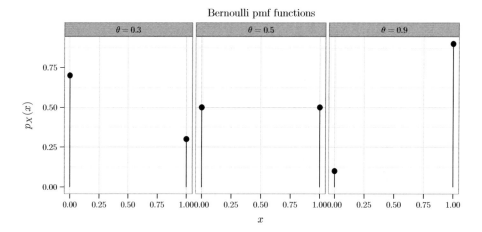

3.2   The Binomial Random Variable

The Binomial RV, $X \sim \text{Bin}(n, \theta)$, where $\theta \in [0, 1]$, $n \in \mathbb{N} \cup \{0\}$, counts the number of successes in n independent Bernoulli experiments with parameter θ, regardless of the ordering of the results of the experiments.

If ordering does matter, the probability of particular sequence of n experiments with x successes and $n - x$ failures is $\theta^x (1 - \theta)^{n-x}$. There are n-choose-x such sequences, implying that

$$p_X(x) = \begin{cases} \dbinom{n}{x} \theta^x (1 - \theta)^{n-x} & x = 0, 1, \ldots, n \\ 0 & \text{otherwise} \end{cases}.$$

The fact that $P(X \in \Omega) = \sum_{x=0}^{n} p_X(x) = 1$ may be ascertained using the binomial theorem (Proposition 1.6.6):

$$1 = 1^n = (\theta + (1 - \theta))^n = \sum_{k=0}^{n} \binom{n}{k} \theta^k (1 - \theta)^{n-k}.$$

The binomial and Bernoulli RVs are closely related. A $\mathsf{Ber}(\theta)$ RV has the same distribution as a $\mathsf{Bin}(1, \theta)$ RV. On the other hand, a $\mathsf{Bin}(n, \theta)$ RV is a sum of n independent Bernoulli RVs $\mathsf{Ber}(\theta)$. In the next chapter we show (Proposition 4.6.3 and Corollary 4.6.2) that if $Z^{(1)}, \ldots, Z^{(n)}$ are RVs corresponding to independent experiments, then $\mathsf{E}(\sum Z^{(i)}) = \sum \mathsf{E} Z^{(i)}$ and $\mathsf{Var}(\sum Z^{(i)}) = \sum \mathsf{Var}(Z^{(i)})$. Using this result and the expectation and variance formulas of the Bernoulli RV, we derive that for $X \sim \mathsf{Bin}(n, \theta)$

$$\mathsf{E}(X) = n\theta$$
$$\mathsf{Var}(X) = n\theta(1 - \theta).$$

The formulas above indicate the following trends: as θ and n increase the expectation increase, and as n increases the variance increases. For a fixed n, the highest variance

$$\max_{\theta \in [0,1]} \mathsf{Var}(X) = \max_{\theta \in [0,1]} n\theta(1 - \theta) = n \max_{\theta \in [0,1]} (\theta - \theta^2)$$

is obtained when $\theta = 0.5$ (this can be verified by solving $0 = (\theta - \theta^2)' = 1 - 2\theta$ for θ). This is in agreement with our intuition that an unbiased coin flip has more uncertainty than a biased coin flip. The R code below graphs the pmf of nine binomial distributions with different n and θ parameters. The trends in the figure are in agreement with the observations above.

```
x = 0:19  # range of values to display in plot
# dbinom(x,n,theta) computes the pmf of
# Bin(n,theta) at x
y1 = dbinom(x, 5, 0.3)
y2 = dbinom(x, 5, 0.5)
y3 = dbinom(x, 5, 0.7)
y4 = dbinom(x, 20, 0.3)
y5 = dbinom(x, 20, 0.5)
y6 = dbinom(x, 20, 0.7)
D = data.frame(mass = c(y1, y2, y3, y4, y5, y6),
    x = x, n = 0, theta = 0)
D$n[1:60] = "$n=5$"
D$n[61:120] = "$n=20$"
D$theta[c(1:20, 61:80)] = "$\\theta=0.3$"
D$theta[c(21:40, 81:100)] = "$\\theta=0.5$"
D$theta[c(41:60, 101:120)] = "$\\theta=0.7$"
qplot(x, mass, data = D, main = "Binomial pmf functions",
    geom = "point", stat = "identity", facets = theta ~
        n, xlab = "$x$", ylab = "$p_X(x)$") +
    geom_linerange(aes(x = x, ymin = 0, ymax = mass))
```

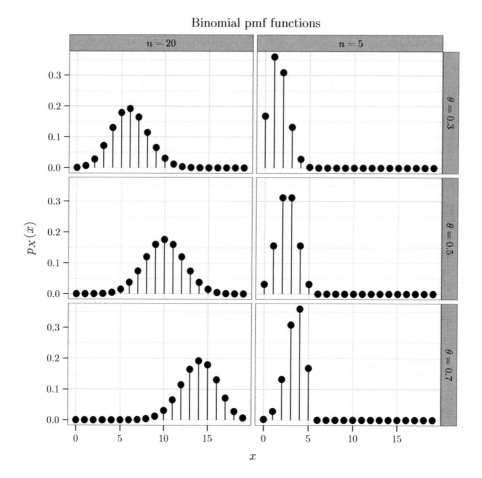

3.3   The Geometric Random Variable

The geometric RV, $X \sim \text{Geom}(\theta)$, where $\theta \in [0, 1]$, is the number of failures we encounter in a sequence of independent Bernoulli experiments with parameter θ before encountering success. The pmf of the geometric RV $X \sim \text{Geom}(\theta)$ is

$$p_X(x) = \begin{cases} \theta(1 - \theta)^x & x \in \mathbb{N} \cup \{0\} \\ 0 & \text{otherwise} \end{cases}.$$

Using the power series formula (D.5) we can ascertain that $P(X \in \Omega) = 1$:

$$\sum_{n=0}^{\infty} p_X(n) = \theta(1 + (1 - \theta) + (1 - \theta)^2 + \cdots) = \theta \frac{1}{1 - (1 - \theta)} = 1.$$

Using (D.6)-(D.7) we derive

$$E(X) = \theta \sum_{n=0}^{\infty} n(1-\theta)^n = \frac{1}{\theta^2} - \frac{1}{\theta} = \theta\frac{1-\theta}{\theta^2} = \frac{1-\theta}{\theta},$$

$$\text{Var}(X) = E(X^2) - (E(X))^2 = (1-\theta)/\theta^2 = \theta \sum_{n=0}^{\infty} n^2(1-\theta)^n - \frac{(1-\theta)^2}{\theta^2}$$

$$= \theta\frac{2(1-\theta)}{\theta^3} - \theta\frac{1-\theta}{\theta^2} - \frac{(1-\theta)^2}{\theta^2} = \frac{2 - 2\theta - \theta + \theta^2 - 1 + 2\theta - \theta^2}{\theta^2}$$

$$= \frac{1-\theta}{\theta^2}.$$

The R code below graphs the pmf of a geometric RV. In accordance with our intuition, it shows that as θ increases, X is less likely to get high values.

```
x = 0:9
D = stack(list(`$\\theta=0.3$` = dgeom(x, 0.3),
    `$\\theta=0.5$` = dgeom(x, 0.5), `$\\theta=0.7$` = dgeom(x,
    0.7)))
names(D) = c("mass", "theta")
D$x = x
qplot(x, mass, data = D, main = "Geometric pmf",
    geom = "point", stat = "identity", facets = theta ~
    ., xlab = "$x$", ylab = "$p_X(x)$") +
    geom_linerange(aes(x = x, ymin = 0, ymax = mass))
```

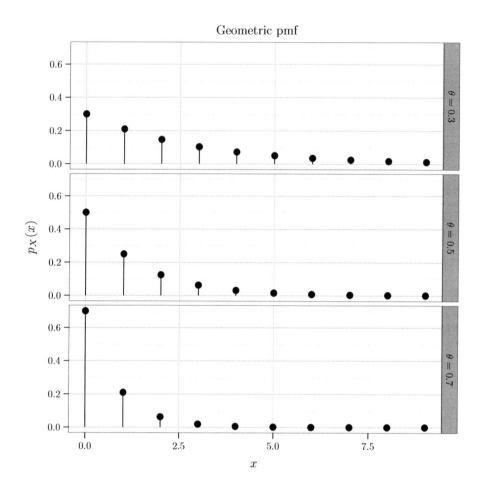

3.4 The Hypergeometric Random Variable

The hypergeometric RV, $X \sim \text{Hyp}(m, n, k)$, where $m, n, k \in \mathbb{N} \cup \{0\}$, describes the number of white balls drawn from an urn with m white balls and n green balls after k draws *without* replacement. It is similar to a binomial RV in that it counts the number of successes in a sequence of experiments, but in the case of the hypergeometric distribution, the probability of success in each experiment or draw changes depending on the previous draws, rather than being constant (the number of white and green balls in the urn changes as balls are drawn from the urn).

The pmf of $X \sim \mathrm{Hyp}(m, n, k)$ is

$$p_X(x) = \begin{cases} \dfrac{\dbinom{m}{x} \dbinom{n}{k-x}}{\dbinom{m+n}{k}} & x \in \{0, 1, \ldots, m\} \\ \\ 0 & \text{otherwise} \end{cases} .$$

The above formula can be derived by noting that the numerator counts the number of possible draws having x white and $k - x$ green balls, and the denominator counts the total number of draws. Assuming the classical model (Chapter 1), the ratio above corresponds to the required probability.

The R code below graphs the pmfs of three hypergeometric RVs with multiple parameter values.

```
m = 15
n = 10
x = 0:18
D = stack(list(`$k=15$` = dhyper(x, m, n, 15),
    `$k=23$` = dhyper(x, m, n, 23), `$k=25$` = dhyper(x,
        m, n, 25)))
names(D) = c("mass", "k")
D$x = x
qplot(x, mass, data = D, geom = "point", stat = "identity",
    facets = k ~ ., xlab = "$x$", ylab = "$p_X(x)$") +
    geom_linerange(aes(x = x, ymin = 0, ymax = mass)) +
    opts(title = "Hypergeometric pmf functions ($m=15, n=10$)")
```

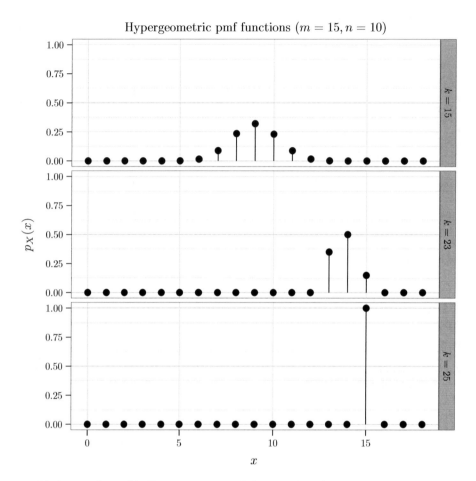

If the number of balls $m + n$ is much larger than k, the hypergeometric distribution converges to the binomial distribution (sampling without replacement is similar to sampling with replacement if the number of drawn balls is much smaller than the number of balls in the urn). This is illustrated by the first row in the figure above, which displays behavior similar to the corresponding binomial. The second and third rows show a strong deviation from the binomial model.

3.5   The Negative Binomial Random Variable

The negative binomial RV, $X \sim NB(r, \theta)$, where $r \in \mathbb{N} \cup \{0\}, \theta \in [0, 1]$, is the number of successes in an infinite sequence of independent Bernoulli experiments before encountering r failures.

The mass function of $X \sim NB(r, \theta)$ is

$$p_X(x) = \begin{cases} \binom{x + r - 1}{x} (1 - \theta)^r \theta^x & x \in \mathbb{N} \cup \{0\} \\ 0 & \text{otherwise} \end{cases}. \tag{3.1}$$

The above pmf can be derived by noting that there are $\binom{x + r - 1}{x}$ arrangements of r failures and x successes with the last failure occurring at the end, and that each of the arrangements occur with probability $(1 - \theta)^r \theta^x$.

The R code below graphs the pmfs of multiple negative binomial RVs with different parameter values.

```
theta = 0.5
x = 0:30
D = stack(list(`$r=1$` = dnbinom(x, 1, theta),
        `$r=4$` = dnbinom(x, 4, theta), `$r=8$` = dnbinom(x,
        8, theta)))
names(D) = c("mass", "r")
D$x = x
qplot(x, mass, data = D, geom = "point", stat = "identity",
        facets = r ~ ., xlab = "$x$", ylab = "$p_X(x)$") +
        geom_linerange(aes(x = x, ymin = 0, ymax = mass)) +
        opts(title = "Negative binomial pmf functions ($\\theta=0.5$)")
```

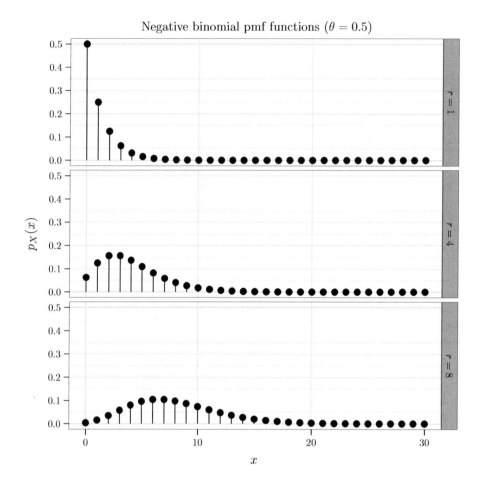

3.6   The Poisson Random Variable

The Poisson RV, $X \sim \text{Pois}(\lambda)$, where $\lambda > 0$, has the following pmf

$$p_X(x) = \begin{cases} \frac{\lambda^x e^{-\lambda}}{x!} & x \in \mathbb{N} \cup \{0\} \\ 0 & \text{otherwise} \end{cases}. \tag{3.2}$$

Using (D.4) we ascertain that $P(X \in \Omega) = 1$:

$$\sum_{x=0}^{\infty} \frac{\lambda^x e^{-\lambda}}{x!} = e^{-\lambda} \sum_{x=0}^{\infty} \frac{\lambda^x}{x!} = e^{-\lambda} e^{\lambda} = 1.$$

We denote the pmf of $X \sim \text{Bin}(n, \theta)$ by $\text{Bin}(x\,;n,\theta)$ and consider the case where $n \to \infty$, $\theta \to 0$ and $n\theta = \lambda \in (0, \infty)$. Using (D.5), we have

$$\log \text{Bin}(x = 0\,; n, \theta) = \log(1 - \theta)^n = \log(1 - \lambda/n)^n = n\log(1 - \lambda/n)$$
$$= -\lambda - \frac{\lambda^2}{2n} - \cdots,$$

implying that

$$\text{Bin}(x = 0\,; n, \theta) \approx \exp(-\lambda).$$

Similarly,

$$\frac{\text{Bin}(x\,; n, \theta)}{\text{Bin}(x - 1\,; n, \theta)} = \frac{n - x + 1}{n}\,\frac{\theta}{1 - \theta} = \frac{\lambda - (x - 1)\theta}{x(1 - \theta)} \approx \frac{\lambda}{x},$$

implying that

$$\text{Bin}(x = 1\,; n, \theta) \approx \frac{\lambda}{x}\text{Bin}(x = 0\,; n, \theta) \approx \frac{\lambda}{1}\exp(-\lambda)$$
$$\text{Bin}(x = 2\,; n, \theta) \approx \frac{\lambda}{x}\text{Bin}(x = 1\,; n, \theta) \approx \frac{\lambda^2}{2 \cdot 1}\exp(-\lambda).$$

Using induction, we observe that the pmf of $\text{Pois}(\lambda)$ approximates the pmf of $\text{Bin}(n, \theta)$ when $n \to \infty$, $\theta \to 0$ and $n\theta = \lambda \in (0, \infty)$. In other words, the $\text{Pois}(\lambda)$ RV is the number of "successes" occurring in a very long sequence of independent Bernoulli experiments $(n \to \infty)$, each having very small success probability $\theta \to 0$, and further assuming that $0 < n\theta < \infty$.

The approximation above explains why many naturally-occurring quantities follow a Poisson distribution. Examples of such quantities are (see [16] for more details):

- the number of particles reaching a counter during the disintegration of radioactive material,

- the number of bombs landing in different regions of London during the Second World War,

- the number of bacteria found in different regions of a Petri dish,

- the number of centenarians (people over the age of 100) dying each year,

- the number of cars arriving at an intersection during a particular time interval, and

- the number of phone calls arriving at a switchboard during a particular time interval.

Using (D.4), we derive the mgf of the Poisson RV

$$m(t) = \sum_{k=0}^{\infty} e^{tk}\frac{\lambda^k e^{-\lambda}}{k!} = e^{-\lambda}\sum_{k=0}^{\infty}\frac{(e^t\lambda)^k}{k!} = e^{\lambda e^t - \lambda}, \tag{3.3}$$

which leads to the following expectation and variance expressions:

$$E(X) = m'(0) = e^{\lambda e^t - \lambda} \lambda e^t \Big|_{t=0} = \lambda$$

$$Var(X) = E(X^2) - (E(X))^2 = m''(0) - \lambda^2 = \left(e^{\lambda e^t - \lambda} \lambda^2 (e^t)^2 + e^{\lambda e^t - \lambda} \lambda e^t \right) \Big|_{t=0} - \lambda^2$$

$$= \lambda.$$

Proposition 4.8.1 states that the mgf of a sum of independent RVs is a product of the corresponding mgfs. Since the product of multiple Poisson mgfs with parameters λ_i is also a Poisson mgf with parameter $\sum_i \lambda_i$, we have the following result: if $Z_i \sim \text{Pois}(\lambda_i)$, $i = 1, \ldots, n$, are independent Poisson RVs then

$$\sum_{i=1}^{n} Z_i \sim \text{Pois}\left(\sum_{i=1}^{n} \lambda_i \right).$$

```
x = 0:19
D = stack(list(`$\\lambda=1$` = dpois(x, 1),
       `$\\lambda=5$` = dpois(x, 5), `$\\lambda=10$` = dpois(x,
           10)))
names(D) = c("mass", "lambda")
D$x = x
qplot(x, mass, data = D, geom = "point", stat = "identity",
       xlab = "$x$", ylab = "$p_X(x)$", facets = lambda ~
           ., main = "Poisson pmf functions") + geom_linerange(aes(x = x,
       ymin = 0, ymax = mass))
```

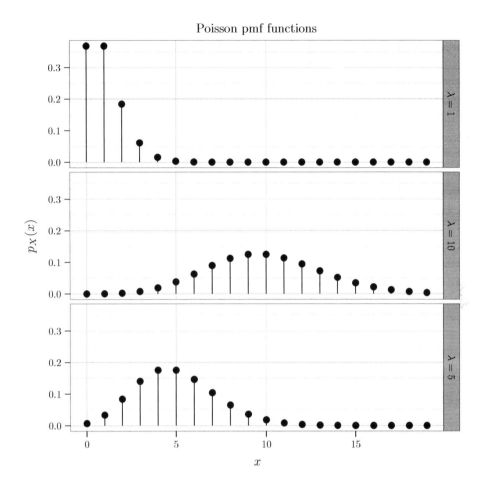

Poisson pmf functions

3.7 The Uniform Random Variable

The uniform RV, $X \sim U(a, b)$, where $a < b$, is the classical model on the interval $[a, b]$ (see Section 1.4). By definition, the pdf of a $U(a, b)$ RV is constant for $x \in [a, b]$ and 0 for $x \notin [a, b]$. The only such constant function that integrates to one is $f(x) = 1/(b - a)$ for $x \in [a, b]$ and 0 otherwise, implying that

$$f_X(x) = \begin{cases} 1/(b - a) & x \in [a, b] \\ 0 & \text{otherwise} \end{cases}.$$

As computed in Example 2.3.2,

$$\mathsf{E}(X) = (a + b)/2$$
$$\mathsf{Var}(X) = (b - a)^2/12$$

implying that the expectation is the mid-point of the interval and the variance increases with the square of the interval width. The R code below graphs the pdf and the cdf of $U(a, b)$ for two different parameter values.

```
x = seq(-1, 2, length = 100)
y1 = dunif(x, 0, 1/2)
y2 = dunif(x, 0, 1)
y3 = punif(x, 0, 1/2)
y4 = punif(x, 0, 1)
D = data.frame(probability = c(y1, y2, y3, y4))
D$parameter[1:100] = "$U(0,1/2)$"
D$parameter[101:200] = "$U(0,1)$"
D$parameter[201:300] = "$U(0,1/2)$"
D$parameter[301:400] = "$U(0,1)$"
D$type[1:200] = "$f_X(x)$"
D$type[201:400] = "$F_X(x)$"
D$x = x
qplot(x, probability, main = "Uniform pdf and cdf",
    facets = parameter ~ type, data = D, geom = "area",
    xlab = "$x$", ylab = "")
```

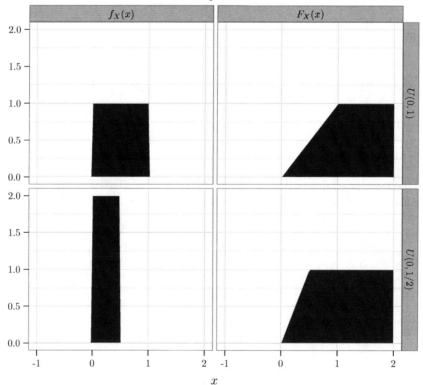

3.8 The Exponential Random Variable

The exponential RV, $X \sim \text{Exp}(\lambda)$, where $\lambda > 0$, has the pdf

$$f_X(x) = \begin{cases} \lambda e^{-\lambda x} & x > 0 \\ 0 & \text{otherwise} \end{cases}.$$

Since the pdf decreases exponentially as x grows (for positive x), it is more probable that X will receive a small positive value than a large positive value. The cdf is

$$F_X(x) = P(X \le x) = \begin{cases} \int_0^x \lambda e^{-\lambda x} = -e^{-\lambda x}\Big|_0^x = 1 - e^{-\lambda x} & x > 0 \\ 0 & \text{otherwise} \end{cases}.$$

The exponential RV is the only continuous RV X with the memoryless property: the probability that X is larger than $s + t$ is the same as the probability that X is larger than s in one experiment and an independent copy of X is larger than t in an independent experiment

$$P(X > s + t) = P(X > s)\, P(X > t). \tag{3.4}$$

(The equation above holds for the exponential RV since $P(X > t) = 1 - F_X(t) = e^{-\lambda x}$.) The term "memoryless" is motivated by noting that $P(X > s + t) = P(X > s)\, P(X > t)$ implies the following lack of memory:

$$P(X > t + h | X > t) = \frac{P(\{X > t + h\} \cap \{X > t\})}{P(X > t)} = \frac{P(\{X > t + h\})}{P(X > t)}$$

$$= \frac{e^{-\lambda(t+h)}}{e^{-\lambda t}} = e^{-\lambda h} = P(X > h).$$

A proof that no other continuous distribution has this property is available for example in [16]. The memoryless property motivates the use of the exponential RV to model times between successive arrivals of customers at a store, cars at an intersection, or phone calls at a switchboard.

The mgf of an exponential RV is

$$m(t) = E(\exp(tX)) = \lambda \int_0^\infty e^{-\lambda x} e^{tx}\, dx = \lambda \int_0^\infty e^{(t-\lambda)x}\, dx = \frac{\lambda}{t - \lambda} e^{(t-\lambda)x}\Big|_0^\infty$$

$$= \frac{\lambda}{\lambda - t}$$

for $t < \lambda$, implying that

$$E(X) = m'(0) = \frac{\lambda}{(\lambda - t)^2}\Big|_{t=0} = \lambda^{-1},$$

$$\text{Var}(X) = m''(0) = \frac{\lambda}{(\lambda - t)^3}\Big|_{t=0} = \lambda^{-2}.$$

The R code below graphs the pdf and cdf of exponential RVs with different parameter values.

```
x = seq(0, 3, length = 100)
y1 = dexp(x, 1)
y2 = dexp(x, 3)
y3 = pexp(x, 1)
y4 = pexp(x, 3)
D = data.frame(probability = c(y1, y2, y3, y4),
    x = x)
D$parameter[1:100] = "$\\lambda=1$"
D$parameter[101:200] = "$\\lambda=2$"
D$parameter[201:300] = "$\\lambda=1$"
D$parameter[301:400] = "$\\lambda=2$"
D$type[1:200] = "$f_X(x)$"
D$type[201:400] = "$F_X(x)$"
qplot(x, probability, main = "Exponential pdf and cdf",
    geom = "area", facets = parameter ~ type,
    data = D, xlab = "$x$", ylab = "")
```

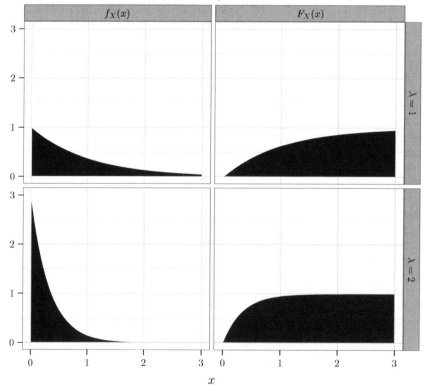

3.9 The Gaussian Random Variable

The Gaussian or normal RV, $X \sim N(\mu, \sigma^2)$, where $\mu \in \mathbb{R}, \sigma^2 > 0$, has the pdf

$$f_X(x) = \frac{1}{\sqrt{2\pi\sigma^2}} \exp\left(-\frac{(x-\mu)^2}{2\sigma^2}\right).$$

If $\mu = 0, \sigma^2 = 1$, we refer to this RV as the standard normal or standard Gaussian RV. We sometimes attach the parameters μ, σ as subscripts, for example $f_{0,1}$ and $F_{0,1}$ denote the pdf and cdf of $X_{0,1}$, the standard normal RV. A common notation for $F_{0,1}(x)$ is $\Phi(x)$.

To verify that the pdf integrates to 1, we use the change of variable $(x - \mu)/\sigma \mapsto y$ to get

$$\int_{-\infty}^{\infty} \frac{1}{\sqrt{2\pi\sigma^2}} \exp\left(-\frac{(x-\mu)^2}{2\sigma^2}\right) dx = \int_{-\infty}^{\infty} \frac{1}{\sqrt{2\pi}} \exp\left(-y^2/2\right) dy \qquad (3.5)$$

$$= \frac{1}{\sqrt{2\pi}} \sqrt{2\pi} = 1 \qquad (3.6)$$

where the first equality follows from Example F.2.3 and the second from Equation (F.25).

Example F.2.3 (substitute $a = -\infty$ and $b = x$) shows the following relations between the cdfs and pdfs of the Gaussian distribution:

$$F_{\mu,\sigma^2}(x) = F_{0,1}\left(\frac{x-\mu}{\sigma}\right) = \Phi\left(\frac{x-\mu}{\sigma}\right), \qquad (3.7)$$

$$f_{\mu,\sigma^2}(x) = \frac{d}{dx} F_{\mu,\sigma^2}(x) = \frac{d}{dx} F_{0,1}\left(\frac{x-\mu}{\sigma}\right) = \frac{1}{\sigma} f_{0,1}\left(\frac{x-\mu}{\sigma}\right). \qquad (3.8)$$

(The last equality follows from the chain rule (Proposition D.1.3).)

Proposition 3.9.1.

$$(X_{\mu,\sigma} - \mu)/\sigma \sim N(0,1)$$
$$\sigma X_{0,1} + \mu \sim N(\mu, \sigma).$$

Proof. Using (3.7) we have $P(X_{\mu,\sigma^2} \leq x) = t = P(X_{0,1} \leq (x-\mu)/\sigma)$, which implies $\sigma X_{0,1} + \mu \sim N(\mu, \sigma^2)$. The second statement follows similarly. ∎

Proposition 3.9.2. *The mgf of $X \sim N(\mu, \sigma^2)$ is*

$$m(t) = \exp(\mu t + t^2\sigma^2/2).$$

Proof. Using the change of variables $u = r - \mu$,

$$
\begin{aligned}
m(t) &= \frac{1}{\sqrt{2\pi\sigma^2}} \int_{\mathbb{R}} e^{tr} e^{-(r-\mu)^2/(2\sigma^2)} dr = \frac{1}{\sqrt{2\pi\sigma^2}} \int_{\mathbb{R}} e^{t(u+\mu)} e^{-u^2/(2\sigma^2)} du \\
&= \frac{1}{\sqrt{2\pi\sigma^2}} e^{t\mu} \int_{\mathbb{R}} e^{tu} e^{-u^2/(2\sigma^2)} du = \frac{1}{\sqrt{2\pi\sigma^2}} e^{t\mu} \int_{\mathbb{R}} e^{-(u^2 - tu2\sigma^2)/(2\sigma^2)} du \\
&= \frac{1}{\sqrt{2\pi\sigma^2}} e^{t\mu} e^{t^2\sigma^2/2} \int_{\mathbb{R}} e^{-(u^2 - tu2\sigma^2 + t^2\sigma^4)/(2\sigma^2)} du \\
&= \frac{1}{\sqrt{2\pi\sigma^2}} e^{t\mu} e^{t^2\sigma^2/2} \int_{\mathbb{R}} e^{-(u - t\sigma^2)^2/(2\sigma^2)} du = e^{t\mu + t^2\sigma^2/2},
\end{aligned}
$$

where the last equality follows from (3.5). ∎

Differentiating the mgf, we have for $X \sim N(\mu, \sigma^2)$,

$$
\begin{aligned}
\mathsf{E}(X) &= \psi'(0) = (\mu + \sigma^2 0)1 = \mu \\
\mathsf{Var}(X) &= \mathsf{E}(X^2) - (\mathsf{E}(X))^2 = \psi''(0) - \mu^2 = \sigma^2 1 + (\mu + 0)^2 1 - \mu^2 = \sigma^2.
\end{aligned}
$$

The Gaussian distribution is one of the most important distributions. The central limit theorem (see Section 8.9 for details) informally states that a sum of many independent random variables is approximately a Gaussian distribution. As a consequence, quantities that are sums of a large number of independent random factors are approximately Gaussian. Examples include performance measures like IQ test results and physiological measurements like height.

The R code below graphs the pdf of the Gaussian distribution for different values of σ and μ.

```
x = seq(-8, 8, length.out = 100)
gf = function(x, s) exp(-x^2/(2 * s^2))/(sqrt(2 *
    pi) * s)
R = stack(list(`$\\sigma=1$` = gf(x, 1), `$\\sigma=2$` = gf(x,
    2), `$\\sigma=3$` = gf(x, 3), `$\\sigma=4$` = gf(x,
    4)))
# below, x is recycled four times
names(R) = c("y", "sigma")
R$x = x
qplot(x, y, color = sigma, lty = sigma, geom = "line",
    xlab = "$x$", ylab = "$f_X(x)$", data = R) +
    opts(title = "Gaussian pdf")
```

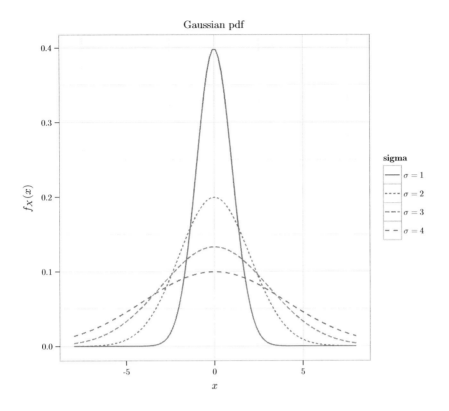

```
x = seq(-4, 4, length = 100)
y1 = dnorm(x, 0, 1)
y2 = dnorm(x, 1, 1)
y3 = dnorm(x, 0, 2)
y4 = pnorm(x, 0, 1)
y5 = pnorm(x, 1, 1)
y6 = pnorm(x, 0, 2)
D = data.frame(probability = c(y1, y2, y3, y4,
    y5, y6))
D$x = x
D$parameter[1:100] = "$N(0,1)$"
D$parameter[301:400] = "$N(0,1)$"
D$parameter[101:200] = "$N(1,1)$"
D$parameter[401:500] = "$N(1,1)$"
D$parameter[201:300] = "$N(0,2)$"
D$parameter[501:600] = "$N(0,2)$"
D$type[1:300] = "$f_X(x)$"
D$type[301:600] = "$F_X(x)$"
qplot(x, probability, data = D, geom = "area",
    facets = parameter ~ type, xlab = "$x$", ylab = "") +
    opts(title = "Gaussian pdf and cdf")
```

Gaussian pdf and cdf

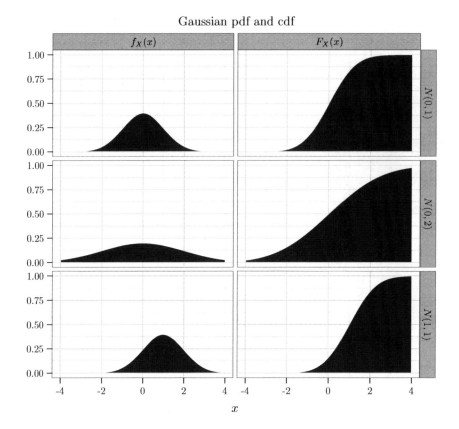

3.10   The Gamma and χ^2 Distributions

Definition 3.10.1. The gamma function is

$$\Gamma(x) = \int_0^\infty t^{x-1} \exp(-t)\, dt, \qquad x > 0.$$

Example F.2.5 shows, using integration by parts, that $\Gamma(n) = (n-1)!$ for positive integers n. The gamma function, however, is also defined for real numbers that are not integers.

Definition 3.10.2. The gamma RV, $X \sim \text{Gam}(k, \theta)$, where $k, \theta \geq 0$, has the following pdf:

$$f_X(x) = \begin{cases} x^{k-1}\theta^{-k}\exp(-x/\theta)/\Gamma(k) & x, k, \theta \geq 0 \\ 0 & \text{otherwise} \end{cases}. \qquad (3.9)$$

The gamma function in the denominator of the pdf ensures that the pdf integrates to 1:

$$\int_0^\infty x^{k-1} \exp(-x/\theta)\, dx = \theta\theta^{k-1} \int_0^\infty z^{k-1} \exp(-z)\, dz = \theta^k \Gamma(k). \qquad (3.10)$$

Proposition 3.10.1. *The mgf of the gamma distribution is* $m(t) = (1 - \theta t)^{-k}$, *for* $t < 1/\theta$.

Proof. Defining $y = (1/\theta - t)x$,

$$m(t) = \mathsf{E}(\exp(tX)) = \int_0^\infty x^{k-1} \frac{\exp(-x/\theta)}{\Gamma(k)\theta^k} \exp(tx)\, dx$$

$$= \frac{1}{\Gamma(k)\theta^k} \int_0^\infty \left(\frac{y}{1/\theta - t}\right)^{k-1} \exp(-y)\, dy \frac{1}{1/\theta - t}$$

$$= \frac{1}{\Gamma(k)(\theta(1/\theta - t)^k} \int_0^\infty y^{k-1} \exp(-y)\, dy = (1 - \theta t)^{-k}.$$

■

Proposition 3.10.2. *If* $X_i \sim Gam(k_i, \theta)$, $i = 1m\dots, n$ *are independent RVs, then*

$$\sum_{i=1}^n X_i \sim Gam\left(\sum_{i=1}^n k_i, \theta\right).$$

Proof. This follows from the fact that the mgf of a sum of RVs is the product of their mgfs: $\prod(1 - \theta t)^{-k_i} = (1 - \theta t)^{\sum k_i}$. ■

Using the mgf, we can compute the expectation and variance of $X \sim Gam(k, \theta)$:

$$\mathsf{E}(X) = m'(0) = k\theta(1 - \theta t)^{-k-1}\Big|_{t=0} = k\theta$$

$$\mathsf{E}(X^2) = m''(0) = k\theta(-k - 1)(1 - \theta t)^{-k-2}\Big|_{t=0}(-\theta) = k\theta^2(k + 1)$$

$$\mathsf{Var}(X) = k\theta^2(k + 1) = k^2\theta^2 = k\theta^2.$$

The non-constant parts of the pdf above (the parts that depend on x) include two multiplied terms: a decaying exponential $\exp(-x/\theta)$ and a polynomial x^{k-1}. As x increases, the polynomial term is monotonic increasing and the exponential term is monotonic decreasing. For large x the exponential decay is stronger than the polynomial growth (see Section B.5), implying that the pdf may be monotonic increasing for a while, but eventually decays to zero as $x \to \infty$. The shape of the pdf is generally unimodal, as in the case of the Gaussian distribution. Since the support is $[0, \infty)$, the pdf is necessarily asymmetric.

The R code below graphs the pdf of multiple gamma RVs with different parameter values.

```
x = seq(0, 30, length = 100)
y1 = dgamma(x, 2, scale = 1)
y2 = dgamma(x, 2, scale = 3)
y3 = dgamma(x, 2, scale = 5)
y4 = dgamma(x, 3, scale = 1)
y5 = dgamma(x, 3, scale = 3)
y6 = dgamma(x, 3, scale = 5)
y7 = dgamma(x, 4, scale = 1)
y8 = dgamma(x, 4, scale = 3)
y9 = dgamma(x, 4, scale = 5)
D = data.frame(probability = c(y1, y2, y3, y4,
    y5, y6, y7, y8, y9))
D$x = x
D$k[1:300] = "$k=2$"
D$k[301:600] = "$k=3$"
D$k[601:900] = "$k=4$"
D$theta[1:100] = "$\\theta=1$"
D$theta[101:200] = "$\\theta=3$"
D$theta[201:300] = "$\\theta=5$"
D$theta[301:400] = "$\\theta=1$"
D$theta[401:500] = "$\\theta=3$"
D$theta[501:600] = "$\\theta=5$"
D$theta[601:700] = "$\\theta=1$"
D$theta[701:800] = "$\\theta=3$"
D$theta[801:900] = "$\\theta=5$"
qplot(x, probability, data = D, main = "Gamma pdf",
    geom = "area", facets = k ~ theta, xlab = "$x$",
    ylab = "$f_X(x)$")
```

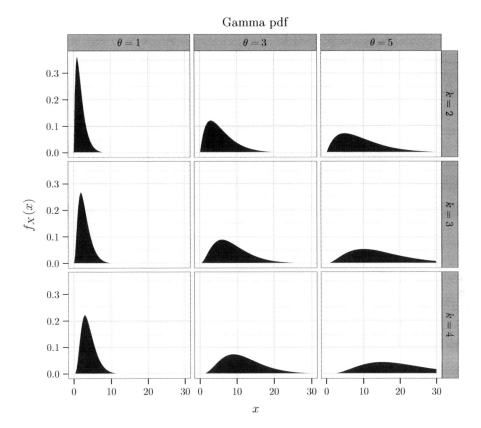

Gamma pdf

Definition 3.10.3. The chi-squared RV, $X \sim \chi_r^2$, where $r \in \mathbb{N}$, is a gamma RV with parameters $k = r/2, \theta = 2$. The parameter r is called the degrees of freedom.

Corollary 3.10.1. *The mgf of the chi-squared distribution χ_r^2 is $m(t) = (1 - 2t)^{-r/2}$.*

The primary motivation of the chi-squared distribution is the following observation.

Proposition 3.10.3. *If $Z_1, \ldots, Z_n$ are independent $N(0, 1)$ RVs then*

$$\sum_{i=1}^{n} Z_i^2 \sim \chi_n^2.$$

Proof. Consider first the case $n = 1$. The mgf of Z_1^2 is

$$\mathsf{E}\left(e^{tZ^2}\right) = \int_{-\infty}^{+\infty} e^{tz^2}(2\pi)^{-1/2} e^{-z^2/2} dz = \int_{-\infty}^{+\infty} (2\pi)^{-1/2} e^{-(1-2t)z^2/2} dz$$

$$= \frac{1}{(1-2t)^{1/2}} \int_{-\infty}^{+\infty} \frac{e^{-z^2/(2(1-2t)^{-1})}}{\sqrt{2\pi}(1-2t)^{-1/2}} dz = \frac{1}{(1-2t)^{1/2}} \cdot 1$$

which is the mgf of χ_1^2.

In the case $n > 1$, the mgf of the sum is the product of n mgfs (see Proposition 4.8.1) of chi-squared RVs with one degree of freedom. This product is $\prod_{i=1}^n (1 - 2t)^{-1/2} = (1 - 2t)^{-n/2}$, which is the mgf of a χ_r^2 RV. ∎

3.11 The t Random Variable

A t_ν RV, where $\nu > 0$ is a parameter called the degrees of freedom, has the following pdf

$$f_X(x) = \frac{\Gamma(\frac{\nu+1}{2})}{\sqrt{\nu\pi}\,\Gamma(\nu/2)} \left(1 + \frac{x^2}{\nu}\right)^{-\frac{\nu+1}{2}}, \quad x \in \mathbb{R}, \nu > 0.$$

It can be shown that if $Z \sim N(0,1), W \sim \chi_\nu^2$ are two independent RVs, then

$$\frac{Z}{\sqrt{W/\nu}} \sim t_\nu.$$

The t distribution with $\nu = 1$ is called the Cauchy distribution. Its pdf is

$$f_X(x) = \frac{1}{\pi(1 + x^2)}.$$

The t distribution with $\nu = 2$ has the following pdf

$$f_X(x) = \frac{1}{(2 + x^2)^{2/3}}.$$

As ν increases, the polynomial decay becomes sharper. At the limit $\nu \to \infty$ the t distribution pdf converges to the pdf of a $N(0,1)$ RV.

The main use of the t-distribution is in statistical tests and confidence intervals. In addition, it is also used to model heavy tailed distributions. A t_ν-RV decays qualitatively slower than any Gaussian RV, regardless of the ν parameter. In fact, the decay of the t_ν-RV pdf is so slow that for $\nu < 1$ the expectation does not exist (the integral diverges). Similarly, the variance of t_ν is ∞ for $1 < \nu \leq 2$ and is undefined for $\nu \leq 1$.

The R code below graphs the pdfs of a Gaussian (shaded) and a t (line curve) RVs. Both have symmetric bell-shaped pdf, but the t pdf decays more slowly and is sharper and the center.

```
x = seq(-6, 6, length.out = 200)
R = data.frame(density = dnorm(x, 0, 1))
R$tdensity = dt(x, 1.5)
R$x = x
G = ggplot(R, aes(x = x, y = density)) + geom_area(fill = I("grey"))
G + geom_line(aes(x = x, y = tdensity), xlab = "$x$",
    ylab = "$f_X(x)$") + opts(title = "Gaussian and t (dof=1.5) pdfs")
```

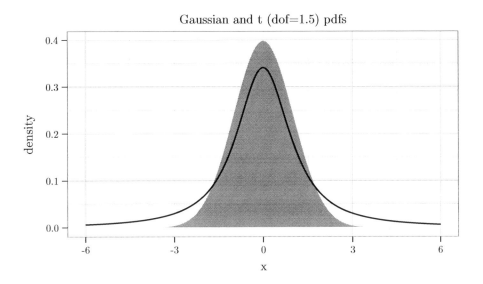

Gaussian and t (dof=1.5) pdfs

3.12 The Beta Random Variable

The beta RV Beta(α, β), where $\alpha, \beta > 0$, has the following pdf

$$f_X(x) = \begin{cases} \frac{1}{B(\alpha,\beta)} x^{\alpha-1}(1-x)^{\beta-1} & x \in [0,1] \\ 0 & x \notin [0,1] \end{cases},$$

where

$$B(\alpha, \beta) = \frac{\Gamma(\alpha)\Gamma(\beta)}{\Gamma(\alpha+\beta)}$$

is the beta function. Example F.6.1 verifies that the pdf integrates to 1 and can be used to compute the expectation and other moments

$$E(X^k) = \frac{1}{B(\alpha,\beta)} \int_0^\infty x^{\alpha+k-1}(1-x)^{\beta-1}\,dx = \frac{B(\alpha+k,\beta)}{B(\alpha,\beta)}$$

$$E(X) = \frac{B(\alpha+1,\beta)}{B(\alpha,\beta)} = \frac{\alpha}{\alpha+\beta}.$$

If $\alpha = \beta = 1$ the beta distribution reduces to the uniform distribution over $[0,1]$. For other values of α, β, however, we get a different behavior. When $\alpha < 1, \beta < 1$ the pdf has a U-shape over $[0,1]$. When $\alpha < 1$ and $\beta \geq 1$ the pdf is strictly decreasing. When $\alpha \geq 1$ and $\beta < 1$ the pdf is strictly increasing. Finally, when $\alpha > 1$ and $\beta > 1$ the pdf is unimodal, with a local maximum in $(0,1)$. If $\alpha = \beta$ the pdf is symmetric around $1/2$ and if $\alpha \neq \beta$ the pdf is asymmetric around $1/2$.

The R code below graphs the pdf of the beta distribution.

```
x = seq(0, 1, length = 100)
y1 = dbeta(x, 1/2, 1/2)
y2 = dbeta(x, 1/2, 1)
y3 = dbeta(x, 1/2, 2)
y4 = dbeta(x, 1, 1/2)
y5 = dbeta(x, 1, 1)
y6 = dbeta(x, 1, 2)
y7 = dbeta(x, 2, 1/2)
y8 = dbeta(x, 2, 1)
y9 = dbeta(x, 2, 2)
D = data.frame(probability = c(y1, y2, y3, y4,
    y5, y6, y7, y8, y9))
D$x = x
D$alpha[1:300] = "$\\alpha=1/2$"
D$alpha[301:600] = "$\\alpha=1$"
D$alpha[601:900] = "$\\alpha=2$"
D$beta[1:100] = "$\\beta=1/2$"
D$beta[101:200] = "$\\beta=1$"
D$beta[201:300] = "$\\beta=2$"
D$beta[301:400] = "$\\beta=1/2$"
D$beta[401:500] = "$\\beta=1$"
D$beta[501:600] = "$\\beta=2$"
D$beta[601:700] = "$\\beta=1/2$"
D$beta[701:800] = "$\\beta=1$"
D$beta[801:900] = "$\\beta=2$"
qplot(x, probability, data = D, geom = "area",
    facets = alpha ~ beta, xlab = "$x$", ylab = "$f_X(x)$",
    main = "Beta pdf") + scale_y_continuous(limits = c(0,
    4))
```

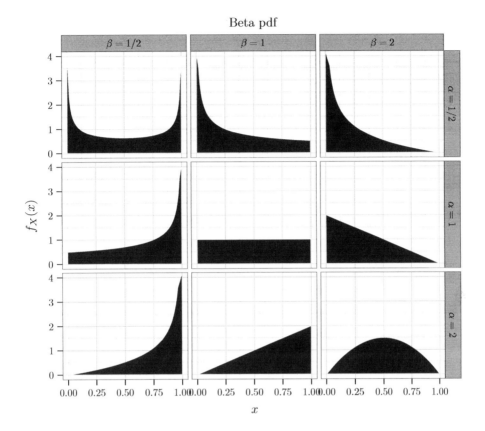

Beta pdf

3.13   Mixture Random Variable

The distributions above have simple shape, in the sense that their pdf or pmf functions are constant, monotonic increasing or decreasing, or unimodal. This section and the next two sections describe methods of constructing distributions that are more complex, potentially having multiple local maxima and minima.

Given k independent RVs $X^{(1)}, \ldots, X^{(k)}$ that are all continuous or all discrete, their mixture is an RV with the following pdf or pmf

$$f_X(x) = \sum_{i=1}^{k} \alpha_i f_{X^{(i)}}(x) \qquad \text{(continuous case)} \qquad (3.11)$$

$$p_X(x) = \sum_{i=1}^{k} \alpha_i p_{X^{(i)}}(x) \qquad \text{(discrete case).} \qquad (3.12)$$

where $\boldsymbol{\alpha} = (\alpha_1, \ldots, \alpha_n)$ is a vector of non-negative numbers that sum to 1. Since the weights sum to one, the above functions are valid pdf or pmf and thus uniquely characterize the mixture distribution.

The moments of a mixture random variables are linear combinations of the corresponding moments of the individual random variables:

$$E(X) = \sum_{i=1}^{k} \alpha_i E(X^{(i)})$$

$$Var(X) = \sum_{i=1}^{k} \alpha_i^2 Var(X^{(i)}).$$

Mixture distributions are able to capture a wide variety of complex distributions. For example, a mixture of k Gaussians can capture a distribution with k modes. Mixture distributions are particularly applicable to situations when the quantity is determined via a two stage experiment: first a mixture component is chosen, and then the value is determined from the appropriate mixture component. For example, consider the situation of measuring with length of fish found in a certain lake containing k species of fish. Assuming fish in different species have significantly different lengths, and that within each of the species the variability in length is limited, we have a mixture model with k components: $X^{(i)}, i = 1, \ldots, k$ represents the distribution of lengths among the species, and w represents the relative frequency of the different species.

The R code below graphs two mixtures of three Gaussians. The first example exhibits a multimodal shape while the second exhibits a asymmetric unimodal shape.

```
x = seq(-3, 6, length = 100)
y1 = dnorm(x, -1, 1/2)
y2 = dnorm(x, 1, 1/2)
y3 = dnorm(x, 3, 1.5)
qplot(x, y1/4 + y2/4 + y3/2, xlab = "$x$", ylab = "$f_X(x)$",
    geom = "area", main = "Mixture of Gaussians pdf")
```

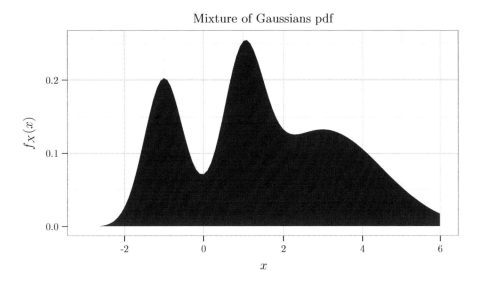

```
x = seq(-6, 4, length = 100)
y1 = dnorm(x, 1, 2)
y2 = dnorm(x, 0, 1/2)
y3 = dnorm(x, -2, 2)
qplot(x, y1/4 + y2/4 + y3/2, xlab = "$x$", ylab = "$f_X(x)$",
    geom = "area", main = "Mixture of Gaussians pdf")
```

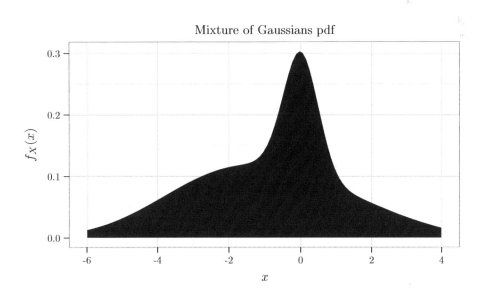

3.14   The Empirical Random Variable

The empirical distribution associated with finite set $A \subset \mathbb{R}$ is a discrete RV with the following pmf

$$p_X(x) = \frac{1}{|A|} \sum_{y \in A} I_{\{y\}}(x),$$

where $I_A(x) = 1$ if $x \in A$ and 0 otherwise. In other words, the empirical distribution places equal mass $(1/|A|)$ on each of the points in A.

In some cases, we associate an empirical distribution with a multiset (a set in which some values may appear more than once). Denoting the multiset as a list $A = (y^{(1)}, \ldots, y^{(n)})$, $y^{(i)} \in \mathbb{R}$ (we may have $y^{(i)} = y^{(j)}$ for $i \neq j$), the empirical distribution pmf is

$$p_X(x) = \frac{1}{n} \sum_{i=1}^{n} I_{\{y^{(i)}\}}(x).$$

The R code below graphs the pmf and cdf of the empirical distribution associated with $(0, 1, 1, 1, 2, 2, 2, 3)$.

```
D = data.frame(x = c(-1, 0, 1, 2, 3, 4), y = c(0,
    1/8, 3/8, 3/8, 1/8, 0))
qplot(x, y, main = "Empirical distribution pmf",
    data = D, xlab = "$x$", ylab = "$p_X(x)$") +
    geom_linerange(aes(x = x, ymin = 0, ymax = y))
```

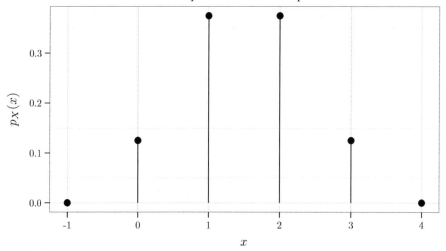

Empirical distribution pmf

```
plot(ecdf(c(0, 1, 1, 1, 2, 2, 2, 3)), verticals = FALSE,
    lwd = 3, xlab = "$x$", ylab = "$F_X(x)$",
    main = "")
title(main = "Empirical distribution cdf", font.main = 1)
```

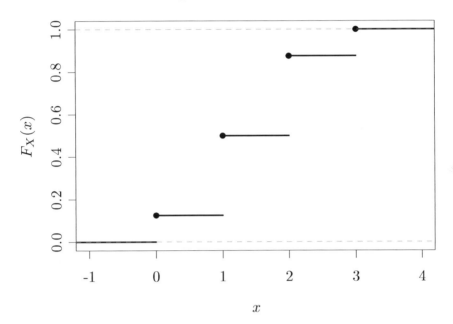

3.15   The Smoothed Empirical Random Variable

One difficulty with the empirical distribution is that it assigns positive probability mass only to a finite number of points, which may be undesirable when modeling continuous variables. The smoothed empirical distribution combines the ideas in the previous two sub-sections. Given a sequence of values $y^{(1)}, \ldots, y^{(n)}$, $y^{(i)} \in \mathbb{R}$, the smoothed empirical distribution is defined as a mixture distribution with the following pdf

$$f_X(x) = \frac{1}{n} \sum_{i=1}^{n} f_\sigma(x - y^{(i)}),$$

where f_σ is usually a symmetric unimodal pdf with expectation 0 and variance σ^2. The mixture pdf is thus a sum of pdfs $f_\sigma(x - y^{(1)}), \ldots, f_\sigma(x - y^{(n)})$ with variance σ^2 and centered at the values $y^{(1)}, \ldots, y^{(n)}$.

A popular choice for f_σ is the Gaussian pdf $N(0, \sigma^2)$. In this case, we have

$$f_X(x) = \frac{1}{n\sqrt{2\pi\sigma^2}} \sum_{i=1}^{n} \exp\left(-(x - y^{(i)})^2/(2\sigma^2)\right).$$

As σ decreases, the smoothed empirical distribution converges to the empirical distribution. As σ increases, the role of the individual mixtures decreases and the distribution approaches a single monolithic component.

The R code below graphs the smoothed version of the empirical distribution corresponding to the sequence $(-1, 0, 0.5, 1, 2, 5, 5.5, 6)$. The graphs show the empirical distribution as a solid line and the mixture components $f_\sigma(x - y^{(i)})$, $i = 1, \ldots, 8$ corresponding to each $y^{(i)}$ with dashed lines (scaled down by a factor of 2 to avoid overlapping solid and dashed lines).

In the first graph below, the σ value is relatively small ($\sigma = 1/6$), resulting in a pdf close to the empirical distribution. In the middle case σ is larger ($\sigma = 1/3$) showing a multimodal shape that is significantly different from the empirical distribution. In the third case above, σ is relatively large ($\sigma = 1$), resulting in a pdf that is resembles two main components. For larger σ, the pdf will resemble a single monolithic component.

```
A = c(-1, 0, 0.5, 1, 2, 5, 5.5, 6)
x = seq(-3, 8, length.out = 100)
D = x %o% rep(1, 8)
f = x * 0
for (i in 1:8) {
    D[, i] = dnorm(x, A[i], 1/6)/8
    f = f + D[, i]
}
plot(x, f, xlab = "$x$", ylab = "$f_X(x)$", type = "l")
for (i in 1:8) lines(x, D[, i]/2, lty = 2)
title("Smoothed empirical distribution ($\\sigma=1/6$)",
    font.main = 1)
```

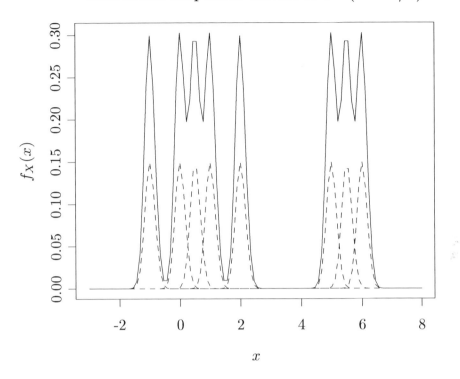

Smoothed empirical distribution ($\sigma = 1/6$)

```
A = c(-1, 0, 0.5, 1, 2, 5, 5.5, 6)
x = seq(-3, 8, length.out = 100)
D = x %o% rep(1, 8)
f = x * 0
for (i in 1:8) {
    D[, i] = dnorm(x, A[i], 1/3)/8
    f = f + D[, i]
}
plot(x, f, xlab = "$x$", ylab = "$f_X(x)$", type = "l")
for (i in 1:8) lines(x, D[, i]/2, lty = 2)
title("Smoothed empirical distribution ($\\sigma=1/3$)",
    font.main = 1)
```

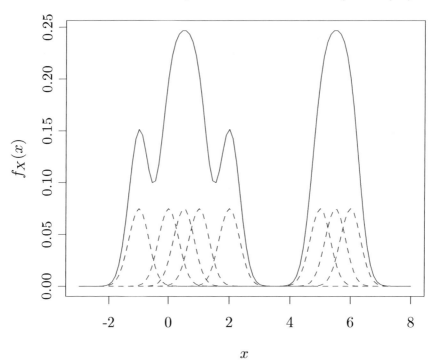

Smoothed empirical distribution ($\sigma = 1/3$)

```
A = c(-1, 0, 0.5, 1, 2, 5, 5.5, 6)
x = seq(-3, 8, length.out = 100)
D = x %o% rep(1, 8)
f = x * 0
for (i in 1:8) {
    D[, i] = dnorm(x, A[i], 1)/8
    f = f + D[, i]
}
plot(x, f, xlab = "$x$", ylab = "$f_X(x)$", type = "l")
for (i in 1:8) lines(x, D[, i]/2, lty = 2)
title("Smoothed empirical distribution ($\\sigma=1$)",
    font.main = 1)
```

Smoothed empirical distribution ($\sigma = 1$)

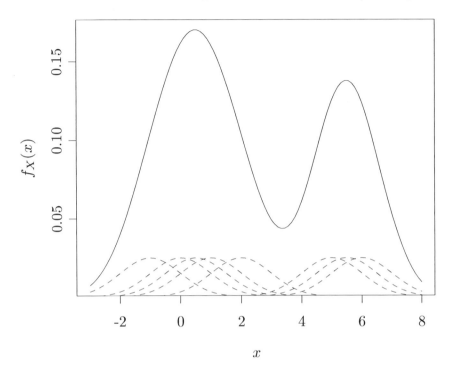

3.16 Notes

In modeling situations, it may seem difficult to select an appropriate distribution among the large variety of possible candidates. The potential candidates may be narrowed by focusing on the domain of the set of values the distribution generates.

The Bernoulli distribution is appropriate for modeling $\{0, 1\}$ values. The binomial and hyper-geometric distributions are appropriate for modeling $\{0, 1, \ldots, n\}$ values. The Poisson and geometric distributions are appropriate for modeling non-negative integers. For continuous data, the beta distribution is appropriate for modeling continuous values in $[0, 1]$ (or some other range if appropriate shifting and scaling are performed). The gamma and exponential distributions are appropriate for modeling continuous non-negative values (exponential for strictly decreasing distributions and gamma for potentially unimodal distributions). The Gaussian and t distributions are appropriate for modeling continuous real-valued data (Gaussian distribution for exponentially decaying distribution and t distribution for polynomially decaying distributions).

The canonical distributions above have pdfs with simple shapes, for example monotonic increasing, monotonic decreasing, or unimodal. In cases where the required distribution exhibits more complex shapes, we can select an appropriate mixture distribution. As the number of mixture components is increased, more flexibility is obtained (at the price of an increase in the number of parameter).

A comprehensive description of these and other distributions is available in manuscripts specializing in distributions, for example [21]. Detailed coverage of modeling mixture distributions is available in [28]. Additional discrete random vectors and their use in modeling discrete data are available in [6].

3.17   Exercises

1. The description of the Poisson distribution contains several examples of real world quantities that are approximately distributed Poisson. After reading the motivation behind the Poisson distribution, reflect on why these quantities approximately follow the Poisson distribution.

2. The description of the exponential distribution states that it is the only continuous RV with the memoryless property. Can you find a discrete distribution that also has this property?

3. Consider the uniform distribution over two disjoint intervals $[a, b] \cup [c, d]$ with $a < b < c < d$. Write down the pdf and cdf of that distribution and derive its expectation and variance.

Chapter 4

Random Vectors

In the previous chapters we considered random variables $X : \Omega \to \mathbb{R}$ and the probabilities associated with them $\mathsf{P}(X \in A)$. Whenever we discussed multiple random variable X, Y we assumed independence: $\mathsf{P}(X \in A, Y \in B) = \mathsf{P}(X \in A) \mathsf{P}(Y \in B)$. In this chapter we extend our discussion to multiple (potentially) dependent random variables. Our exposition is general, and applies to n random variables, but it is useful to keep in mind the intuitive $n = 2$ case. We continue in this chapter our approach of considering random vectors that are either discrete or continuous. This allows us to avoid using measure theory in most of the proofs.

4.1 Basic Definitions

Definition 4.1.1. A random vector $\boldsymbol{X} = (X_1, \ldots, X_n)$ is a collection of n random variables $X_i : \Omega \to \mathbb{R}$ that together may be considered a mapping $\boldsymbol{X} : \Omega \to \mathbb{R}^n$. We denote

$$\{\boldsymbol{X} \in E\} \stackrel{\text{def}}{=} \{\omega \in \Omega : (X_1(\omega), \ldots, X_n(\omega)) \in E\}, \qquad E \subset \mathbb{R}^n,$$
$$\mathsf{P}(\boldsymbol{X} \in E) = \mathsf{P}(\{\omega \in \Omega : \boldsymbol{X}(\omega) = (X_1(\omega), \ldots, X_n(\omega)) \in E\}).$$

Figure 4.1 illustrates the definition above. Note that we denote random vectors in bold-face and its components using subscripts, for example $\boldsymbol{X} = (X_1, \ldots, X_n)$. This mirrors our notation of vectors in $\mathbb{R}^n$ using lower-case bold-face letters, for example $\boldsymbol{x} = (x_1, \ldots, x_n) \in \mathbb{R}^n$.

Definition 4.1.2. If a set $E \subset \mathbb{R}^n$ is of the form $E = \{\boldsymbol{x} \in \mathbb{R}^n : x_1 \in S_1, \ldots, x_n \in S_n\}$, we call it a product set and denote it as $E = S_1 \times \cdots \times S_n$.

Figure 4.2 illustrates the definition above.

Definition 4.1.3. Two RVs, X_1, X_2, are independent if for all product sets, $S_1 \times S_2$

$$\mathsf{P}((X_1, X_2) \in S_1 \times S_2) = \mathsf{P}(X_1 \in S_1) \mathsf{P}(X_2 \in S_2).$$

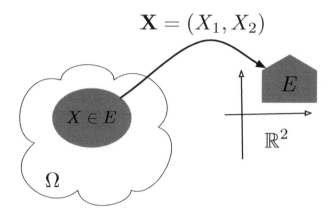

Figure 4.1: A random vector $\boldsymbol{X} = (X_1, X_2)$ is a mapping from Ω to $\mathbb{R}^2$. The set $\{\boldsymbol{X} \in E\}$ is a subset of Ω corresponding to all $\omega \in \Omega$ that are mapped to E.

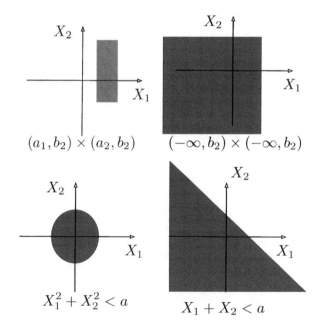

Figure 4.2: Two product sets (top) and two non-product sets (bottom) for $\boldsymbol{X} = (X_1, X_2)$.

The components of a random vector are pairwise independent if every pair of components is independent. The components of a random vector $\boldsymbol{X} = (X_1, \ldots, X_n)$ are independent if for all product sets $S_1 \times \cdots \times S_n$,

$$\mathsf{P}(\boldsymbol{X} \in S_1 \times \cdots \times S_n) = \mathsf{P}(X_1 \in S_1) \cdots \mathsf{P}(X_n \in S_n).$$

As in the case of independence of events (Definition 1.5.2), mutual independence implies pairwise independence but not necessarily vice verse. The R code below graphs an illustration of several independent and dependent random vectors (see Figure 4.3 for the graph itself).

```
X = runif(200)
Y = runif(200)
W = rnorm(300, sd = 0.2)
Z = rnorm(300, sd = 0.2)
Q = runif(300) * 2 - 1
P = runif(300) * 2 - 1
qplot(X, Y, xlab = "$X_1$", ylab = "$X_2$")
qplot(W, Z, xlab = "$X_1$", ylab = "$X_2$")
qplot(X, X + runif(200)/10, xlab = "$X_1$", ylab = "$X_2$")
qplot(X, -X + runif(200)/10, xlab = "$X_1$", ylab = "$X_2$")
qplot(cos(seq(0, 2 * pi, length = 30)), sin(seq(0,
    2 * pi, length = 30)), xlab = "$X_1$", ylab = "$X_2$")
qplot(Q[(Q^2 + P^2) < 1], P[(Q^2 + P^2) < 1],
    xlab = "$X_1$", ylab = "$X_2$")
```

Example 4.1.1. *In a random experiment describing drawing a phone number from a phone book, the sample space Ω is the collection of phone numbers and an outcome $\omega \in \Omega$ corresponds to a specific phone number. Let $X_1, X_2, X_3 : \Omega \to \mathbb{R}$ be the weight, height, and IQ of the phone number's owner. The event*

$$\{X_1 \in (a_1, b_1)\} = \{\omega \in \Omega : X_1(\omega) \in (a_1, b_1)\}$$

corresponds to "the weight of the selected phone number's owner lies in the range (a_1, b_1)". The event $(X_1, X_2) \in (a_1, b_1) \times (a_2, b_2)$ corresponds to "the weight is in the range (a_1, b_1) and the height is in the range (a_2, b_2)". The set $(a_1, b_1) \times (a_2, b_2)$ is a product set, but since we do not expect height and weight to be independent RVs, we normally have

$$\mathsf{P}((X_1, X_2) \in (a_1, b_1) \times (a_2, b_2)) \neq \mathsf{P}(X_1 \in (a_1, b_1)) \mathsf{P}(X_2 \in (a_2, b_2)).$$

The set $(a_2, b_2) \times (a_3, b_3)$ is also a product event, and assuming that height and IQ are independent, we have

$$\mathsf{P}((X_2, X_3) \in (a_2, b_3) \times (a_3, b_3)) = \mathsf{P}(X_2 \in (a_2, b_2)) \mathsf{P}(X_3 \in (a_3, b_3)).$$

Two examples of non-product sets are $X_2/X_1^2 \le a$ (body-mass index less than a) and $X_1 \le X_2 + a$.

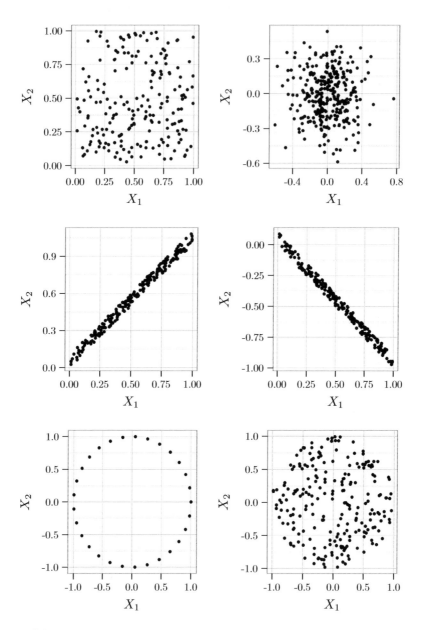

Figure 4.3: Samples from six different distributions over (X_1, X_2). The cases in the top row exhibit independence (uniform distribution in the unit square and spherical symmetric distribution with more samples falling closer to the origin). The middle row and bottom row exhibit a lack of independence (knowing the value of X_2 influences the probability of X_1: $P(X_1|X_2) \neq P(X_1)$). The top and bottom rows exhibit zero covariance while the middle row exhibits positive covariance (left) and negative covariance (right).

A random vector $\boldsymbol{X}$ with corresponding Ω, P defines a new distribution P' on $\Omega' = \mathbb{R}^n$:

$$\mathsf{P}'(E) = \mathsf{P}(\boldsymbol{X} \in E).$$

It is straightforward to verify that P' satisfies the three probability axioms. In many cases the probability space $(\mathbb{R}^n, \mathsf{P}')$ is conceptually simpler than (Ω, P).

Example 4.1.2. *In the dice-throwing experiment (Example 1.3.1), for $X_1(a, b) = a + b$ (an RV measuring the sum of the outcomes of two dice) and $X_2(a, b) = a - b$ (an RV measuring the difference of the outcomes of the two dice) we have*

$$\mathsf{P}(X_1 = 2, X_2 = 2) = \mathsf{P}(X_1 \in \{2\}, X_2 \in \{2\}) = \mathsf{P}(\emptyset) = 0$$
$$\mathsf{P}(X_1 = 6, X_2 = 2) = \mathsf{P}(\{(4, 2)\}) = 1/36$$
$$\mathsf{P}(X_1 > 4, X_2 < 0) = \mathsf{P}(\{(1, 4), (1, 5), (1, 6), (2, 3), (2, 4), (2, 5),$$
$$(2, 6), (3, 4), (3, 5), (3, 6), (4, 5), (4, 6), (5, 6)\}) = \frac{13}{36}$$
$$\mathsf{P}(X_1 > 4) = 1 - \mathsf{P}(\{(1, 1), (1, 2), (2, 1), (1, 3), (3, 1), (2, 2)\}) = 30/36$$
$$\mathsf{P}(X_2 < 0) = 15/36.$$

Since $\mathsf{P}(X_1 > 4, X_2 < 0) \neq \mathsf{P}(X_1 > 4) \, \mathsf{P}(X_2 < 0)$, X_1, X_2 are not independent.

4.2 Joint Pmf, Pdf, and Cdf Functions

Definition 4.2.1. The random vector $\boldsymbol{X}$ is discrete if there exists a finite or countable set $K \subset \mathbb{R}^n$ such that $\mathsf{P}(\boldsymbol{X} \in K) = 1$. The random vector $\boldsymbol{X}$ is continuous if $\mathsf{P}(\boldsymbol{X} = \boldsymbol{x}) = 0$ for all $\boldsymbol{x} \in \mathbb{R}^n$.

The random vector in Example 4.1.1 may be continuous (assuming weight, height and IQ are measured with infinite precision) while the random vector in Example 4.1.2 is clearly discrete. As with one dimensional random variables, we define the cdf, the pmf, and the pdf below.

Definition 4.2.2. For any random vector $\boldsymbol{X} = (X_1, \ldots, X_n)$ we define the cdf as $F_{\boldsymbol{X}} : \mathbb{R}^n \to \mathbb{R}$

$$F_{\boldsymbol{X}}(\boldsymbol{x}) = \mathsf{P}(X_1 \leq x_1, X_2 \leq x_2, \ldots, X_n \leq x_n).$$

For continuous $\boldsymbol{X}$ we define the pdf as

$$f_{\boldsymbol{X}}(\boldsymbol{x}) = \frac{\partial^n}{\partial x_1 \cdots \partial x_n} F_{\boldsymbol{X}}(x_1, \ldots, x_n)$$

and 0 if the derivative does not exist. For discrete $\boldsymbol{X}$ we define the pmf as

$$p_{\boldsymbol{X}}(\boldsymbol{x}) = \mathsf{P}(X_1 = x_1, \ldots, X_n = x_n).$$

The cdf, pdf, and pmf of a random vector have similar properties to the properties of the cdf, pdf, and pmf of a random variable derived in Chapter 2 (with obvious modification). Most proofs carry over from the random variable case with little modifications. Examples appear below.

- The pdf is non-negative and integrates to one

$$\int \cdots \int_{\mathbb{R}^n} f_{\boldsymbol{X}}(\boldsymbol{x}) \, dx_1 \cdots dx_n = 1.$$

- The pmf is non-negative and sums to one

$$\sum_{x_1} \cdots \sum_{x_2} p_{\boldsymbol{X}}(\boldsymbol{x}) = 1.$$

- The cdf is monotonic increasing: if $a_i \le b_i$ for all $i = 1, \ldots, n$, then

$$F_{\boldsymbol{X}}(a_1, \cdots, a_n) \le F_{\boldsymbol{X}}(b_1, \cdots, b_n).$$

-

$$0 = \lim_{x_1 \to -\infty} \cdots \lim_{x_n \to -\infty} F_{\boldsymbol{X}}(x_1, \cdots, x_n)$$
$$1 = \lim_{x_1 \to \infty} \cdots \lim_{x_n \to \infty} F_{\boldsymbol{X}}(x_1, \cdots, x_n).$$

- For continuous random vectors

$$F_{\boldsymbol{X}}(\boldsymbol{x}) = \int_{-\infty}^{x_1} \int_{-\infty}^{x_2} \cdots \int_{-\infty}^{x_n} f_{\boldsymbol{X}}(t_1, t_2, \ldots, t_n) \, dt_1 dt_2 \cdots dt_n.$$

-

$$\mathsf{P}(\boldsymbol{X} \in A) = \begin{cases} \sum_{\boldsymbol{x} \in A} p_{\boldsymbol{X}}(\boldsymbol{x}) & \boldsymbol{X} \text{ is a discrete random vector} \\ \int_A f_{\boldsymbol{X}}(\boldsymbol{x}) \, d\boldsymbol{x} & \boldsymbol{X} \text{ is a continuous random vector} \end{cases}. \quad (4.1)$$

4.3   Marginal Random Vectors

We can relate probabilities involving $\boldsymbol{X} \in A$ to probabilities involving the one dimensional components X_i or, more generally, to probabilities involving a subset of the dimensions of $\boldsymbol{X}$. The most convenient way to do so is to relate the corresponding pmfs, cdfs, and pdfs.

Proposition 4.3.1.

$$p_{X_i}(x_i) = \sum_{x_1} \cdots \sum_{x_{i-1}} \sum_{x_{i+1}} \cdots \sum_{x_n} p_{\boldsymbol{X}}(x_1, \ldots, x_n)$$

$$f_{X_i}(x_i) = \int \cdots \int f_{\boldsymbol{X}}(x_1, \ldots, x_n) \, dx_1 \cdots dx_{i-1} dx_{i+1} dx_n$$

$$F_{X_i}(x_i) = \lim_{x_1 \to \infty} \cdots \lim_{x_i \to \infty} \lim_{x_{i+1} \to \infty} \cdots \lim_{x_n \to \infty} F_{\boldsymbol{X}}(x_1, \cdots, x_n).$$

Proof. The proof follows from the fact that the following two sets are equal

$$\{X_1 \in S_1\} = \{(X_1, X_2, \ldots, X_n) \in S_1 \times \mathbb{R} \times \cdots \times \mathbb{R}\}.$$

∎

Example 4.3.1. *In the two-dimensional case*

$$p_{X_1}(x_1) = \sum_{x_2} p_{X_1, X_2}(x_1, x_2)$$

$$f_{X_1}(x_1) = \int_{-\infty}^{\infty} f_{X_1, X_2}(x_1, x_2)\, dx_2$$

$$F_{X_1}(x_1) = \lim_{x_2 \to \infty} F_{X_1, X_2}(x_1, x_2).$$

A result similar to Proposition 4.3.1 (with obvious modifications) applies to higher-order marginals. For example,

$$F_{X_1, X_4}(x_1, x_4) = \lim_{x_2 \to \infty} \lim_{x_3 \to \infty} F_{X_1, X_2, X_3, X_4}(x_1, x_2, x_3, x_4)$$

$$f_{X_1, X_4}(x_1, x_4) = \int_{-\infty}^{\infty} \int_{-\infty}^{\infty} f_{X_1, X_2, X_3, X_4}(x_1, x_2, x_3, x_4)\, dx_2 dx_3.$$

Proposition 4.3.2. *The vector $\boldsymbol{X}$ is independent if and only if*

$$F_{\boldsymbol{X}}(x_1, \ldots, x_n) = F_{X_1}(x_1) \cdots F_{X_n}(x_n)$$

which occurs for discrete random vectors if and only if

$$p_{\boldsymbol{X}}(x_1, \ldots, x_n) = p_{X_1}(x_1) \cdots p_{X_n}(x_n)$$

and for continuous vectors if and only if

$$f_{\boldsymbol{X}}(x_1, \ldots, x_n) = f_{X_1}(x_1) \cdots f_{X_n}(x_n).$$

Proof*. The proof uses results from measure theory. Formally, the equations in the proposition hold almost everywhere (see Definition F.3.6). If the cdf factorizes then by definition the pdf or the pmf factorize as well. Similarly, if the pmf or the pdf factorize it is easy to see that the cdf factorizes as well. If the components of $\boldsymbol{X}$ are independent, the cdf factorizes.

It remains to show that if the cdf factorizes, the components of the random vector $\boldsymbol{X}$ are independent. From a measure theory perspective, factorization of the cdf is equivalent to independence of

$$\mathcal{A}_k = \{\mathbb{R}^{k-1} \times (-\infty, a] \times \mathbb{R}^{n-k} : a \in \mathbb{R}\}, \qquad k = 1, \ldots, n \qquad (*)$$

under $(\mathbb{R}^n, \mathcal{B}(\mathbb{R}^n), \mathsf{P}')$. Similarly, independence of the components of $\boldsymbol{X}$ corresponds to independence of the sets

$$\mathcal{C}_k = \{\mathbb{R}^{k-1} \times B \times \mathbb{R}^{n-k} : B \in \mathcal{B}(\mathbb{R})\}, \qquad i = 1, \ldots, n \qquad (**)$$

under $(\mathbb{R}^n, \mathcal{B}(\mathbb{R}^n), \mathsf{P}')$. See Definition 4.9.1 at the end of this chapter for a definition of the concept of independence of sets.

The equivalence between (*) and (**) is asserted by Proposition 4.9.1, which
states that the independence of $\mathcal{A}_1, \ldots, \mathcal{A}_n$ implies the independence of $\sigma(\mathcal{A}_1) =$
$\mathcal{C}_1, \ldots, \sigma(\mathcal{A}_n) = \mathcal{C}_n$.   ∎

Example 4.3.2. *Consider the following pdf*

$$f_{X,Y}(x,y) = \frac{1}{2\pi} e^{-(x^2+y^2)/2},$$

*which factorizes as $f_X(x)f_Y(y)$ where f_X and f_Y are the pdfs of $N(0,1)$ Gaus-
sian RVs. This implies that X,Y are independent $N(0,1)$ Gaussian RVs. The
factorization leads to simplified calculation of probabilities, for example*

$$\mathsf{P}(3 < X < 6, Y < 9) = \int_3^6 \int_{-\infty}^9 f_{X,Y}(x,y)\,dx dy = \int_3^6 \int_{-\infty}^9 f_X(x)f_Y(y)\,dx dy$$

$$= \int_3^6 f_X(x)dx \int_\infty^9 f_Y(y)dy.$$

Example 4.3.3. *For a random vector (X,Y) where X,Y are independent exponential random variables with parameters λ_1, λ_2 we have*

$$f_{X,Y}(x,y) = f_X(x)f_Y(y) = \lambda_1\lambda_2 e^{-\lambda_1 x}e^{-\lambda_2 y} = \lambda_1\lambda_2 e^{-\lambda_1 x - \lambda_2 y}.$$

The cdf is

$$F_{X,Y}(x,y) = \int_{-\infty}^x \int_{-\infty}^y \lambda_1\lambda_2 e^{-\lambda_1 x - \lambda_2 y} dx dy = \int_{-\infty}^x \lambda_1 e^{-\lambda_1 x} \int_{-\infty}^y \lambda_2 e^{-\lambda_2 y}$$

$$= (1 - e^{-\lambda_1 x})(1 - e^{-\lambda_1 x}) = F_X(x)F_Y(y).$$

4.4   Functions of a Random Vector

Recall that when X is a random variable and $g : \mathbb{R} \to \mathbb{R}$ is a real valued function
then $g(X)$ is also a random variable and its cdf, pdf or pmf are directly related
to the corresponding functions of X (Chapter 2). The same holds for a random
vector. Specifically, for a random vector $\boldsymbol{X} = (X_1, \ldots, X_n)$ and

$$g = (g_1, \ldots, g_k) : \mathbb{R}^n \to \mathbb{R}^k, \qquad g_i : \mathbb{R}^n \to \mathbb{R}, \quad i = 1, \ldots, k,$$

we have a new k-dimensional random vector $\boldsymbol{Y} = g(\boldsymbol{X})$ with

$$Y_i = g_i(X_1, \ldots, X_n), \quad i = 1, \ldots, k.$$

Figure 4.4 illustrates this concept.

As in the case of random variables, we consider several techniques for relating
the cdf, pmf, or pdf of $g(\boldsymbol{X})$ to that of $\boldsymbol{X}$.

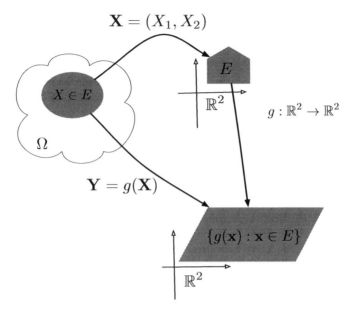

Figure 4.4: A random vector $\boldsymbol{X} = (X_1, X_2)$ and $g : \mathbb{R}^2 \to \mathbb{R}^2$ define a new random vector $\boldsymbol{Y} = g(\boldsymbol{X})$ that is a mapping from Ω to $\mathbb{R}^2$.

1. The first technique is to compute the cdf $F_{\boldsymbol{Y}}(\boldsymbol{y})$, for all $\boldsymbol{y} \in \mathbb{R}^k$. When $k = 1$, $\boldsymbol{Y} = Y = g(X_1, \ldots, X_n)$ and

$$F_Y(y) = \mathsf{P}(g(X_1, \ldots, X_n) \leq y) = \begin{cases} \int_{\boldsymbol{x}:g(\boldsymbol{x})\leq y} f_{\boldsymbol{X}}(\boldsymbol{x}) d\boldsymbol{x} & \boldsymbol{X} \text{ is continuous} \\ \sum_{\boldsymbol{x}:g(\boldsymbol{x})\leq y} p_{\boldsymbol{X}}(\boldsymbol{x}) & \boldsymbol{X} \text{ is discrete} \end{cases}.$$

When $k > 2$,

$$F_{Y_1,\ldots,Y_k}(y_1, \ldots, y_k) = \mathsf{P}(g_1(\boldsymbol{X}) \leq y_1, \ldots, g_k(\boldsymbol{X}) \leq y_1)$$
$$= \begin{cases} \int_A f_{\boldsymbol{X}}(\boldsymbol{x})\, d\boldsymbol{x} & \boldsymbol{X} \text{ is continuous} \\ \sum_{\boldsymbol{x}\in A} p_{\boldsymbol{X}}(\boldsymbol{x}) & \boldsymbol{X} \text{ is discrete} \end{cases}$$

where

$$A = \{\boldsymbol{x} \in \mathbb{R}^n : g_j(\boldsymbol{x}) \leq y_j \text{ for } j = 1, \ldots, k\}.$$

2. If $\boldsymbol{Y}$ is discrete we can find its pmf by

$$p_{\boldsymbol{Y}}(y_1, \ldots, y_k) = \mathsf{P}(g_1(\boldsymbol{X}) = y_1, \ldots, g_k(\boldsymbol{X}) = y_k)$$
$$= \begin{cases} \int_A f_{\boldsymbol{X}}(\boldsymbol{x})\, d\boldsymbol{x} & \boldsymbol{X} \text{ is continuous} \\ \sum_{\boldsymbol{x}\in A} p_{\boldsymbol{X}}(\boldsymbol{x}) & \boldsymbol{X} \text{ is discrete} \end{cases}$$

where

$$A = \{\boldsymbol{x} \in \mathbb{R}^n : g_j(\boldsymbol{x}) = y_j \text{ for all } j = 1, \ldots, k\}.$$

Note that if g is one-to-one the set A consists of a single element.

3. If Y is continuous we can obtain the pdf f_Y by differentiating the joint cdf (if it is available)

$$f_Y(y) = \frac{\partial^k}{\partial y_1 \cdots \partial y_k} F_Y(y)$$

or using the change of variable technique assuming that $k = n$ (see below).

Proposition 4.4.1. *For an n-dimensional continuous random vector X and $Y = g(X)$ with an invertible and differentiable $g : \mathbb{R}^n \to \mathbb{R}^n$ (in the range of X),*

$$f_Y(y) \cdot |\det J(g^{-1}(y))| = f_X(g^{-1}(y))$$

where $J(g^{-1}(y))$ is the Jacobian matrix at $x = g^{-1}(y)$:

$$J(x) = \begin{pmatrix} \frac{\partial g_1}{\partial x_1}(x) & \cdots & \frac{\partial g_1}{\partial x_n}(x) \\ \vdots & \vdots & \vdots \\ \frac{\partial g_n}{\partial x_1}(x) & \cdots & \frac{\partial g_n}{\partial x_n}(x) \end{pmatrix}.$$

Proof. See Proposition F.6.3.   ∎

4. The third technique uses the moment generating function. See the corresponding technique in Chapter 2 and the generalization of the moment generating function to random vectors at the end of this chapter.

The following example illustrates the second method above in the important special case of an invertible linear transformation (see Chapter C).

Example 4.4.1. *If $g : \mathbb{R}^n \to \mathbb{R}^n$ is a linear transformation expressed by an invertible matrix T of size $n \times n$, then $g(x) = Tx$ and $g^{-1}(y) = T^{-1}y$. Since the Jacobian $J(g^{-1}(y)) = T$,*

$$f_Y(y) = |\det T|^{-1} f_X(T^{-1}y).$$

The example above leads to a general technique for finding the pdf or pmf of a sum of independent random variables. This is explored in the example below. Proposition 4.4.3 shows how to obtain the same result using the convolution operator.

Example 4.4.2. *Consider an independent random vector $X = (X_1, X_2)$ and the mapping $X \mapsto Y$ defined by $Y_1 = X_1$, $Y_2 = X_1 + X_2$ or in matrix notation $g(x) = Ax$ for $A = \begin{pmatrix} 1 & 0 \\ 1 & 1 \end{pmatrix}$. Note that the determinant of A is 1 and that the inverse mapping is $X_1 = Y_1$ and $X_2 = Y_2 - Y_1$.*
We therefore have

$$f_{Y_1, Y_2}(y_1, y_2) = \frac{1}{|1|} f_{X_1, X_2}(y_1, y_2 - y_1) = f_{X_1}(y_1) f_{X_2}(y_2 - y_1).$$

Integrating both sides of the equations above with respect to y_1 gives

$$f_{X_1+X_2}(z) = \int f_{X_1}(t) f_{X_2}(z - t) \, dt.$$

The example below illustrates the second method of finding the distribution of $g(X)$ in the case of a non-linear transformation g.

Example 4.4.3. *For two independent exponential RVs X_1, X_2 (with parameter λ) and $Y = g(X)$ with $g_1(x_1, x_2) = x_1/(x_1+x_2)$, $g_2(x_1, x_2) = x_1+x_2$ (the inverse transformation is $g_1^{-1}(y_1, y_2) = y_1 y_2$, $g_2^{-1}(y_1, y_2) = (1 - y_1)y_2$) the Jacobian is*

$$J(x_1, x_2) = \begin{pmatrix} \frac{x_2}{(x_1+x_2)^2} & -\frac{x_1}{(x_1+x_2)^2} \\ 1 & 1 \end{pmatrix}$$

$$\det J(x_1, x_2) = \frac{x_2}{(x_1 + x_2)^2} + \frac{x_1}{(x_1 + x_2)^2} = \frac{1}{x_1 + x_2},$$

and

$$f_{Y_1,Y_2}(y_1, y_2) = |(y_1 y_2 + (1 - y_1)y_2)| \, \lambda e^{-\lambda y_1 y_2} \lambda e^{-\lambda(1-y_1)y_2} = y_2 \lambda^2 e^{-\lambda y_2}$$

for $0 < y_1 < 1$, $0 < y_2$ and 0 otherwise.

Definition 4.4.1. The convolution of two functions $f, g : \mathbb{R} \to \mathbb{R}$ is the function

$$(f * g)(z) = \begin{cases} \sum_t f(t)g(z - t) & f, g \text{ are discrete functions} \\ \int_{-\infty}^{\infty} f(t)g(z - t) \, dt & f, g \text{ otherwise} \end{cases}.$$

(a function f is discrete if $f(x) = 0$ except on a finite or countable set of x values.)

The parenthesis are often omitted resulting in the notation $f * g$.

Proposition 4.4.2. *The convolution is commutative: $(f * g)(t) = (g * f)(t)$ and associative: $((f * g) * h)(t) = (f * (g * h))(t)$.*

Proof. In the continuous case we have

$$(f * g)(z) = \int_{-\infty}^{\infty} f(t)g(z - t) \, dt = -\int_{\infty}^{-\infty} f(z - t')g(t') \, dt'$$

$$= \int_{-\infty}^{\infty} g(t')f(z - t') \, dt' = (g * f)(z).$$

The proof in the discrete case is similar. The proof of associativity is along similar lines but more tedious. ∎

The convolution's associativity justifies leaving out the parenthesis, for example we write $(f * g * h)(t)$ instead of $(f * (g * h))(t)$.

Proposition 4.4.3. *If* $X = (X_1, \ldots, X_n)$ *is an independent random vector, then*

$$f_{\sum_{i=1}^n X_i}(y) = f_{X_1} * \cdots * f_{X_n}(y) \qquad \boldsymbol{X} \text{ is continuous}$$

$$p_{\sum_{i=1}^n X_i}(y) = p_{X_1} * \cdots * p_{X_n}(y) \qquad \boldsymbol{X} \text{ is discrete.}$$

Proof. When $n = 2$ and $\boldsymbol{X}$ is continuous, the cdf of $Y = X_1 + X_2$ is

$$F_Y(y) = \mathsf{P}(X_1 + X_2 \le y) = \int_{-\infty}^{\infty} \int_{-\infty}^{y-w} f_{X_1, X_2}(w, z) \, dz dw$$

(since $y = z + w$) and by the fundamental theorem of calculus (Proposition F.2.2)

$$f_Y(y) = \frac{d}{dy} F_Y(y) = \int_{-\infty}^{\infty} f_{X_1, X_2}(w, y - w) \, dw = f_{X_1} * f_{X_2}(y),$$

where the last equality follows from the independence of X_1 and X_2. By induction we have that $f_{X_1 + \ldots + X_n}(y) = (f_{X_1} * \cdots * f_{X_n})(y)$. The proof in the case of a discrete $\boldsymbol{X}$ is similar. ∎

4.5 Conditional Probabilities and Random Vectors

Conditional probabilities for random vectors are defined similarly to the scalar case. Considering a joint distribution over the random vector $\boldsymbol{Z} = (\boldsymbol{X}, \boldsymbol{Y})$, the conditional probability $\mathsf{P}(\boldsymbol{X} \in A | \boldsymbol{Y} = \boldsymbol{y})$ reflects an updated likelihood for the event $\boldsymbol{X} \in A$ given that $\boldsymbol{Y} = \boldsymbol{y}$.

The conditional cdf, pdf, and pmf are defined as follows

$$F_{\boldsymbol{X}|\boldsymbol{Y}=\boldsymbol{y}}(\boldsymbol{x}) = \begin{cases} \mathsf{P}(\boldsymbol{X} \le \boldsymbol{x}, \boldsymbol{Y} = \boldsymbol{y})/p_{\boldsymbol{Y}}(\boldsymbol{y}) & \boldsymbol{Y} \text{ is discrete} \\ \mathsf{P}(\boldsymbol{X} \le \boldsymbol{x}, \boldsymbol{Y} = \boldsymbol{y})/f_{\boldsymbol{Y}}(\boldsymbol{y}) & \boldsymbol{Y} \text{ is continuous} \end{cases} \tag{4.2}$$

$$f_{\boldsymbol{X}|\boldsymbol{Y}=\boldsymbol{y}}(\boldsymbol{x}) = \frac{\partial^n}{\partial x_1 \cdots \partial x_n} F_{\boldsymbol{X}|\boldsymbol{Y}=\boldsymbol{y}}(\boldsymbol{x}) \tag{4.3}$$

$$p_{\boldsymbol{X}|\boldsymbol{Y}=\boldsymbol{y}}(\boldsymbol{x}) = \frac{\mathsf{P}(\boldsymbol{X} = \boldsymbol{x}, \boldsymbol{Y} = \boldsymbol{y})}{\mathsf{P}(\boldsymbol{Y} = \boldsymbol{y})} = \frac{p_{\boldsymbol{X}, \boldsymbol{Y}}((\boldsymbol{x}, \boldsymbol{y}))}{p_{\boldsymbol{Y}}(\boldsymbol{y})}. \tag{4.4}$$

Note that we assume above that $f_{\boldsymbol{Y}}(\boldsymbol{y})$ and $p_{\boldsymbol{Y}}(\boldsymbol{y})$ are not zero.

When both $\boldsymbol{X}$ and $\boldsymbol{Y}$ are continuous their joint cdf is differentiable and

$$f_{\boldsymbol{X}|\boldsymbol{Y}=\boldsymbol{y}}(\boldsymbol{x}) = \frac{f_{\boldsymbol{X}, \boldsymbol{Y}}(\boldsymbol{x}, \boldsymbol{y})}{f_{\boldsymbol{Y}}(\boldsymbol{y})}.$$

Computing conditional probabilities from the conditional pdf and pmf proceeds as in the non-conditional case, by integrating over the corresponding pdf or summing over the corresponding pmf. The proof is similar to the scalar case (see Chapter 2).

$$P(\boldsymbol{X} \in A | \boldsymbol{Y} = \boldsymbol{y}) = \begin{cases} \int_A f_{\boldsymbol{X}|\boldsymbol{Y}=\boldsymbol{y}}(\boldsymbol{x}) d\boldsymbol{x} & \boldsymbol{X} \text{ is continuous} \\ \sum_{\boldsymbol{x} \in A} p_{\boldsymbol{X}|\boldsymbol{Y}=\boldsymbol{y}}(\boldsymbol{x}) & \boldsymbol{X} \text{ is discrete} \end{cases}.$$

Example 4.5.1.

$$F_{X_2|X_1=x_1,X_3=x_3}(x_2) = \begin{cases} \frac{P(X_1=x_1, X_2 \le x_2, X_3=x_3)}{\sum_{x_2} p_{\boldsymbol{X}}(X_1=x_1, X_2=x_2, X_3=x_3)} & \boldsymbol{X} \text{ is discrete} \\ \frac{P(X_1=x_1, X_2 \le x_2, X_3=x_3)}{\int f_{\boldsymbol{X}}(X_1=x_1, X_2=x_2, X_3=x_3) dx_2} & \boldsymbol{X} \text{ is continuous} \end{cases}$$

Example 4.5.2.

$$f_{X_i|\{X_j=x_j:j\ne i\}}(x_i) = \frac{\frac{d}{dx_i} \int_{-\infty}^{x_i} f_{X_1,\ldots,X_n}(x_1,\ldots,x_n) dx_i}{f_{X_1,\ldots,X_{i-1},X_{i+1},\ldots,X_n}(x_1,\ldots,x_{i-1},x_{i+1},\ldots,x_n)}$$

$$= \frac{f_{X_1,\ldots,X_n}(x_1,\ldots,x_n)}{\int_{-\infty}^{\infty} f_{X_1,\ldots,X_n}(x_1,\ldots,x_n) dx_i}.$$

In the case of $n = 3$, we have

$$f_{X_2|X_1=x_1,X_3=x_3}(x_2) = \frac{d}{dx_2} F_{X_2|X_1=x_1,X_3=x_3}(x_2)$$

$$= \frac{f_{X_1,X_2,X_3}(x_1,x_2,x_3)}{\int_{-\infty}^{\infty} f_{X_1,X_2,X_3}(x_1,x_2,x_3) dx_2}.$$

The above formulas lead to the following generalization of the Bayes rule for events $P(A|B) = \frac{P(B|A)\,P(A)}{P(B)}$ (Proposition 1.5.2).

Proposition 4.5.1 (Bayes Theorem).

$$f_{\boldsymbol{X}}(\boldsymbol{X}) = f_{X_i|\{X_j=x_j:j\ne i\}}(x_i) f_{X_1,\ldots,X_{i-1},X_{i+1},\ldots,X_n}(x_1,\ldots,x_{i-1},x_{i+1},\ldots,x_n)$$

$$p_{\boldsymbol{X}}(\boldsymbol{X}) = p_{X_i|\{X_j=x_j:j\ne i\}}(x_i) p_{X_1,\ldots,X_{i-1},X_{i+1},\ldots,X_n}(x_1,\ldots,x_{i-1},x_{i+1},\ldots,x_n)$$

Proof. The pdf formula follows from Example 4.5.2. The derivation of the pmf formula is similar. ∎

Corollary 4.5.1.

$$f_{X_1,\ldots,X_n}(x_1,\ldots,x_n) = f_{X_1}(x_1) f_{X_2|X_1=x_1}(x_2) f_{X_3|X_1=x_1,X_2=x_2}(x_3) \cdots$$
$$f_{X_n|X_1=x_1,\ldots,X_{n-1}=x_{n-1}}(x_n). \quad (4.5)$$

$$p_{X_1,\ldots,X_n}(x_1,\ldots,x_n) = p_{X_1}(x_1) p_{X_2|X_1=x_1}(x_2) p_{X_3|X_1=x_1,X_2=x_2}(x_3) \cdots$$
$$p_{X_n|X_1=x_1,\ldots,X_{n-1}=x_{n-1}}(x_n).$$

Proof. Repeated use of Proposition 4.5.1 establishes (4.5). ∎

The ordering of $X_1, \ldots, X_n$ in the decomposition above is arbitrary and similar formulas hold when the variables in (4.5) are relabeled (for example, replace in (4.5) X_1 with X_2, X_2 with X_3, and X_3 with X_1, or any other arbitrary relabeling[1]). For example the following two equations hold.

$$f_{X_1,X_2,X_3}(\boldsymbol{x}) = f_{X_1}(x_1)f_{X_2|X_1=x_1}(x_2)f_{X_3|X_1=x_1,X_2=x_2}(x_3)$$
$$f_{X_1,X_2,X_3}(\boldsymbol{x}) = f_{X_2}(x_2)f_{X_3|X_2=x_2}(x_3)f_{X_1|X_2=x_2,X_3=x_3}(x_1).$$

Example 4.5.3. *Suppose that a point X is chosen from a uniform distribution in the interval $[0,1]$ and that after $X = x$ is observed a point Y is drawn from a uniform distribution on the interval $[x, 1]$. We have*

$$f_{X,Y}(x,y) = f_X(x)f_{Y|X=x}(x) = \begin{cases} 1 \cdot \frac{1}{1-x} & 0 < x < y < 1 \\ 0 & \text{otherwise} \end{cases}$$

$$f_Y(y) = \begin{cases} \int_{-\infty}^{\infty} f_{X,Y}(x,y)dx = \int_0^y \frac{1}{1-x}dx = -\log(1-y) & 0 < y < 1 \\ 0 & \text{otherwise} \end{cases}$$

$$f_{X|Y=y}(x) = f_{X,Y}(x,y)/f_Y(y) = \begin{cases} \frac{-1}{(1-x)\log(1-y)} & 0 < x < y < 1 \\ 0 & \text{otherwise} \end{cases}.$$

4.6   Moments

Definition 4.6.1. The expectation of a random vector $\boldsymbol{X} = (X_1, \ldots, X_n)$ is the vector of expectations of the corresponding random variables

$$\mathsf{E}(\boldsymbol{X}) \stackrel{\text{def}}{=} (\mathsf{E}(X_1), \ldots, \mathsf{E}(X_n)) \in \mathbb{R}^n.$$

In some cases we arrange the components of a random vector $\boldsymbol{X} = (X_1, \ldots, X_n)$ in a matrix form. In this case the expectation is defined similarly as a matrix whose entries are the expectations of the corresponding RVs. It is common to use vector or matrix equations when dealing with random vectors or matrices in this cases. For example,

$$1 + \mathsf{E}\begin{pmatrix} X_1 & X_2 \\ X_3 & X_4 \end{pmatrix} = \begin{pmatrix} 1 + \mathsf{E}\,X_1 & 1 + \mathsf{E}\,X_2 \\ 1 + \mathsf{E}\,X_3 & 1 + \mathsf{E}\,X_4 \end{pmatrix}.$$

The following example, shows that the linearity property of expectation from Chapter 2 extends to random vectors and random matrices. It may be verified by straightforward application of the linearity properties of expectation (see Chapter 2) and vector and matrix addition and multiplication.

Example 4.6.1. *For a vector of RVs $\boldsymbol{X}$ we have the following vector equality*

$$\mathsf{E}(a\boldsymbol{X} + \boldsymbol{b}) = a\,\mathsf{E}(\boldsymbol{X}) + \boldsymbol{b}.$$

---

[1] Formally, given a permutation function $\pi : \{1, \ldots, n\} \to \{1, \ldots, n\}$, which is a one-to-one and onto function, a relabeling of the vector $(X_1, \ldots, X_n)$ is the vector $(X_{\pi(1)}, \ldots, X_{\pi(n)})$.

For a matrix of RVs $\boldsymbol{Y}$, we have the following matrix equality

$$\mathsf{E}(A\boldsymbol{Y}B + C) = A\,\mathsf{E}(\boldsymbol{Y})B + C.$$

Above, we assume that $\boldsymbol{b}$ is a constant vector, a is a constant scalar, and A, B, C are constant matrices.

The following proposition generalizes Proposition 2.3.1.

Proposition 4.6.1. *For a random vector $\boldsymbol{X} = (X_1, \ldots, X_n)$ and a function $g : \mathbb{R}^n \to \mathbb{R}$,*

$$\mathsf{E}(g(\boldsymbol{X})) = \begin{cases} \int_{\mathbb{R}^n} g(\boldsymbol{x})f_{\boldsymbol{X}}(\boldsymbol{x})\,d\boldsymbol{x} & \boldsymbol{X} \text{ is continuous} \\ \sum_{\boldsymbol{x} \in \mathbb{R}^n} g(\boldsymbol{x})p_{\boldsymbol{X}}(\boldsymbol{x}) & \boldsymbol{X} \text{ is discrete} \end{cases}.$$

Proof. The proof is identical to the proof of Proposition 2.3.1. ∎

If $g(\boldsymbol{X})$ is a function of only a subset of the component RVs $X_1, \ldots, X_n$ (for example $g(\boldsymbol{X}) = (X_1 + X_3)/2$), the expression in Proposition 4.6.1 may be simplified by taking the integration or summation only with respect to the components of $\boldsymbol{X}$ present in $g(\boldsymbol{X})$. In this case $p_{\boldsymbol{X}}$ or $f_{\boldsymbol{X}}$ may be replaced with the joint pmf or pdf over the reduced set of variables. The following two examples demonstrate this principle and justify its application.

Example 4.6.2. *For a continuous random vector $\boldsymbol{X} = (X_1, X_2, X_3)$ and $g(\boldsymbol{X}) = X_1/X_2$,*

$$\begin{aligned}
\mathsf{E}(X_1/X_2) &= \iiint f_{X_1,X_2,X_3}(x_1, x_2, x_3)x_1/x_2 \, dx_1 dx_2 dx_3 \\
&= \iiint f_{X_3|X_1=x_1,X_2=x_2}(x_3)f_{X_1,X_2}(x_1, x_2)x_1/x_2 \, dx_1 dx_2 dx_3 \\
&= \iiint f_{X_3|X_1=x_1,X_2=x_2}(x_3) \, dx_3 f_{X_1,X_2}(x_1, x_2)x_1/x_2 \, dx_1 dx_2 \\
&= 1 \cdot \iint f_{X_1,X_2}(x_1, x_2)x_1/x_2 \, dx_1 dx_2.
\end{aligned}$$

A similar expression holds in the discrete case (replace pdfs with pmfs and integrals with sums).

Example 4.6.3. *Taking expectation of X_j with respect to the marginal f_{X_j} or with respect to the joint $f_{\boldsymbol{X}}$ produces identical results:*

$$\begin{aligned}
&\mathsf{E}(X_j) \\
&= \int \cdots \int x_j f_{\boldsymbol{X}}(x_1, \ldots, x_n) \, dx_1 \cdots dx_n \\
&= \int x_j f_{X_j}(x_j) \, dx_j \int \cdots \int f_{\{X_i : i \neq j\}|X_j = x_j}(\{x_i : i \neq j\}) \, dx_1 \cdots dx_{j-1} dx_{j+1} \cdots dx_n \\
&= \int x_j f_{X_j}(x_j) \, dx_j \cdot 1.
\end{aligned}$$

A similar expression holds in the discrete case (replace pdfs with pmfs and integrals with sums).

Definition 4.6.2. The covariance of two random variables X, Y is

$$\mathsf{Cov}(X, Y) \stackrel{\text{def}}{=} \mathsf{E}((X - \mathsf{E}(X))(Y - \mathsf{E}(Y))).$$

The covariance matrix of two random vectors $\boldsymbol{X} = (X_1, \ldots, X_n)$, $\boldsymbol{Y} = (Y_1, \ldots, Y_m)$ is the $n \times m$ matrix defined by

$$[\mathsf{Cov}(\boldsymbol{X}, \boldsymbol{Y})]_{ij} = \mathsf{Cov}(X_i, Y_j),$$

or using matrix notation (assuming $\boldsymbol{X}, \boldsymbol{Y}$ are row vectors)

$$\mathsf{Cov}(\boldsymbol{X}, \boldsymbol{Y}) = \mathsf{E}((\boldsymbol{X} - \mathsf{E}(\boldsymbol{X}))(\boldsymbol{Y} - \mathsf{E}(\boldsymbol{Y}))^\top)$$

(where $A^\top$ is the transpose of the matrix A). The variance matrix of the random vector $\boldsymbol{X} = (X_1, \ldots, X_n)$ is the $n \times n$ matrix defined by

$$[\mathsf{Var}(\boldsymbol{X})]_{ij} = \mathsf{Cov}(X_i, X_j),$$

or using matrix notation (assuming $\boldsymbol{X}, \boldsymbol{Y}$ are row vectors)

$$\mathsf{Var}(\boldsymbol{X}) = \mathsf{Cov}(\boldsymbol{X}, \boldsymbol{X}) = \mathsf{E}((\boldsymbol{X} - \mathsf{E}(\boldsymbol{X}))(\boldsymbol{X} - \mathsf{E}(\boldsymbol{X}))^\top).$$

Note that the variance of a random variable $\mathsf{Var}(X)$ is in agreement with the 1×1 variance matrix of a random vector with one component $\boldsymbol{X} = (X)$.

Proposition 4.6.2.

$$\mathsf{Cov}(X_i, X_j) = \mathsf{E}(X_i X_j) - \mathsf{E}(X_i) \mathsf{E}(X_j) \tag{4.6}$$
$$\mathsf{Cov}(\boldsymbol{X}, \boldsymbol{Y}) = \mathsf{E}(\boldsymbol{X} \boldsymbol{Y}^\top) - \mathsf{E}(\boldsymbol{X}) \mathsf{E}(\boldsymbol{Y})^\top. \tag{4.7}$$

Proof. The following derivation proves the first statement

$$\begin{aligned}
\mathsf{Cov}(X_i, X_j) &= \mathsf{E}(X_i X_j - X_i \mathsf{E}(X_j) - X_j \mathsf{E}(X_i) + \mathsf{E}(X_i) \mathsf{E}(X_j)) \\
&= \mathsf{E}(X_i X_j) - 2 \mathsf{E}(X_i) \mathsf{E}(X_j) + \mathsf{E}(X_i) \mathsf{E}(X_j) \\
&= \mathsf{E}(X_i X_j) - \mathsf{E}(X_i) \mathsf{E}(X_j).
\end{aligned}$$

The second statement follows by applying the first statement to each element of the relevant matrices. ∎

Proposition 4.6.3 (Linearity Property of Expectation).

$$\mathsf{E}\left(\sum_{i=1}^n a_i X_i\right) = \sum_{i=1}^n a_i \mathsf{E}(X_i)$$

Proof. Using the argument in Example 4.6.3 we have

$$\mathsf{E}(aX + bY) = \iint_{\mathbb{R}^2} (ax + by)f_{X,Y}(x,y)\,dxdy$$

$$= \int_{-\infty}^{\infty} ax \int_{-\infty}^{\infty} f_{X,Y}(x,y)\,dydx + \int_{-\infty}^{\infty} by \int f_{X,Y}(x,y)\,dxdy$$

$$= a\,\mathsf{E}(X) + b\,\mathsf{E}(Y).$$

(If X, Y are discrete replace integrals with sums and pdf functions with pmf functions). Using induction completes the proof. ∎

We emphasize that the linearity of expectation in Proposition 4.6.3 above holds for any random vector $\boldsymbol{X}$, including the case where the variables X_i are dependent.

Proposition 4.6.4. *For a random vector $\boldsymbol{X}$ with independent components and any functions $g_i : \mathbb{R} \to \mathbb{R}$, $i = 1, \ldots, n$ we have*

$$E\left(\prod_{i=1}^{n} g_i(X_i)\right) = \prod_{i=1}^{n} \mathsf{E}(g_i(X_i)).$$

In particular for independent X, Y we have $\mathsf{E}(XY) = \mathsf{E}(X)\,\mathsf{E}(Y)$.

Proof.

$$\mathsf{E}(g_1(X)g_2(Y)) = \iint_{\mathbb{R}^2} g_1(x)g_2(y)f_{X,Y}(x,y)\,dxdy$$

$$= \iint_{\mathbb{R}^2} g_1(x)g_2(y)f_X(x)f_Y(y)\,dxdy$$

$$= \left(\int_{-\infty}^{\infty} g_1(x)f_X(x)dx\right)\left(\int_{-\infty}^{\infty} g_2(y)f_Y(y)dy\right)$$

$$= \mathsf{E}(g_1(X))\,\mathsf{E}(g_2(Y)).$$

The proof for discrete RVs is similar (replace integrals with sums and pdf functions with pmf functions). Using induction completes the proof. ∎

Corollary 4.6.1. *If $\boldsymbol{X}$ is an independent random vector, then $\mathsf{Var}(\boldsymbol{X})$ is a diagonal matrix, or in other words $i \neq j$ implies $\mathsf{Cov}(X_i, X_j) = 0$.*

Proof. According to Definition 4.6.2

$$\mathsf{Cov}(X, Y) = E(XY) - E(X)E(Y),$$

which by Proposition 4.6.4 is zero for independent X, Y. ∎

While independence implies zero covariance, the converse is not necessarily true: two dependent RVs may have zero or non-zero covariance. Intuitively,

$\mathsf{Cov}(X, Y)$ measures the extent to which there exists a linear relationship between X and Y: $X = \alpha Y + \beta$, $\alpha, \beta \in \mathbb{R}$. If there is no linear relationship, the covariance is zero but the variables may still be dependent. The bottom row of Figure 4.3 displays examples of dependent RVs with zero covariance.

Proposition 4.6.5.

$$\mathsf{Var}\left(\sum_{i=1}^{n} X_i\right) = \sum_{i=1}^{n}\sum_{j=1}^{n} \mathsf{Cov}(X_i, X_j) = \sum_{i=1}^{n} \mathsf{Var}(X_i) + 2\sum_{i=1}^{n}\sum_{j>i} \mathsf{Cov}(X_i, X_j).$$

In particular for any two RVs X, Y,

$$\mathsf{Var}(X + Y) = \mathsf{Var}(X) + \mathsf{Var}(Y) + 2\,\mathsf{Cov}(X, Y)$$
$$\mathsf{Var}(X - Y) = \mathsf{Var}(X) + \mathsf{Var}(Y) - 2\,\mathsf{Cov}(X, Y).$$

Proof. For two RVs,

$$\mathsf{Var}(X + Y)$$
$$= \mathsf{E}(((X + Y) - \mathsf{E}(X + Y))^2)$$
$$= \mathsf{E}(X^2 + Y^2 + 2XY + (\mathsf{E}(X))^2 + (\mathsf{E}(Y))^2 + 2\,\mathsf{E}(X)\,\mathsf{E}(Y) - 2(X + Y)\,\mathsf{E}(X + Y))$$
$$= \mathsf{E}(X^2) + \mathsf{E}(Y^2) + 2\,\mathsf{E}(XY) + (\mathsf{E}(X))^2 + (\mathsf{E}(Y))^2 + 2\,\mathsf{E}(X)\,\mathsf{E}(Y) - 2(\mathsf{E}(X) + \mathsf{E}(Y))^2$$
$$= \mathsf{E}(X^2) - (\mathsf{E}(X))^2 + \mathsf{E}(Y^2) - (\mathsf{E}(Y))^2 + 2\,\mathsf{E}(XY) - 2\,\mathsf{E}(X)\,\mathsf{E}(Y)$$
$$= \mathsf{Var}(X) + \mathsf{Var}(Y) + 2\,\mathsf{Cov}(X, Y).$$

The case of $X - Y$ reduces to $X + (-Y)$, causing the covariance to receive a minus sign. The more general case follows by induction. ∎

Corollary 4.6.2. *If $X = (X_1, \ldots, X_n)$ is an independent random vector*

$$\mathsf{Var}\left(\sum_{i=1}^{n} X_i\right) = \sum_{i=1}^{n} \mathsf{Var}(X_i).$$

Proof. The proof follows the proposition above and the fact that independent variables have zero covariance. ∎

Proposition 4.6.6.

$$\mathsf{E}(X^{\top} A X) = \mathsf{tr}(A\,\mathsf{Var}(X)) + (\mathsf{E}(X))^{\top} A\,\mathsf{E}(X).$$

Proof. Denoting $\mathsf{E}(\boldsymbol{X}) = \boldsymbol{\mu}$ and $\mathsf{Var}(\boldsymbol{X}) = \Sigma$,

$$
\begin{aligned}
\mathsf{E}(\boldsymbol{X}^\top A \boldsymbol{X}) &= \mathsf{tr}(\mathsf{E}(\boldsymbol{X}^\top A \boldsymbol{X}) \\
&= \mathsf{E}(\mathsf{tr}(\boldsymbol{X}^\top A \boldsymbol{X}) \\
&= \mathsf{E}(\mathsf{tr}(A \boldsymbol{X} \boldsymbol{X}^\top)) \\
&= \mathsf{tr}(\mathsf{E}(A \boldsymbol{X} \boldsymbol{X}^\top)) \\
&= \mathsf{tr}(A \mathsf{E}(\boldsymbol{X} \boldsymbol{X}^\top)) \\
&= \mathsf{tr}(A(\mathsf{Var}(\boldsymbol{X}) + \boldsymbol{\mu}\boldsymbol{\mu}^\top) \\
&= \mathsf{tr}(A\Sigma) + \mathsf{tr}(A\boldsymbol{\mu}\boldsymbol{\mu}^\top) \\
&= \mathsf{tr}(A\Sigma) + \boldsymbol{\mu}^\top A \boldsymbol{\mu}.
\end{aligned}
$$

∎

Proposition 4.6.7.

$$
\mathsf{Cov}(A\boldsymbol{X}, B\boldsymbol{Y}) = A\,\mathsf{Cov}(\boldsymbol{X}, \boldsymbol{Y})B^\top
$$

Proof. Using the linearity property of the expectation (Proposition 4.6.3),

$$
\begin{aligned}
\mathsf{Cov}(A\boldsymbol{X}, B\boldsymbol{Y}) &= \mathsf{E}((A\boldsymbol{X} - A\mathsf{E}(X))(B\boldsymbol{Y} - B\mathsf{E}(Y))^\top) \\
&= \mathsf{E}(A(\boldsymbol{X} - \mathsf{E}(X))((\boldsymbol{Y} - \mathsf{E}(Y)^\top B^\top))) \\
&= A\,\mathsf{E}((\boldsymbol{X} - \mathsf{E}(X))((\boldsymbol{Y} - \mathsf{E}(Y)^\top)))B^\top \\
&= A\,\mathsf{Cov}(\boldsymbol{X}, \boldsymbol{Y})B^\top.
\end{aligned}
$$

∎

Corollary 4.6.3. *For any matrix A and random vector $\boldsymbol{X}$*

$$
\mathsf{Var}(A\boldsymbol{X}) = A\,\mathsf{Var}(\boldsymbol{X})A^\top. \tag{4.8}
$$

Proposition 4.6.8. *The matrix $\mathsf{Var}(\boldsymbol{X})$ is positive semi-definite. It is positive definite if no component of $\boldsymbol{X}$ is a linear combination of the other components.*

Proof. Recall that the variance is an expectation of a non-negative (squared) random variable. Using Corollary 4.6.3 we have for all column vectors $\boldsymbol{v}$,

$$
0 \le \mathsf{Var}(\boldsymbol{v}^\top \boldsymbol{X}) = \boldsymbol{v}^\top \mathsf{Var}(\boldsymbol{X})\boldsymbol{v},
$$

which shows that the variance matrix is positive semi-definite. The inequality above holds with equality if and only if $\boldsymbol{v}$ is the zero vector or $\boldsymbol{v}^\top \boldsymbol{X}$ is a deterministic RV taking a constant value with probability 1. This can only happen if one component of $\boldsymbol{X}$ is a linear combination of the other components. ∎

Definition 4.6.3. The correlation coefficient of two RVs X, Y is defined as

$$\mathsf{Cor}(X, Y) = \frac{\mathsf{Cov}(X, Y)}{\sqrt{\mathsf{Var}(X)}\sqrt{\mathsf{Var}(Y)}}.$$

Proposition 4.6.9. *For any random variables X, Y we have*

$$-1 \le \mathsf{Cor}(X, Y) \le 1,$$

where the inequality above holds with equality if and only if there is a linear relationship between X and Y, specifically $X = aY + b$ or $Y = aX + b$ for some $a, b \in \mathbb{R}$.

Proof. Since expectation of non-negative RVs is non-negative

$$0 \le E\left(\left(\frac{X - E(X)}{\sqrt{\mathsf{Var}(X)}} \pm \frac{Y - E(Y)}{\sqrt{\mathsf{Var}(Y)}}\right)^2\right)$$

$$= \frac{E((X - E(X))^2)}{\mathsf{Var}\, X} + \frac{E((Y - E(Y))^2)}{\mathsf{Var}\, Y} \pm 2\,\mathsf{Cor}(X, Y)$$

$$= 2(1 \pm \mathsf{Cor}(X, Y)).$$

(The notation $\pm$ above means that the entire derivation may be repeated alternatively with a plus and a minus sign.) The proposition follows. ∎

4.7   Conditional Expectations

Definition 4.7.1. The conditional expectation of the RV Y conditioned on $X = x$ is

$$E(Y|X = x) = \begin{cases} \int_{-\infty}^{\infty} y f_{Y|X=x}(y)\, dy & Y|X = x \text{ is a continuous RV} \\ \sum_y y p_{Y|X=x}(y) & Y|X = x \text{ is a discrete RV} \end{cases}$$

Intuitively, $E(Y|X = x)$ represents the average or expected value of Y if we know that $X = x$. Definition 4.7.1 extends naturally to conditioning on multiple random variables, for example $E(X_i|\{X_j = x_j : j \ne i\})$ (replace in Definition 4.7.1 the appropriate conditional pdf or pmf).

For a given x, the conditional expectation $E(Y|X = x)$ is a real number. We can also consider the conditional expectation $E(Y|X = x)$ as a function of x: $g(x) = E(Y|X = x)$. The mapping $x \mapsto g(x) = E(Y|X = x)$ corresponds to the random variable $E(Y|X)$.

Definition 4.7.2. The conditional expectation $E(Y|X)$ is a random variable $E(Y|X) : \Omega \to \mathbb{R}$ defined as follows:

$$E(Y|X)(\omega) = E(Y|X = X(\omega)).$$

In other words, for every value $\omega \in \Omega$ we obtain a value $X(\omega) \in \mathbb{R}$, which we may denote as x, and this in turn leads to $E(Y|X = x)$. Note that $E(Y|X)$ is a RV that is a function of the random variable X, or in other words $E(Y|X) = g(X)$ for some function g.

Since $E(Y|X)$ is a random variable, we can compute its expectation. The following proposition discovers an interesting relationship between the expectation of $E(Y|X)$ and the expectation of Y: for any X, we have $E(E(Y|X)) = E(Y)$.

Proposition 4.7.1.
$$E(E(Y|X)) = E(Y).$$

Proof.

$$E(E(Y|X)) = \int_{-\infty}^{\infty} E(Y|X = x) f_X(x)\, dx$$
$$= \int_{-\infty}^{\infty} \int_{-\infty}^{\infty} y f_{Y|X=x}(y)\, dy f_X(x)\, dx$$
$$= \int_{-\infty}^{\infty} y \int_{-\infty}^{\infty} f_{X,Y}(x, y)\, dx dy$$
$$= \int_{-\infty}^{\infty} y f_Y(y)\, dy$$
$$= E(Y)$$

where the first equality holds since $E(Y|X)$ is a function of X and $E(g(X)) = \int g(x) f_X(x) dx$. The proof in the discrete case is similar. ∎

The proposition above is sometimes useful for simplifying the calculation of $E X$. An example appears below.

Example 4.7.1. *Recalling Example 4.5.3, where*

$$X \sim U([0, 1])$$
$$\{Y|X = x\} \sim U([x, 1]) \quad x \in (0, 1),$$

we have $E(Y|X = x) = (x + 1)/2$ *(as shown in Section 3.7 it is the middle point of the interval* $[x, 1]$*). It follows that* $E(Y|X) = (X + 1)/2$ *and by the linearity of the expectation,*

$$E(Y) = E(E(Y|X)) = (E(X) + 1)/2 = \left(\frac{1}{2} + 1\right)/2 = 3/4.$$

4.8 Moment Generating Function

Proposition 4.8.1. *If* $X_1, \ldots, X_n$ *are independent RVs with mgfs* $m_{X_i}, \ldots, m_{X_n}$*, then the mgf of their sum is*

$$m_{\sum_{i=1}^{n} X_i}(t) = \prod_{i=1}^{n} m_{X_i}(t)$$

Proof.

$$m_{\sum_{i=1}^{n} X_i}(t) = \mathsf{E}\left(\exp\left(t\sum_{i=1}^{n} X_i\right)\right)$$

$$= \mathsf{E}\left(\prod_{i=1}^{n} \exp(tX_i)\right)$$

$$= \prod_{i=1}^{n} \mathsf{E}(\exp(tX_i))$$

$$= \prod_{i=1}^{n} m_{X_i}(t).$$

$\blacksquare$

Definition 4.8.1. We define the moment generating function of a random vector X as follows

$$m_X : \mathbb{R}^n \to \mathbb{R}, \qquad m_X(t) = \mathsf{E}(\exp(t^{\top} X)).$$

As in the one dimensional case the mgf uniquely characterizes the distribution of the random vector.

Proposition 4.8.2. *Suppose that X, Y are two random vectors whose mgfs $m_X(t)$, $m_Y(t)$ exist for all $t \in B_\epsilon(0)$ for some $\epsilon > 0$. If $m_X(t) = m_Y(t)$ for all $t \in B_\epsilon(0)$, then the cdfs of X and Y are identical (and consequentially also the pdf or pmf functions are identical).*

Proof. The proof is similar to the proof of the one dimensional case (Proposition 2.4.2). $\blacksquare$

4.9 Random Vectors and Independent σ-Algebras*

As mentioned in Section 2.5, a random variable is a measurable function from the measurable space $(\Omega, \mathcal{F})$ to the measurable space $(\mathbb{R}, \mathcal{B}(\mathbb{R}))$. Similarly, a random vector is a measurable function from $(\Omega, \mathcal{F})$ to $(\mathbb{R}^d, \mathcal{B}(\mathbb{R}^d))$. This ensures that probabilities $\mathsf{P}(X \in A)$ are defined for all $A \in \mathcal{B}(\mathbb{R}^d)$.

Definition 4.9.1. The sets $\mathcal{A}_1, \ldots, \mathcal{A}_n$ of events are independent if $A_1 \in \mathcal{A}_1, \ldots, A_n \in \mathcal{A}_n$ implies that $A_1, \ldots, A_n$ are independent (see Definition 1.5.2).

Proposition 4.9.1. *If $\mathcal{A}_1, \ldots, \mathcal{A}_n$ are independent π-systems, then $\sigma(\mathcal{A}_1), \ldots, \sigma(\mathcal{A}_n)$ are independent.*

Both the definition and proposition above extend from a finite n to a potentially infinite number of sets.

Proof. We denote the probability measure space by $(\Omega, \mathcal{F}, \mathsf{P})$, note that $\mathcal{A}'_i = \mathcal{A} \cup \{\Omega\}$, $i = 1, \ldots, n$ are π-systems, and denote $\mathcal{L} \subset \mathcal{F}$ to be the set of all sets B for which

$$\mathsf{P}(B \cap A_2 \cdots \cap A_n) = \mathsf{P}(B) \, \mathsf{P}(A_2) \cdots \mathsf{P}(A_n)$$

for some fixed sets $A_2 \in \mathcal{A}_2, \ldots, A_n \in \mathcal{A}_n$.

We verify below that $\mathcal{L}$ is a λ-system. We have $\Omega \in \mathcal{L}$ since

$$\mathsf{P}(\Omega \cap A_2 \cap \cdots A_n) = \mathsf{P}(A_2 \cap \cdots \cap A_n) = \mathsf{P}(A_2) \cdots \mathsf{P}(A_n).$$

The set $\mathcal{L}$ is closed under complement since $B \in \mathcal{L}$ implies that

$$
\begin{aligned}
\mathsf{P}(B^c \cap A_2 \cap \cdots A_n) &= \mathsf{P}((\Omega \setminus B) \cap A_2 \cap \cdots A_n) \\
&= \mathsf{P}(\Omega \cap A_2 \cap \cdots \cap A_n \setminus B \cap A_2 \cdots \cap A_n) \\
&= \mathsf{P}(\Omega \cap A_2 \cap \cdots \cap A_n) - \mathsf{P}(B \cap A_2 \cdots \cap A_n) \\
&= \mathsf{P}(A_2) \cdots \mathsf{P}(A_n) - \mathsf{P}(B) \, \mathsf{P}(A_2) \cdots \mathsf{P}(A_n) \\
&= (1 - \mathsf{P}(B)) \, \mathsf{P}(A_2) \cdots \mathsf{P}(A_n).
\end{aligned}
$$

The set $\mathcal{L}$ is closed under disjoint union: if $B_k \in \mathcal{L}, k \in \mathbb{N}$ are pairwise disjoint sets whose union is B, then

$$
\begin{aligned}
\mathsf{P}((\cup_{k \in \mathbb{N}} B_k) \cap A_2 \cap \cdots \cap A_n) &= \mathsf{P}(\cup_{k \in \mathbb{N}} (B_k \cap A_2 \cap \cdots \cap A_n)) \\
&= \sum_{k \in \mathbb{N}} \mathsf{P}(B_k \cap A_2 \cap \cdots \cap A_n) \\
&= \sum_{k \in \mathbb{N}} \mathsf{P}(B_k) \, \mathsf{P}(A_2) \cdots \mathsf{P}(A_n) \\
&= \mathsf{P}(A_2) \cdots \mathsf{P}(A_n) \sum_{k \in \mathbb{N}} \mathsf{P}(B_i).
\end{aligned}
$$

Since $\mathcal{L}$ is a λ-system that contains the π-system $\mathcal{A}'_1$, it follows from Dynkin's theorem (Proposition E.3.4) that $\sigma(\mathcal{A}_1), \mathcal{A}_2, \ldots, \mathcal{A}_n$ are independent. Repeating the argument multiple times concludes the proof. ∎

4.10   Notes

More information on random vectors is available in nearly any probability text-book. Elementary exposition that avoids measure theory is available in most undergraduate probability textbooks, for example [48, 14]. A rigorous measure-theoretic description is available in [17, 10, 5, 1, 33, 25]. Convolutions are usually described in detail in signal processing books, for example [31].

4.11   Exercises

1. Consider two independent discrete RVs X, Y that are uniform over $\{1, 2, 3, 4\}$. Derive the pmf of the RV $X + Y$ using the convolution technique.

2. Repeat (1) for $X - Y$.

3. Specify formally a pair of RVs X, Y for which $\text{Cov}(X, Y) = 0$.

Chapter 5

Important Random Vectors

The two most important random vectors are the multinomial (discrete) and the multivariate Gaussian (continuous). The first generalizes the binomial random variable and the second generalizes the Gaussian random variable. This chapter describes both distributions, and concludes with a discussion of three more important random vectors: Dirichlet, mixtures, and the exponential family random vectors.

5.1 The Multinomial Random Vector

Definition 5.1.1. A unit vector $x \in \mathbb{R}^k$ is a vector whose components are all zeros except for one component that is one.

Definition 5.1.2. The multinomial simplex, or simply the simplex, is the set of all non-negative vectors that sum to one:

$$\mathbb{P}_k = \left\{ z \in \mathbb{R}^k \ : \ z_i \geq 0, \ i = 1, \dots, k, \ \sum_{i=1}^{k} z_i = 1 \right\}.$$

Since $\mathbb{P}_k$ is a set of points a linear constraint $\sum z_i = 1$, $\mathbb{P}_k$ is a part of a $k - 1$ dimensional linear hyperplane in $\mathbb{R}^k$. Figure 5.1 illustrates the spatial relationship between $\mathbb{P}_k$ and $\mathbb{R}^k$ for $k = 3$.

As described in Section 1.2, a probability function over a finite sample space $\Omega = \{\omega_1, \dots, \omega_k\}$ is characterized by a vector of probability values $\boldsymbol{\theta} = (\theta_1, \dots, \theta_k)$ satisfying $\theta_i \geq 0, i = 1, \dots, k$ and $\sum \theta_i = 1$. The probability function is defined by the following equation

$$\mathsf{P}(A) = \sum_{i \in A} \theta_i.$$

This implies that the simplex $\mathbb{P}_k$ parameterizes the set of all distributions on a sample size Ω of size k. There exists a one-to-one and onto function mapping each elements of the simplex to distinct distributions.

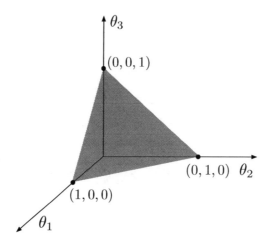

Figure 5.1: The simplex $\mathbb{P}_3$ as a subset in $\mathbb{R}^3$ and the three unit vectors corresponding to its corners.

Definition 5.1.3. The multivariate Bernoulli random vector $\boldsymbol{X} \sim \mathrm{Ber}(\boldsymbol{\theta})$, where $\boldsymbol{\theta} \in \mathbb{P}_k$, is a k dimensional discrete random vector with the following pmf

$$p_{\boldsymbol{X}}(\boldsymbol{x}) = \begin{cases} \theta_i & x_i = 1 \text{ and } x_j = 0 \text{ for } j \neq i \\ 0 & \boldsymbol{x} \text{ is not a unit vector} \end{cases}.$$

We make the following comments.

- The multivariate Bernoulli random vector is a unit vector describing the outcome of a random experiment with k possible outcome – each outcome occurring with probability θ_k.

- Since $X_i = 1$ implies $X_j = 0$ for $j \neq i$, the multivariate Bernoulli random vector does not have independent components.

- The pmf of a Bernoulli random vector may be written as

$$p_{\boldsymbol{X}}(\boldsymbol{x}) = \begin{cases} \prod_{i=1}^{k} \theta_i^{x_i} & \boldsymbol{x} \text{ is a unit vector} \\ 0 & \text{otherwise} \end{cases}.$$

The expectation vector and variance matrix (see Chapter 4) of the multivariate Bernoulli random vector are

$$\mathsf{E}(\boldsymbol{X}) = \boldsymbol{\theta} \tag{5.1}$$

$$\mathsf{Var}(\boldsymbol{X}) = \mathsf{E}(\boldsymbol{X}\boldsymbol{X}^{\top}) - \mathsf{E}(\boldsymbol{X})\,\mathsf{E}(\boldsymbol{X})^{\top} = \mathrm{diag}(\boldsymbol{\theta}) - \boldsymbol{\theta}\boldsymbol{\theta}^{\top} \tag{5.2}$$

where $\mathrm{diag}(\boldsymbol{\theta})$ is a diagonal matrix with $(\theta_1, \ldots, \theta_k)$ as its diagonal (and 0 elsewhere). The derivations above follow from the fact that an expectation of a binary random variable is the probability of it being one.

Definition 5.1.4. The multinomial random vector, $\boldsymbol{X} = (X_1, \ldots, X_k) \sim \text{Mult}(N, \boldsymbol{\theta})$, where $N \in \mathbb{N}$ and $\boldsymbol{\theta} \in \mathbb{P}_k$, is a sum of N independent mutlivariate Bernoulli random vectors with parameter $\boldsymbol{\theta}$.

We make the following observations.

- Recall from Proposition 1.6.7 that the number of possible sequences of N multivariate Bernoulli experiments resulting in x_i successes in experiment i for $i = 1, \ldots, k$ is $N!/(x_1! \cdots x_k!)$ (where $N = \sum x_i$). It follows that the pmf of the multinomial random vector is

$$p_{\boldsymbol{X}}(\boldsymbol{x}) = \begin{cases} \frac{N!}{x_1! \cdots x_k!} \theta_1^{x_1} \cdots \theta_k^{x_k} & x_i \geq 0, \ i = 1, \ldots, k, \ \sum_{i=1}^{k} x_i = N \\ 0 & \text{otherwise} \end{cases}.$$

- As in the case of the multivariate Bernoulli vector, the multinomial vector does not have independent components.

- Since the multinomial random vector is a sum of independent multivariate Bernoulli random vectors, its expectation vector and variance matrix are (see Proposition 4.6.3 and Corollary 4.6.2)

$$\mathsf{E}(\boldsymbol{X}) = N\boldsymbol{\theta}$$
$$\mathsf{Var}(\boldsymbol{X}) = N\text{diag}(\boldsymbol{\theta}) - N\boldsymbol{\theta}\boldsymbol{\theta}^\top.$$

- The support of the multinomial pmf (the set of vectors $\boldsymbol{x}$ for which $p_{\boldsymbol{X}}(\boldsymbol{x}) > 0$) satisfies a linear constraint $\sum x_i = N$, indicating that it lies on a $k - 1$ dimensional linear hyperplane in $\mathbb{R}^k$.

Example 5.1.1. *An American Roulette wheel has 38 possible outcomes: 18 red, 18 black and 2 green outcomes. Playing a fair American Roulette (all outcomes are equally likely) is a multivariate Bernoulli experiment with $\theta_1 = \theta_2 = 18/38$ and $\theta_3 = 2/38$. The distribution of the outcomes over multiple games follows a multinomial distribution. For example, we play the roulette 10 times, the probability that we get 4 red outcomes, 2 black outcomes and 4 green is*

$$p_{X_1, X_2, X_3}(4, 2, 4) = \frac{10!}{4!2!4!}(18/38)^4(18/38)^2(2/38)^4$$

The multinomial coefficient is present since there are 10!/(4!2!4!) ways to play 10 times and obtain 4 red 2 black and 4 green outcomes.

The following two graphs illustrate the multinomial pmf corresponding to $N = 12$, and $\boldsymbol{\theta} = (0.5, 0.3, 0.2)$. The first graphs the pmf as a three dimensional bar chart, where the height of each bar indicates the probability or pmf value, and the two dimensional position of each bar corresponds to the vector (x_1, x_2). The third dimension is redundant since the three components of $\boldsymbol{x}$ have to sum to one. It appears that the multinomial pmf has a unimodal shape with the peak around the expectation vector $(6, 3.6, 2.4)$ (the peak in the graph is located at the first two components of the expectation vector.)

```
library(scatterplot3d)
R = expand.grid(0:12, 0:12)
R[, 3] = 12 - R[, 1] - R[, 2]
names(R) = c("x1", "x2", "x3")
for (i in 1:nrow(R)) {
    v = R[i, 1:3]
    if (sum(v) == 12 & all(v >= 0)) {
        R$p[i] = dmultinom(v, prob = c(0.5, 0.3,
            0.2))
    } else R$p[i] = 0
}
scatterplot3d(x = R$x1, y = R$x2, z = R$p, type = "h",
    lwd = 5, pch = " ", xlab = "$x_1$", ylab = "$x_2$",
    zlab = "$p_{X_1,X_2,X_3}(x_1,x_2,1-x_1-x_2)$",
    color = grey(1/2), col.axis = "blue", col.grid = "lightblue")
title(list("Multinomial pmf (third dimension suppressed)",
    cex = 0.7))
```

Multinomial pmf (third dimension suppressed)

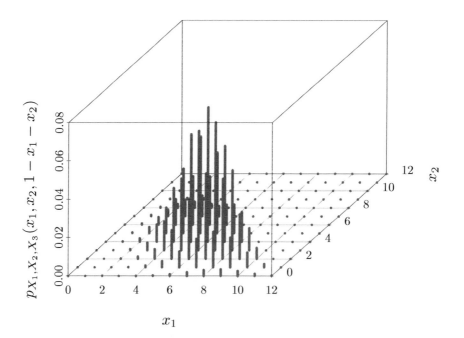

The second graph shows the non-negative values of the pmf using a three dimensional scatter plot. The three dimensions of each point correspond to its components (x_1, x_2, x_3) and its gray level corresponds to its pmf $p_X(\boldsymbol{x})$. The colors are assigned in the following manner: light gray colors correspond to high probabilities while dark gray colors correspond to low probabilities. It appears that the multinomial pmf is non-zero over a linear hyperplane, and has unimodal shape with the peak around the expectation vector $(6, 3.6, 2.4)$.

```
scatterplot3d (x = R$x1, y = R$x2, z = R$x3, color = gray(0.9 -
    R$p * 10), type = "p", xlab = "$x_1$", pch = 16,
    ylab = "$x_2$", zlab = "$x_3$", cex.symbols = 1.5,
    col.axis = "blue", col.grid = "lightblue")
title(list("Multinomial pmf (dark indicates high probability)",
    cex = 0.7))
```

Multinomial pmf (dark indicates high probability)

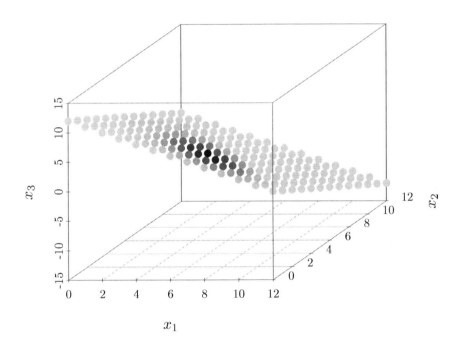

5.2   The Multivariate Normal Random Vector

Definition 5.2.1. The multivariate Gaussian random vector, $\boldsymbol{X} = (X_1, \ldots, X_n) \sim N(\boldsymbol{\mu}, \Sigma)$, where $\boldsymbol{\mu} \in \mathbb{R}^n$ and $\Sigma \in \mathbb{R}^{n \times n}$ is a symmetric positive definite matrix, is a continuous random vector with following pdf

$$f_X(x) = \frac{1}{(2\pi)^{n/2}\sqrt{\det \Sigma}} \exp\left(-\frac{1}{2}(x - \mu)^\top \Sigma^{-1}(x - \mu)\right).$$

We make the following observations.

- As Example C.1.2 shows, we have

$$f_X(x) = C(\mu, \Sigma) \exp\left(-\frac{1}{2}\sum_{i=1}^{n}\sum_{j=1}^{n}(x_i - \mu_i)[\Sigma^{-1}]_{ij}(x_j - \mu_j)\right)$$

  where $C^{-1} = (2\pi)^{n/2}\sqrt{\det \Sigma}$ is constant in x. This implies that the pdf decreases as the value of the quadratic form $\sum_{i=1}^{n}\sum_{j=1}^{n}(x_i - \mu_i)[\Sigma^{-1}]_{ij}(x_j - \mu_j)$ increases.

- In the one dimensional case $n = 1$, Σ is a scalar, $\Sigma^{-1} = 1/\Sigma$, and the pdf above reduces to the Gaussian RV pdf defined in Chapter 3.

- Since the determinant of a symmetric positive definite matrix is positive (see Section C.4), the square root of the determinant is a well defined positive real number.

- The fact that the density integrates to 1 can be ascertained by the following change of variables

$$y = \Sigma^{-1/2}(x - \mu), \qquad x = \Sigma^{1/2}y + \mu.$$

(see Proposition C.4.6 for a definition of the square root matrix $A^{1/2}$ of a given symmetric positive definite A) with a Jacobian determinant $|J| = \det(\Sigma^{1/2}) = (\det \Sigma)^{1/2}$. Using Proposition F.6.3, we have

$$\int \exp\left(-\frac{1}{2}(x - \mu)^\top \Sigma^{-1}(x - \mu)\right) dx = \int \exp\left(-\frac{1}{2}y^\top \Sigma^{1/2}\Sigma^{-1}\Sigma^{1/2}y\right)|J|dy$$

$$= \int \exp\left(-\frac{1}{2}y^\top y\right)|J|dy$$

$$= |J|\prod_{j=1}^{n}\int \exp\left(-\frac{1}{2}y_j^2\right)dy_j$$

$$= |J|\left((2\pi)^{1/2}\right)^n$$

$$= (2\pi)^{n/2}(\det \Sigma)^{1/2} \qquad (5.3)$$

where the second-to-last equality follows from (F.25).

Proposition 5.2.1. *If $X \sim N(\boldsymbol{\mu}, \Sigma)$ then*

$$\Sigma^{-1/2}(X - \boldsymbol{\mu}) \sim N(\boldsymbol{0}, I)$$
$$\mathsf{E}(X) = \boldsymbol{\mu}$$
$$\mathsf{Var}(X) = \Sigma$$

Proof. Using the change of variables $Y = \Sigma^{-1/2}(X - \boldsymbol{\mu})$ and derivations similar to (5.3), we have

$$f_Y(\boldsymbol{y}) = f_X(\Sigma^{1/2}\boldsymbol{y} + \boldsymbol{\mu})\,|J|$$
$$= \frac{\sqrt{\det(\Sigma)}}{(2\pi)^{n/2}\sqrt{\det(\Sigma)}} \exp(-\boldsymbol{y}^\top \Sigma^{1/2}\Sigma^{-1}\Sigma^{1/2}\boldsymbol{y})$$
$$= \prod_{j=1}^{n} \frac{1}{\sqrt{2\pi}} e^{-y_i^2}$$

implying that $(Y_1, \ldots, Y_n)$ are mutually independent $N(0,1)$ RVs.

$$\mathsf{E}(X) = \mathsf{E}(\Sigma^{1/2}Y + \boldsymbol{\mu}) = \Sigma^{1/2}\,\mathsf{E}(Y) + \boldsymbol{\mu} = \boldsymbol{0} + \boldsymbol{\mu}$$
$$\mathsf{Var}(X) = \mathsf{Var}(\Sigma^{1/2}Y + \boldsymbol{\mu}) = \mathsf{Var}(\Sigma^{1/2}Y) = (\Sigma^{1/2})^\top\,\mathsf{Var}(Y)\Sigma^{1/2} = (\Sigma^{1/2})^\top I\Sigma^{1/2}$$
$$= \Sigma.$$

Above, we used the fact that $\Sigma^{1/2}$ is symmetric and Corollary 4.6.3. ∎

The transformation $Y = \Sigma^{-1/2}(X - \boldsymbol{\mu})$ implies that if $X_1, \ldots, X_n$ are Gaussian RVs that are uncorrelated, or in other words $\mathsf{Var}(X) = I$, they are also independent. This is in contrast to the general case where a lack of correlation or $\mathsf{Var}(X) = I$ does not necessarily imply independence.

Proposition 5.2.2. *If $X_i \sim N(\mu_i, \sigma_i), i = 1, \ldots, n$ are independent RVs then*

$$\boldsymbol{a}^\top X = \sum_{i=1}^{n} a_i X_i \sim N\left(\sum_{i=1}^{n} a_i \mu_i, \sum_{i=1}^{n} a_i^2 \sigma_i^2\right), \qquad \boldsymbol{a} \in \mathbb{R}^n.$$

Proof. Since the mgf of $N(\mu, \sigma^2)$ (see Section 3.9) is $e^{\mu t + \sigma^2 t^2/2}$, the mgf of $a_i X_i$ is $\mathsf{E}(\exp(ta_i X_i))$, which is the mgf of X_i at ta_i: $e^{\mu t a_i + \sigma_i^2 t^2 a_i^2/2}$. This is also the mgf of $N(a_i\mu, a_i^2\sigma_i^2)$, and by Proposition 2.4.2 this means that $a_i X_i \sim N(a_i\mu, a_i^2\sigma_i^2)$. Finally, from Proposition 4.8.1,

$$\prod_i e^{\mu t a_i + \sigma^2 t^2 a_i^2/2} = e^{\sum_i \mu t a_i + \sigma_i^2 t^2 a_i^2/2} = e^{t(\sum_i \mu a_i) + t^2(\sum \sigma_i^2 a_i^2)/2}$$

which is the mgf of $N(\sum a_i \mu_i, \sum a_i^2 \sigma_i^2)$. Using Proposition 2.4.2 again concludes the proof. ∎

Proposition 5.2.3. *The multivariate mgf of* $X \sim N(\mu, \Sigma)$ *is*

$$\mathsf{E}(\exp(t^\top X)) = \exp\left(t^\top \mu + t^\top \Sigma t/2\right).$$

Proof. The mgf of $Y \sim N(0, I)$ is

$$\mathsf{E}(\exp(t^\top Y)) = \mathsf{E}\left(\exp\left(\sum_{i=1}^n t_i Y_i\right)\right)$$

$$= \mathsf{E}\left(\prod_{i=1}^n \exp(t_i Y_i)\right)$$

$$= \prod_{i=1}^n \mathsf{E}(\exp(t_i Y_i))$$

$$= \exp\left(t^\top t/2\right).$$

Since $X = \Sigma^{1/2} Y + \mu$, we use the transformation $u = \Sigma^{1/2} t$ to get

$$\mathsf{E}(\exp(t^\top X)) = \mathsf{E}\left(\exp\left(t^\top(\Sigma^{1/2} Y + \mu)\right)\right)$$

$$= \mathsf{E}\left(\exp((\Sigma^{1/2} t)^\top Y)\right) \exp(t^\top \mu)$$

$$= \mathsf{E}\left(\exp(u^\top Y)\right) \mathsf{E}\exp(t^\top \mu)$$

$$= \exp\left(u^\top u/2\right) \exp(t^\top \mu)$$

$$= \exp\left((\Sigma^{1/2} t)^\top(\Sigma^{1/2} t)/2\right) \exp(t^\top \mu)$$

$$= \exp(t^\top \Sigma t/2 + t^\top \mu).$$

∎

The proposition below shows that any affine transformation of a Gaussian random vector is also a Gaussian vector.

Proposition 5.2.4. *If* $X \sim N(\mu, \Sigma)$ *and A is a matrix of full rank then*

$$AX + \tau \sim N(A\mu + \tau, A^\top \Sigma A).$$

Proof. Using the transformation $u = A^\top t$,

$$\mathsf{E}\exp(t^\top(AX + \tau)) = \left(\mathsf{E}\exp((A^\top t)^\top X)\exp(t^\top \tau)\right)$$

$$= \mathsf{E}(\exp(u^\top X))\exp(t^\top \tau)$$

$$= \exp(u^\top \Sigma u/2 + u^\top \mu)\exp(t^\top \tau)$$

$$= \exp((A^\top t)^\top \Sigma (A^\top t)/2 + (A^\top t)^\top \mu + t^\top \tau)$$

$$= \exp(t^\top(A\Sigma A^\top)t)/2 + t^\top(A\mu + \tau)),$$

which is the mgf of the $N(A\mu + \tau, A\Sigma A^\top)$ distribution. The fact that A is full rank is needed to ensure that $A\Sigma A^\top$ is positive definite (this is a required property for the variance of the multivariate Gaussian distribution). ∎

As shown in Chapter 3, Gaussian RVs have pdfs that resemble a bell-shaped curve centered at the expectation and whose spread corresponds to the variance. The Gaussian random vector has an n-dimension pdf whose contour lines resemble ellipsoids centered at μ and whose shape is determined by Σ. Recall that the leading term in the Gaussian pdf is constant in x, implying that it is sufficient to study the exponent in order to investigate the shape of the equal height contours of f_X.

1. If Σ is the identity matrix, its determinant is 1, its inverse is the identity as well, and the exponent of the Gaussian pdf becomes $-\sum_{i=1}^{n}(x_i - \mu_i)^2/2$. This indicates that the pdf factors into the product of n pdf functions of normal RVs, with means μ_i and variance $\sigma_i^2 = 1$. In this case, the contours have spherical symmetry.

2. If Σ is a diagonal matrix with elements $[\Sigma]_{ij} = \sigma_i^2$, then its inverse is a diagonal matrix with elements $[\Sigma^{-1}]_{ij} = 1/\sigma_i^2$ and $\det\Sigma = \prod_i \sigma_i^2$. The exponent of the pdf factors into a sum, indicating that the pdf factors into a product of independent Gaussian RVs $X_i \sim N(\mu_i, \sigma_i^2), i = 1, \ldots, n$. The equal height contours are axis-aligned ellipses, whose spread along each dimension depends on σ_i^2.

3. In the general case, the exponent is $-\sum_i \sum_j (x_i - \mu_i)\Sigma_{ij}^{-1}(x_j - \mu_j)$, which is a quadratic form in x. As a result, the equal height contours of the pdf will be elliptical with a center determined by μ and shape determined by Σ. The precise relationship is investigated in Section C.3 (see also exercise 3 at the end of this chapter).

The R code below graphs the pdf of a two dimensional Gaussian random vector with expectation $(0, 0)$ and variance matrix $\begin{pmatrix} 1 & 0 \\ 0 & 6 \end{pmatrix}$.

```
library(lattice)
R = expand.grid(seq(-8, 8, length = 30), seq(-8,
    8, length = 30))
names(R) = c("x1", "x2")
R$p = dmvnorm(R, sigma = matrix(c(1, 0, 0, 6),
    nrow = 2, ncol = 2))
wireframe(p ~ x1 + x2, data = R, xlab = "$x_1$",
    ylab = "$x_2$", zlab = "", main = "Bivariate Gaussian pdf",
    scales = list(arrows = FALSE), drape = TRUE)
```

Bivariate Gaussian pdf

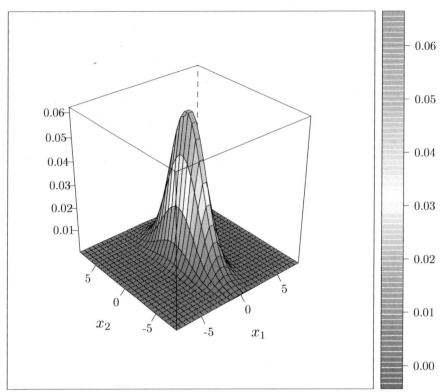

The R code below graphs (see Figure 5.2) the equal height contours of eight two dimensional Gaussian pdfs with different μ and Σ parameters. Note how modifying the μ parameter shifts the pdf and modifying Σ changes the shape of the contour lines as described above.

```
R = expand.grid(seq(-2, 2, length = 100), seq(-2,
    2, length = 100))
z1 = dmvnorm(R, c(0, 0), diag(2))
z2 = dmvnorm(R, c(0, 0), diag(2)/2)
z3 = dmvnorm(R, c(0, 0), matrix(c(3, 0, 0, 1),
    nrow = 2, ncol = 2))
z4 = dmvnorm(R, c(0, 0), matrix(c(3, 1, 1, 3),
    nrow = 2, ncol = 2))
z5 = dmvnorm(R, c(1, 1), diag(2))
z6 = dmvnorm(R, c(1, 1), diag(2)/2)
z7 = dmvnorm(R, c(1, 1), matrix(c(3, 0, 0, 1),
    nrow = 2, ncol = 2))
```

```
z8 = dmvnorm(R, c(1, 1), matrix(c(3, 1, 1, 3),
    nrow = 2, ncol = 2))
S1 = stack(list(`$\\Sigma=I$` = z1, `$\\Sigma=I/2$` = z2,
    `$\\Sigma=(3,0;0,1)$` = z3, `$\\Sigma=(3,1;1,3)$` = z4,
    `$\\Sigma=I$` = z5, `$\\Sigma=I/2$` = z6,
    `$\\Sigma=(3,0;0,1)$` = z7, `$\\Sigma=(3,1;1,3)$` = z8))
S1$mu = "$\\mu=(0,0)$"
S1$x = R[, 1]
S1$y = R[, 2]
S1$mu[1:40000] = "$\\mu=(0,0)$"
S1$mu[40001:80000] = "$\\mu=(1,1)$"
ggplot(S1, aes(x = x, y = y, z = values)) + xlab("$x$") +
    ylab("$y$") + facet_grid(ind ~ mu) + stat_contour()
```

5.3   The Dirichlet Random Vector

Definition 5.3.1. The Dirichlet random vector $\boldsymbol{X} \sim \mathrm{Dir}(\boldsymbol{\alpha})$, where $\alpha \in \mathbb{R}^k$ with $\alpha_i \geq 0$ for all $i = 1, \ldots, k$, is a continuous random vector, defined by

$$X_i = \frac{W_i}{\sum_{i=1}^{k} W_i}, \quad i = 1, \ldots, k,$$

where $W_i \sim \mathrm{Gam}(\alpha_i, 1)$, $i = 1, \ldots, k$ are independent Gamma RVs (see Section 3.10).

Proposition 5.3.1. *The Dirichlet random vector* $\boldsymbol{X} \sim \mathrm{Dir}(\boldsymbol{\alpha})$ *has the following pdf*

$$f_{\boldsymbol{X}}(\boldsymbol{x}) = \frac{\Gamma\left(\sum_{i=1}^{k} \alpha_i\right)}{\prod_{i=1}^{k} \Gamma(\alpha_i)} \prod_{i=1}^{k} x_i^{\alpha_i - 1}$$

if $\boldsymbol{x} \in \mathbb{P}_k$ *and* $f_{\boldsymbol{X}}(\boldsymbol{x}) = 0$ *otherwise.*

Proof. We can obtain the pdf by applying the change of variable technique to the following pdf of the random vector $\boldsymbol{W}$

$$f_{\boldsymbol{W}}(\boldsymbol{w}) = \prod_{i=1}^{k} e^{-w_i} w_i^{\alpha_i - 1} / \Gamma(\alpha_i),$$

and the invertible transformation $\boldsymbol{W} \mapsto (X_1, \ldots, X_{k-1}, Z)$ where $X_i = \frac{W_i}{\sum_{i=1}^{k} W_i}$ and $Z = \sum_{i=1}^{k} W_i$. We can express the transformation as $W_i = X_i Z$ for $i = 1, \ldots, k - 1$ and $W_k = Z X_k = Z(1 - X_1 - \cdots - X_{k-1})$. It follows that the

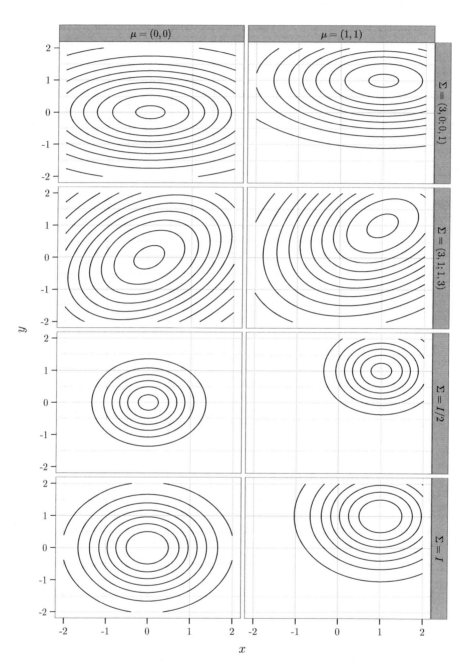

Figure 5.2: Equal height contours of eight two dimensional Gaussian pdfs $f_{X,Y}(x,y)$ with different $\boldsymbol{\mu}$ and Σ parameters.

Jacobian of the transformation is the following matrix

$$J = \begin{pmatrix} Z & 0 & \cdots & X_1 \\ 0 & Z & \cdots & X_2 \\ \vdots & \vdots & \ddots & \vdots \\ -Z & \cdots & -Z & X_k \end{pmatrix}.$$

To use the change of variables technique, we need to compute the determinant of the matrix above. Recalling that adding one of the rows of the matrix to another row does not change the determinant, we can add the first $k-1$ rows to the last row, and then compute the determinant of the resulting matrix. That matrix is an upper diagonal matrix (all elements below the diagonal are zero) with diagonal elements $Z, Z, Z, \cdots, Z, 1$. Since the determinant of an upper diagonal matrix is the product of the diagonal terms (all other terms in the product defining the determinant expression zero out), the determinant is Z^{k-1}. It follows that the pdf of the transformed variables $(X_1, \ldots, X_{k-1}, Z)$ is

$$f_{X_1,\ldots,X_{k-1},Z}(x_1,\ldots,x_{k-1},z) = z^{k-1} \prod_{i=1}^{k} e^{-x_i z}(x_i z)^{\alpha_i - 1}/\Gamma(\alpha_i)$$

$$= z^{\sum \alpha_i - 1} e^{-1 \cdot z} \prod_{i=1}^{k} x_i^{\alpha_i - 1}/\Gamma(\alpha_i).$$

We integrate the above expression with respect to z in order to obtain the pdf of $X_1, \ldots, X_{k-1}$.

$$f_{X_1,\ldots,X_{k-1}}(x_1,\ldots,x_{k-1}) = \prod_{i=1}^{k} x_i^{\alpha_i - 1}/\Gamma(\alpha_i) \cdot \int_0^\infty z^{\sum \alpha_i - 1} e^{-1 \cdot z}\, dz$$

$$= \prod_{i=1}^{k} x_i^{\alpha_i - 1}/\Gamma(\alpha_i) \cdot \Gamma\left(\sum_{i=1}^{k} \alpha_i\right)$$

where the last equation follows from the fact that Gamma pdf integrates to one (see Section 3.10). Since $\sum_{i=1}^{k} X_i = 1$, the RV X_k must equal $1 - \sum_{i=1}^{k-1} X_i$. The pdf above therefore remains the same if we add the RV X_k taking value $1 - \sum_{i=1}^{k-1} X_i$ (the pdf becomes zero for all other values of X_k). ∎

Recall that the simplex $\mathbb{P}_k$ represents the set of all probability functions on a sample space Ω of size k. The Dirichlet distribution may thus be viewed as a distribution over distributions over a finite sample space.

Proposition 5.3.2. *Let* $W_i \sim Gam(\alpha_i, 1)$, $i = 1,\ldots,k$ *be independent Gamma RVs. The random vector* $(X_1,\ldots,X_{k-1})$, *where* $X_i = W_i/\sum_{i=1}^{k} W_i$ *for* $i = 1,\ldots,k-2$ *and* $X_{k-1} = (W_{k-1}+W_k)/\sum_{i=1}^{k} W_i$ *is a Dirichlet random vector with parameter* $(\alpha_1,\ldots,\alpha_{k-2},\alpha_{k-1}+\alpha_k)$.

Proof. Recall from Section 3.10 that a sum of independent Gamma RVs with the same second parameter is a Gamma RV whose parameter is the sum of the corresponding parameters. The rest of the proof follows from the definition of the Dirichlet distribution above.   ∎

It is straightforward to generalize the proposition above and show that whenever we aggregate some of the Gamma W_i RVs in categories, and then form the normalization construction that defines the Dirichlet random vector, we obtain a Dirichlet random vector with a parameter vector that is similarly aggregated.

Proposition 5.3.3. *Let X be a Dirichlet random vector with parameter vector α. The marginal distribution of X_i is $Beta(\alpha_i, \sum_{j=1}^{k} \alpha_j - \alpha_i)$, and we have*

$$\mathsf{E}(X_i) = \frac{\alpha_i}{\sum_{j=1}^{k} \alpha_j}$$

$$\mathsf{Var}(X_i) = \frac{\alpha_i (\sum_{j=1}^{k} \alpha_j - \alpha_i)}{(\sum_{j=1}^{k} \alpha_j)^3 - (\sum_{j=1}^{k} \alpha_j)^2}.$$

Proof. Let $W_i \sim \text{Gam}(\alpha_i, 1)$, $i = 1, \ldots, k$ be the Gamma RVs underlying the Dirichlet construction of X. The proposition states that the Dirichlet construction corresponding to the Gamma RVs W_i and $\sum_{j:j\neq i} W_j$ results in a random vector $(X_i, 1 - X_i)$ that have a Dirichlet distribution with a parameter vector $(\alpha_i, \sum_{j:j\neq i} \alpha_j)$. Comparing the pdf of $\text{Dir}(\alpha_i, \sum_{j:j\neq i} \alpha_j)$ to the beta distribution pdf, and noting that the distribution of X_i is directly obtained from the distribution of $(X_i, 1 - X_i)$ (the second RV is a function of the first and therefore it is a deterministic quantity) concludes the first part of the proof. The second part of the proof follows from the derivations of the expectation and variance of the beta distribution (see Section 3.12).   ∎

The R code below graphs the pdf of several Dirichlet distributions. Dark shades correspond to high probability and light shades correspond to low probabilities. Note that the pdf assigns non-zero values only to the simplex $\mathbb{P}_k$ (in this case $k = 2$).

```
library(MCMCpack)
R = expand.grid(seq(0.001, 0.999, length.out = 80),
    seq(0.001, 0.999, length.out = 80))
R[, 3] = 1 - R[, 1] - R[, 2]
names(R) = c("x1", "x2", "x3")
R = R[R[, 3] > 0, ]
for (i in 1:nrow(R)) R$p[i] = ddirichlet(R[i,
    1:3], alpha = c(3, 3, 3))
cloud(x3 ~ x1 * x2, data = R, col = gray(1 - R$p/(max(R$p) +
    3) - 0.1), pch = 20, cex = 0.8, main = "Dirichlet (3,3,3) pdf",
    scales = list(arrows = FALSE), xlab = "$x_1$",
    ylab = "$x_2$", zlab = "$x_3$")
```

Dirichlet (3,3,3) pdf

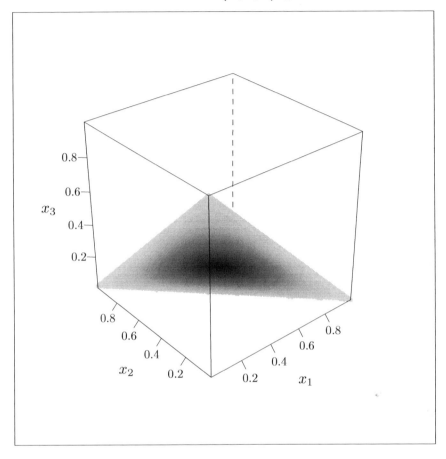

```
for (i in 1:nrow(R)) R$p[i] = ddirichlet(R[i,
    1:3], alpha = c(10, 10, 10))
cloud(x3 ~ x1 * x2, data = R, col = gray(1 - R$p/(max(R$p) +
    3) - 0.1), pch = 20, cex = 0.8, main = "Dirichlet (10,10,10) pdf",
    scales = list(arrows = FALSE), xlab = "$x_1$",
    ylab = "$x_2$", zlab = "$x_3$")
```

Dirichlet (10,10,10) pdf

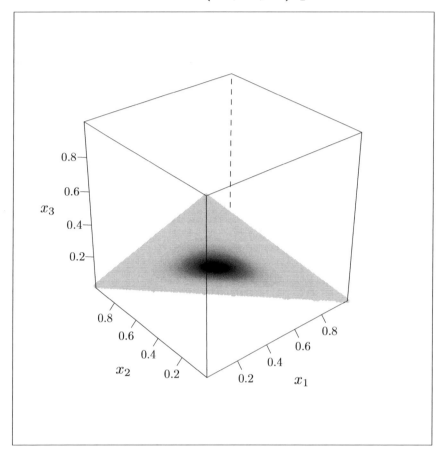

```
for (i in 1:nrow(R)) R$p[i] = ddirichlet(R[i,
    1:3], alpha = c(1, 3, 6))
cloud(x3 ~ x1 * x2, data = R, col = gray(1 - R$p/(max(R$p) +
    3) - 0.1), pch = 20, cex = 0.8, main = "Dirichlet (1,3,6) pdf",
    scales = list(arrows = FALSE), xlab = "$x_1$",
    ylab = "$x_2$", zlab = "$x_3$")
```

Dirichlet (1,3,6) pdf

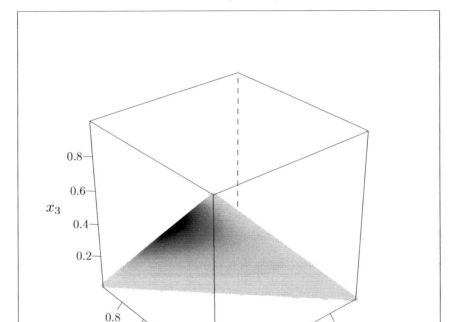

Much of the usefulness of the Dirichlet distribution stems from it being the conjugate prior to the multinomial distribution. More details on conjugate distributions and Bayesian modeling are available in [2]. The Dirichlet random vector also serves as the basis for latent Dirichlet allocation [8], a recently proposed model for text documents, and for the the Dirichlet process [18, 7], a random process that is often used in non-parametric Bayesian statistics.

5.4 Mixture Random Vectors

Mixture random vectors are the direct multivariate analog of mixture random variables (see Section 3.13).

Definition 5.4.1. Given k independent random vectors $X^{(1)}, \ldots, X^{(k)}$ that are all continuous or all discrete, their mixture is a random vector with the following pdf or pmf

$$f_X(x) = \sum_{i=1}^{k} \alpha_i f_{X^{(i)}}(x) \qquad \text{(continuous case)} \qquad (5.4)$$

$$p_X(x) = \sum_{i=1}^{k} \alpha_i p_{X^{(i)}}(x) \qquad \text{(discrete case)}. \qquad (5.5)$$

where $\alpha = (\alpha_1, \ldots, \alpha_n)$ is a vector of non-negative numbers that sum to 1.

Since the weights sum to one, the above functions are valid pdf or pmf and thus uniquely characterize the mixture distribution. The moments of a mixture random vector are linear combinations of the corresponding moments of the individual random vectors (see Section 3.13).

As in the univariate case of Section 3.13, mixture distributions are able to capture a wide variety of complex distributions. For example, a mixture of k multivariate Gaussians can capture a distribution with k modes. Mixture distributions are particularly applicable to situations when the vector is determined via a two stage experiment: first a mixture component is chosen, and then the vector value is determined from the appropriate mixture component. For example, consider the situation of taking multiple physiological measurements (e.g., height, weight) of a person, drawn at random from a population containing multiple ethnicities such as African, Asian, and Caucasian. If the disparities in the physiological measurements between the ethnicities are high relative to the disparities within each ethnicity, a mixture distribution will be appropriate where α corresponds to the proportions of different ethnicities in the population and $X^{(i)}$, $i = 1, \ldots, k$ corresponds to the measurement vector distribution within the different ethnicities. If, on the other hand, the measurement disparity between ethnicities is low relative to the disparities within each ethnicity a simpler non-mixture model may be preferred.

The R code below graphs the pdf of a mixture of two dimensional Gaussian random vectors.

```
R = expand.grid(seq(-4, 4, length = 30), seq(-4,
    4, length = 30))
names(R) = c("x1", "x2")
M1 = matrix(c(0.5, 0.2, 0.2, 1), nrow = 2, ncol = 2)
M2 = matrix(c(2, 0.7, 0.7, 0.5), nrow = 2, ncol = 2)
M3 = matrix(c(1, 0.5, 0.5, 1), nrow = 2, ncol = 2)
R$p = dmvnorm(R, mean = c(2, 1), sigma = M1)/2 +
    dmvnorm(R, mean = c(-1, 2), sigma = M2)/4 +
    dmvnorm(R, mean = c(0, -2), sigma = M3)/4
wireframe(p ~ x1 + x2, data = R, xlab = "$x_1$",
    ylab = "$x_2$", zlab = "", main = "Mixture random vector pdf",
    scales = list(arrows = FALSE), drape = TRUE)
```

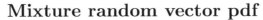

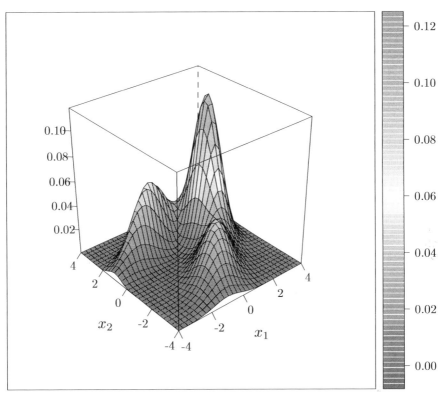

5.5 The Exponential Family Random Vector

Definition 5.5.1. The continuous exponential family random vector is a random vector whose pdf is

$$f_{\boldsymbol{X}}(\boldsymbol{x}) = \frac{\alpha(\boldsymbol{x})}{Z(\boldsymbol{\theta})} \exp\left(\sum_{i=1}^{k} h_i(\boldsymbol{\theta}) g_i(\boldsymbol{x})\right), \quad \boldsymbol{x} \in \mathcal{X} \subset \mathbb{R}^n, \boldsymbol{\theta} \in \Theta \subset \mathbb{R}^d,$$

where

$$
\begin{aligned}
g_i &: \mathcal{X} \to \mathbb{R}, & i &= 1, \ldots, k \\
h_i &: \Theta \to \mathbb{R}, & i &= 1, \ldots, k \\
\alpha &: \mathcal{X} \to \mathbb{R}
\end{aligned}
$$

and Z ensures that the pdf normalizes to one

$$Z(\boldsymbol{\theta}) = \int_{\mathcal{X}} \alpha(\boldsymbol{x}) \exp\left(\sum_{i=1}^{k} h_i(\boldsymbol{\theta}) g_i(\boldsymbol{x})\right) d\boldsymbol{x}.$$

The discrete analog is defined similarly by replacing the pdf by a pmf and the integral by a sum. If $h_i(\boldsymbol{\theta}) = \theta_i$ we say that the exponential family has a canonical form.

We make the following observations.

- In order for the definition above to make sense we need the integral or sum in Z to converge.

- The function h needs to be discrete ($h(\boldsymbol{x}) > 0$ on a countable set of $\boldsymbol{x}$ values) in order for $\boldsymbol{X}$ to be a discrete random vector and continuous ($h(\boldsymbol{x}) > 0$ on non-countable sets of $\boldsymbol{x}$ values) in order for $\boldsymbol{X}$ to be a continuous random vector.

- When the exponential family is in canonical form, we have the following simple connection between $\log Z$ and the expectation and variance of $\boldsymbol{X}$

$$E(g_j(\boldsymbol{X})) = \int g_j(\boldsymbol{x}) \frac{\alpha(\boldsymbol{x})}{Z(\boldsymbol{\theta})} \exp\left(\sum_{i=1}^{k} h_i(\boldsymbol{\theta}) g_i(\boldsymbol{x})\right) d\boldsymbol{x}$$

$$= \frac{\int g_j(\boldsymbol{x}) \alpha(\boldsymbol{x}) \exp\left(\sum_{i=1}^{k} h_i(\boldsymbol{\theta}) g_i(\boldsymbol{x})\right) d\boldsymbol{x}}{Z(\boldsymbol{\theta})}$$

$$= -\frac{\partial \log Z(\boldsymbol{\theta})}{\partial \theta_j}$$

$$\text{Cov}(g_j(X), g_k(X)) = E(g_j(X) g_k(X)) - E(g_j(X)) E(g_k(X))$$

$$= \frac{\partial^2 Z}{\partial \theta_j \partial \theta_k} / Z - \left(\frac{\partial Z}{\partial \theta_j} / Z\right)\left(\frac{\partial Z}{\partial \theta_k} / Z\right)$$

$$= -\left(\frac{\partial Z}{\partial \theta_j} \frac{\partial Z}{\partial \theta_k} - \frac{\partial^2 Z}{\partial \theta_j \partial \theta_k} Z\right) / Z^2$$

$$= -\frac{\partial^2 \log Z(\boldsymbol{\theta})}{\partial \theta_j \partial \theta_k}.$$

(Replace integrals with sums in the discrete case.) Or in vector notation

$$E(\boldsymbol{g}(\boldsymbol{X})) = -\nabla \log Z(\boldsymbol{\theta})$$

$$\text{Var}(\boldsymbol{g}(\boldsymbol{X})) = -\nabla^2 \log Z(\boldsymbol{\theta}).$$

Note that since the variance matrix is non-negative definite, the last equation implies that the functions $\log f_{\boldsymbol{X}}(\boldsymbol{x})$ or $\log p_{\boldsymbol{X}}(\boldsymbol{x})$, when viewed as functions of $\boldsymbol{\theta}$ for an arbitrary fixed $\boldsymbol{x}$, are concave functions. This motivates the use of exponential family in statistical estimation using the maximum likelihood procedure.

Many of the important random variables and random vectors that we have seen so far can be re-expressed in a way that complies with Definition 5.5.1 and thus are exponential family random vectors (or random variables). Specific cases include the binomial, Gaussian, exponential, Poisson, beta, and gamma random variables and the multinomial, multivariate normal, and Dirichlet random vectors. Notable exceptions are uniform, t, and mixture random variables and the corresponding random vectors.

5.6 Notes

Additional information on multinomial, multivariate Gaussian, and Dirichlet random vectors is available in most manuscripts specializing in distributions, for example [21]. A good source of additional information on multivariate Gaussian distribution is linear regression manuscripts, for example [38]. Exponential family distributions are typically explored in statistics textbooks. A detailed description is available in [11].

5.7 Exercises

1. Justify the derivations in (5.1)-(5.2).

2. Consider a multinomial vector $\boldsymbol{X} = (X_1, \ldots, X_n)$ and a mapping $\boldsymbol{X} \to Y$ defined by $Y = \sum_{i \in A} X_i$ for some $A \subset \{1, \ldots, n\}$. What is the distribution of Y? Write down the pmf in a compact form.

3. Characterize the elliptical contours of the multivariate Gaussian pdf with non-diagonal Σ in terms of the eigenvalues and eigenvectors of Σ. Hint: use spectral decomposition (Proposition C.3.8) and the relationship in Proposition 5.2.4.

4. Express the exponential, Poisson RVs and the multinomial random vector as exponential family random vectors.

Chapter 6

Random Processes

Random processes are a natural generalization of random vectors. We describe in this chapter some basic definitions and a few examples. We avoid measure theoretic issues for simplicity, with the exception of Section 6.5, which contains a proof of Kolmogorov's extension theorem for a countable index set. The next chapter will explore in more detail several important random processes.

6.1 Basic Definitions

Definition 6.1.1. A random process (RP) indexed by a set $J \subset \mathbb{R}^k$ is a collection of random variables $\mathcal{X} = \{X_t : t \in J\}$ where $X_t : \Omega \to \mathbb{R}$. If J is finite or countably infinite, for example $J = \mathbb{N}^d$ or $J = \mathbb{Z}^d$, the process has discrete time. If J is non-countably infinite, for example $J = \mathbb{R}^d$ or $J = [0, \infty)^d$, the process has continuous time. The process has discrete state if all RVs X_t are discrete RVs, and has continuous state if all the RVs X_t are continuous RVs.

We make the following comments.

- If $|J|$ is finite $\mathcal{X}$ reduces to a random vector, and if $|J| = 1$, $\mathcal{X}$ reduces to an random variable.

- For a RP $\mathcal{X} = \{X_t : t \in J\}$, the random variables X_t have a common sample space Ω. Given $\omega \in \Omega$, we denote by $\mathcal{X}(\omega)$ the following function from J to $\mathbb{R}$:

$$\mathcal{X}(\omega) : J \to \mathbb{R}, \qquad (\mathcal{X}(\omega))(t) = X_t(\omega) \in \mathbb{R}, \qquad t \in J.$$

The function $\mathcal{X}(\omega)$ is called a realization or sample path of the process. We can thus view an RP $\mathcal{X}$ as a mapping from the sample space Ω to a space $\mathcal{H}$ of functions from J to $\mathbb{R}$.

- In accordance with the definition of $X \in A, A \subset \mathbb{R}$ (for a random variable X) and $\boldsymbol{X} \in A, A \subset \mathbb{R}^n$ (for a random vector $\boldsymbol{X}$), we define the set

$$\{\mathcal{X} \in \mathcal{F}\} = \{\omega \in \Omega : \mathcal{X}(\omega) \in \mathcal{F}\} \subset \Omega,$$

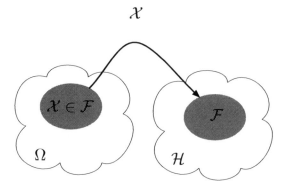

Figure 6.1: An RP maps the sample space Ω to a space $\mathcal{H}$ of functions from J to $\mathbb{R}$. Given a set of functions $\mathcal{F} \subset \mathcal{H}$, the set $\mathcal{X} \in \mathcal{F}$ is the event $\{\omega \in \Omega : \mathcal{X}(\omega) \in \mathcal{F}\} \subset \Omega$.

where $\mathcal{F}$ is a set of functions from J to $\mathbb{R}$ (Figure 6.1). Note that the set $\{\mathcal{X} \in \mathcal{F}\}$ is a subset of Ω, associated with the following probabilities

$$P(\mathcal{X} \in \mathcal{F}) = P(\{\omega \in \Omega : \mathcal{X}(\omega) \in \mathcal{F}\}), \quad \text{or}$$
$$P(\mathcal{X} = f) = P(\{\omega \in \Omega : \mathcal{X}(\omega) = f\}).$$

Example 6.1.1. *The process $\{X_t : t \in \mathbb{R}\}$ with $\Omega = \mathbb{R}$ and $X_t(\omega) = \omega \sin(2\pi t)$ is a continuous-time RP whose sample paths are sinusoidal curves amplified by $\omega \in \mathbb{R}$. If $\omega \sim N(0, \sigma^2)$ then sample paths with small amplitude are more probable than sample paths with large amplitude. The variables $X_t, X_{t'}$ are highly dependent and are in fact functions of each other: $X_t = f(X_{t'})$. For example, we have*

$$P\left(\mathcal{X} \in \left\{f : \sup_{t \in \mathbb{R}} |f(t)| \le 3\right\}\right) = P(|\omega| \le 3) = \frac{1}{\sqrt{2\pi}} \int_{-3}^{3} e^{-z^2/2} \, dz.$$

(Note that the probability on the right hand side above is the probability of the RP taking values in set of functions.) The following R code displays 10 and then 200 sample paths. The latter graph uses semi-transparent curves to demonstrate the higher density of sample paths near the $y = 0$ axis.

```
x = seq(-1, 1, length = 100)
A = sin(2 * pi * x) %*% t(rnorm(200))
plot(c(-1, 1), c(-2, 2), type = "n", xlab = "$J$",
    ylab = "$\\mathcal{X}(\\omega)$")
for (i in 1:10) lines(x, A[, i])
```

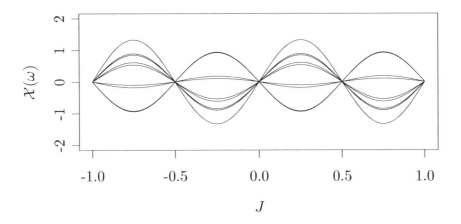

```
plot(c(-1, 1), c(-2, 2), type = "n", xlab = "$J$",
    ylab = "$\\mathcal{X}(\\omega)$")
for (i in 1:200) lines(x, A[, i], col = rgb(0,
    0, 0, alpha = 0.5))
```

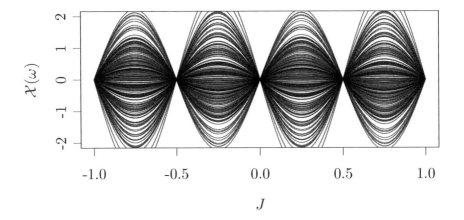

6.2   Random Processes and Marginal Distributions

Given a random process we can consider its finite dimensional marginals, namely the distribution of $X_{t_1}, \ldots, X_{t_k}$ for some $k \in \mathbb{N}$ and some $t_1, \ldots, t_k \in J$. The marginals distributions are characterized by the corresponding joint cdf functions

$$F_{X_{t_1}, \ldots, X_{t_k}}(r_1, \ldots, r_k) = \mathsf{P}(X_{t_1} \leq r_1, \ldots, X_{t_k} \leq r_k)$$

where $k \geq 1$, $t_1, \ldots, t_k \in J$ and $r_1, \ldots, r_k \in \mathbb{R}$. If the process RVs are discrete, we can consider instead the joint pmf functions, and if the process RVs are continuous, we can consider instead the joint pdf functions.

Example 6.2.1. *Consider a random process similar to the one in Example 6.1.1, with $J = \mathbb{R}$ and $X_t = Y \cos(2\pi t)$ for some discrete RV Y with pmf p_Y. Since each X_t is a function of any other $X_{t'}$*

$$p_{X_{t_1}, \ldots, X_{t_k}}(r_1, \ldots, r_k) = \begin{cases} p_Y(r_j/\cos(2\pi t_j)) & r_j/\cos(2\pi t_j) = r_i/\cos(2\pi t_i) \; \forall i, j \\ 0 & \text{otherwise} \end{cases}.$$

Given an RP $\mathcal{X} = \{X_t : t \in J\}$, all finite dimensional marginals are defined. Kolmogorov's extension theorem below states that the reverse also holds: a collection of finite dimensional marginal distributions uniquely defines a random process, as long as the marginals do not contradict each other. Consequentially, we now have a systematic way to define random processes by defining a collection of all finite dimensional marginals (that are consistent with each other).

Definition 6.2.1. A collection of finite dimensional marginal distributions $\mathcal{L}$ are said to be consistent if the following two conditions apply:

1. If $F_{X_{t_1}, \ldots, X_{t_k}} \in \mathcal{L}$ and $F_{X_{t_1}, \ldots, X_{t_k}, X_{t_{k+1}}} \in \mathcal{L}$ then for all $r_1, \ldots, r_{k+1} \in \mathbb{R}$

$$F_{X_{t_1}, \ldots, X_{t_k}}(r_1, \ldots, r_k) = \lim_{r_{k+1} \to +\infty} F_{X_{t_1}, \ldots, X_{t_k}, X_{t_{k+1}}}(r_1, \ldots, r_k, r_{k+1}).$$

2. If $\mathcal{L}$ has two finite dimensional marginal cdfs over the same RVs, the order in which the RVs appear is immaterial; for example,

$$F_{X_{t_1}, X_{t_2}}(r_1, r_2) = F_{X_{t_2}, X_{t_1}}(r_2, r_1).$$

The finite dimensional marginals arising from a random process satisfy the consistency definition above since they are derived from the same joint distribution P over the common sample space Ω. Kolmogorov's extension theorem below states that the converse also holds. The proof for discrete time RPs is available in Section 6.5. A proof for continuous time RPs is available in [5].

Proposition 6.2.1 (Kolmogorov's Extension Theorem). *Given a collection $\mathcal{L}$ of all possible finite dimensional marginal cdfs that is consistent in the sense of Definition 6.2.1, there exists a unique random process whose finite dimensional marginals coincide with $\mathcal{L}$.*

Instead of specifying a process by its finite dimensional marginal cdfs, we may do so using all finite dimensional marginal pdfs (if X_t, $t \in J$ are continuous) or all finite dimensional marginal pmfs (if X_t, $t \in J$ are discrete). In the former case, the first consistency condition in Definition 6.2.1 becomes

$$f_{X_{t_1},\ldots,X_{t_k}}(r_1,\ldots,r_k) = \int_{\mathbb{R}} f_{X_{t_1},X_{t_k},X_{t_{k+1}}}(r_1,\ldots,r_k,r_{k+1})\,dr_{k+1}$$

and in the latter case, the condition becomes

$$p_{X_{t_1},\ldots,X_{t_k}}(r_1,\ldots,r_k) = \sum_{r_{k+1}} p_{X_{t_1},\ldots,X_{t_k},X_{t_{k+1}}}(r_1,\ldots,r_k,r_{k+1}).$$

Definition 6.2.2. An RP $\mathcal{X} = \{X_t : t \in J\}$ for which all finite dimensional cdfs factor with the same univariate cdf

$$F_{X_{t_1},\ldots,X_{t_k}}(r_1,\ldots,r_k) = \prod_{i=1}^{k} F(r_i), \quad \forall k \in \mathbb{N}, \quad \forall t_1,\ldots,t_k \in J$$

is called an independent identically distributed process (iid) with base distribution F. We denote this by $X_t \overset{\text{iid}}{\sim} F$.

Note that the iid process above satisfies the consistency conditions, and as a result of Kolmogorov's extension theorem it characterizes a unique RP.

Example 6.2.2. *The iid process $\mathcal{X} = \{X_t : t \in \mathbb{N}\}$ with a $Ber(1/2)$ univariate marginal distribution is a discrete-time discrete-state process representing an infinite sequence of independent fair coin flips. For all $k \geq 1$, $t_1 < t_2 < \cdots < t_k$, and $r_1,\ldots,r_k \in \{0,1\}$,*

$$\mathsf{P}(X_{t_1} = r_1,\ldots,X_{t_k} = r_k) = 2^{-k}.$$

We also have, for example,

$$\mathsf{P}\left(\mathcal{X} \in \left\{f : \sum_{i=1}^{10} f(i) \leq 5\right\}\right) = \sum_{j=0}^{5} \frac{10!}{j!(10-j)!} 2^{-10}.$$

The following R code generates one sample path $(t = 1,2,\ldots,20)$ from this RP.

```
J = 0:20
X = rbinom(n = 21, size = 1, prob = 0.5)
qplot(x = J, y = X, geom = "point", size = I(5),
    xlab = "$J$", ylab = "$\\mathcal{X}(\\omega)$")
```

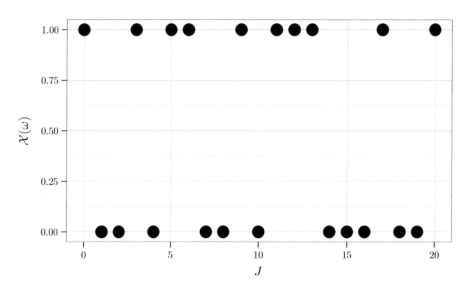

The examples we have considered thus far represent the two extreme cases: total independence in Example 6.2.2 and total dependence in Example 6.2.1 (X_t is a function of $X_{t'}$ for all $t, t' \in J$). In general, the RVs X_t, $X_{t'}$ may be neither independent nor functions of each other.

Two important classes of random processes are stationary processes and Markov processes. They are defined below.

Definition 6.2.3. An RP $\mathcal{X}$ with $J = \mathbb{R}$ or $J = \mathbb{Z}$ is stationary if for all k and for all $r_1, \ldots, r_k, \tau \in \mathbb{R}$,

$$f_{X_{t_1}, \ldots, X_{t_k}}(r_1, \ldots, r_k) = f_{X_{t_1+\tau}, \ldots, X_{t_k+\tau}}(r_1, \ldots, r_k) \quad \text{for continuous RP } \mathcal{X}, \text{ or}$$

$$p_{X_{t_1}, \ldots, X_{t_k}}(r_1, \ldots, r_k) = p_{X_{t_1+\tau}, \ldots, X_{t_k+\tau}}(r_1, \ldots, r_k) \quad \text{for discrete RP } \mathcal{X}.$$

Definition 6.2.4. An RP with $J = \mathbb{R}$ or $J = \mathbb{Z}$ has independent increments if for all $k \geq 2$ and for all $t_1 < t_2 < \cdots < t_k$, the RVs $X_{t_2} - X_{t_1}$, $X_{t_3} - X_{t_2}$, $\ldots$, $X_{t_k} - X_{t_{k-1}}$ are independent.

Definition 6.2.5. A process is a Markov process if for any $k \geq 1$ and for all $t_1 < \cdots < t_k$,

$$f_{X_{t_k}|X_{t_1}=x_{t_1}, \ldots, X_{t_{k-1}}=x_{t_{k-1}}}(x_{t_k}) = f_{X_{t_k}|X_{t_{k-1}}=x_{t_{k-1}}}(x_{t_k}) \text{ for continuous RP}, or$$

$$p_{X_{t_k}|X_{t_1}=x_{t_1}, \ldots, X_{t_{k-1}}=x_{t_{k-1}}}(x_{t_k}) = p_{X_{t_k}|X_{t_{k-1}}=x_{t_{k-1}}}(x_{t_k}) \text{ for discrete RP}.$$

A process with independent increments is necessarily Markov, but the converse does not hold in general.

6.3 Moments

In the case of random vectors, the expectation is a vector and the variance is a matrix. In the case of random processes, the expectation and variance become functions.

Definition 6.3.1. The expectation of a random process $\mathcal{X} = \{X_t : t \in J\}$ is the function $m : J \to \mathbb{R}$ defined by $m(t) = \mathsf{E}(X_t)$.

Definition 6.3.2. The variance of a random process $\mathcal{X} = \{X_t : t \in J\}$ is the function $v : J \to \mathbb{R}$ defined by $v(t) = \mathsf{Var}(X_t)$.

Definition 6.3.3. The autocorrelation function of a random process $\mathcal{X} = \{X_t : t \in J\}$ is the function $R : J \times J \to \mathbb{R}$ defined by $R(t_1, t_2) = \mathsf{E}(X_{t_1} X_{t_2})$.

Definition 6.3.4. The auto-covariance function of a random process $\mathcal{X} = \{X_t : t \in J\}$ is the function $C : J \times J \to \mathbb{R}$ defined by

$$C(t_1, t_2) = \mathsf{E}((X_{t_1} - m(t_1))(X_{t_2} - m(t_2)) = R(t_1, t_2) - m(t_1)m(t_2)$$

where the second equality follows from (4.6).

Definition 6.3.5. The correlation-coefficient function of a random process $\mathcal{X} = \{X_t : t \in J\}$ is the function $\rho : J \times J \to \mathbb{R}$ defined by

$$\rho(t_1, t_2) = \frac{C(t_1, t_2)}{\sqrt{v(t_1)v(t_2)}}.$$

Example 6.3.1. *For the random process $X_t = Y \cos(2\pi t)$ in Example 6.2.1,*

$$m(t) = \mathsf{E}(Y \cos(2\pi t)) = \mathsf{E}(Y) \cos(2\pi t)$$
$$R(t_1, t_2) = \mathsf{E}(Y \cos(2\pi t_1) Y \cos(2\pi t_2)) = \mathsf{E}(Y^2) \cos(2\pi t_1) \cos(2\pi t_2)$$
$$C(t_1, t_2) = R(t_1, t_2) - m(t_1)m(t_2) = (\mathsf{E}(Y^2) - (\mathsf{E}(Y))^2) \cos(2\pi t_1) \cos(2\pi t_2).$$

Example 6.3.2. *For the iid RP (see Definition 6.2.2),*

$$m(t) = \mathsf{E}(X_t) = \mathsf{E}(X_t) = \mu$$
$$C(t_1, t_2) = \mathsf{E}((X_{t_1} - \mu)(X_{t_2} - \mu)) = \begin{cases} 0 & t_1 \neq t_2 \\ \mathsf{Var}(X_{t_1}) = \sigma^2 & t_1 = t_2 \end{cases}$$
$$R(t_1, t_2) = C(t_1, t_2) + m(t_1)m(t_2) = \delta_{t_1, t_2} \sigma^2 + \mu^2$$

where μ, σ^2 are the expectation and variance associated with the cdf F, and $\delta_{ij} = 1$ if $i = j$ and 0 otherwise.

Example 6.3.3. *For the iid RP with $F = Ber(\theta)$ (see Chapter 3), we have*

$$m(t) = \theta$$
$$v(t) = \theta(1 - \theta)$$
$$C(t_1, t_2) = \delta_{t_1, t_2} \theta(1 - \theta),$$
$$R(t_1, t_2) = \delta_{t_1, t_2} \theta(1 - \theta) + \theta^2.$$

Recall that given a random vector $\boldsymbol{X}$ we can define a new random vector $\boldsymbol{Y}$ that is a function of it. The same also holds for random processes.

Example 6.3.4. *Consider the iid Process $\mathcal{X} = \{X_t : t \in J\}$ with $F = Ber(\theta)$ and define the iid process $\mathcal{Y} = \{Y_t : t \in J\}$, $Y_t = 2X_t - 1$. The RVs Y_t takes on value 1 with probability θ and value -1 with probability $1 - \theta$, resulting in*

$$m(t) = \mathsf{E}(2X - 1) = 2\theta - 1$$
$$\mathsf{Var}(Y_t) = \mathsf{Var}(2X_t - 1) = 4\,\mathsf{Var}(X_t) = 4\theta(1 - \theta)$$
$$C(t_1, t_2) = \delta_{t_1, t_2} 4\theta(1 - \theta)$$
$$R(t_1, t_2) = \delta_{t_1, t_2} 4\theta(1 - \theta) + (2\theta - 1)^2.$$

Definition 6.3.6. Two processes $\mathcal{X}, \mathcal{Y}$ are independent if for all $k, l \in \mathbb{N}$ and for all $t_1, \ldots, t_k$ and $t'_1, \ldots, t'_l$

$$F_{X_{t_1}, \ldots, X_{t_k}, Y_{t'_1}, \ldots, Y_{t'_l}}(r_1, \ldots, r_k, s_1, \ldots, s_l)$$
$$= F_{X_{t_1}, \ldots, X_{t_k}}(r_1, \ldots, r_k) F_{Y_{t'_1}, \ldots, Y_{t'_l}}(s_1, \ldots, s_l).$$

Definition 6.3.7. The cross-correlation of the processes $\mathcal{X}, \mathcal{Y}$ is $R_{\mathcal{X}, \mathcal{Y}}(t_1, t_2) = \mathsf{E}(X_{t_1} Y_{t_2})$. If it is always zero, the two processes are orthogonal. The cross-covariance of the two processes is

$$C_{\mathcal{X}, \mathcal{Y}}(t_1, t_2) = \mathsf{E}((X_{t_1} - m_{\mathcal{X}}(t_1))(Y_{t_2} - m_{\mathcal{Y}}(t_2))).$$

If it is always zero, the two processes are uncorrelated.

Definition 6.3.8. An RP $\mathcal{X}$ with $J = \mathbb{R}$ or $J = \mathbb{N}$ is wide sense stationary (WSS) if its mean function $m(t)$ is constant and its autocorrelation $R(t, s)$ is a function only of $|s - t|$, or in other words $R(t, s) = R(t + \tau, s + \tau)$. In this case, we can characterize the auto-correlation function using the function $R : \mathbb{R} \to \mathbb{R}$:

$$R(\tau) \stackrel{\text{def}}{=} R(t, t + \tau).$$

In the case of WSS processes, the $R(\cdot)$ function satisfies the following properties.

1. We have $R(0) = \mathsf{E}(X_t^2) \geq 0$, implying that the second moment function is constant for all t.

2. The function $R(\cdot)$ is even:

$$R(\tau) = \mathsf{E}(X_t X_{t+\tau}) = \mathsf{E}(X_{t+\tau} X_t) = R(-\tau).$$

3. Using the Cauchy-Schwartz inequality (Proposition B.4.1) for the inner product $g(h_1, h_2) = \mathsf{E}(h_2 h_2)$, we have

$$(R(\tau))^2 = (\mathsf{E}(X_{t+\tau} X_t))^2 \leq \mathsf{E}((X_{t+\tau})^2)\,\mathsf{E}((X_t))^2 = (R(0))^2,$$

implying that $R(\tau)$ attains its maximum at $\tau = 0$.

6.4 One Dimensional Random Walk

Thus far, we examined in detail two simple processes: the sinusoidal process where every random variable is a function of any other random variable and the iid process where every random variable is independent of any other random variable. We consider in this section the random walk process, an RP that is neither independent nor completely dependent. In the next chapter, we will see in detail four additional RPs: the Poisson process, the Markov chain, the Gaussian process, and the Dirichlet process.

Recall from Example 6.3.4 the RP $\mathcal{Y} = \{Y_t : t \in \mathbb{N}\}$ where Y_t are iid random variables taking value 1 with probability θ and -1 with probability $1 - \theta$. The random walk process is the discrete-time discrete-state random process $\mathcal{Z} = \{Z_t : t = 1, 2, \ldots\}$ where

$$Z_t = \sum_{n=1}^{t} Y_n.$$

Note that $\mathcal{Z}$ is neither iid nor completely dependent, and it has independent increments; for example, $Z_5 - Z_3 = Y_4 + Y_5$ and $Z_9 - Z_7 = Y_8 + Y_9$ are independent RVs, as they are sums of iid RVs.

The term random walk comes from the following interpretation. Consider a walker who chooses each step at random. The walker chooses whether to walk forward with probability θ or backward with probability $1 - \theta$. The position of the walk at time t is precisely Z_t (assuming the walker starts from 0). An alternative interpretation is related to gambles that win \$1 with probability θ and lose \$1 with probability $1 - \theta$. The RV Z_t measures the total profit or losses of a persistent gambler at time t, assuming the gambler can continue gambling forever (more realistically a gambler with a fixed fortune will not be able to continue gambling after losing all of the money).

The moment functions for $\mathcal{Z}$ are the following functions: for all $t, t_1, t_2 \in \mathbb{N}$

$$m(t) = \mathsf{E}\left(\sum_{i=1}^{t} Y_i\right) = \sum_{i=1}^{t} \mathsf{E}(Y_i) = t(2\theta - 1)$$

$$v(t) = \mathsf{Var}\left(\sum_{i=1}^{t} Y_i\right) = t\,\mathsf{Var}(Y_1) = 4t\theta(1-\theta)$$

$$C(t_1, t_2) = \mathsf{E}\left(\left(\sum_{i=1}^{t_1}(Y_i - \mathsf{E}(Y_1))\right)\left(\sum_{j=1}^{t_2}(Y_j - \mathsf{E}(Y_1))\right)\right)$$

$$= \mathsf{E}\left(\sum_{i=1}^{t_1}\sum_{j=1}^{t_2}(Y_i - \mathsf{E}(Y_1))(Y_j - \mathsf{E}(Y_1))\right)$$

$$= \sum_{i=1}^{t_1}\sum_{i=1}^{t_2}\mathsf{E}((Y_i - \mathsf{E}(Y_1))(Y_j - \mathsf{E}(Y_1))) = \sum_{i=1}^{t_1}\sum_{i=1}^{t_2}\delta_{t_1, t_2} 4\theta(1-\theta)$$

$$= \min(t_1, t_2)4\theta(1-\theta)$$

$$\rho(t_1, t_2) = \frac{C(t_1, t_2)}{\sqrt{v(t_1)v(t_2)}} = \frac{\min(t_1, t_2)}{\sqrt{t_1 t_2}}.$$

The expectation function m increases linearly if $\theta > 0.5$, decreases linearly if $\theta < 0.5$ and is 0 if $\theta = 0.5$. This is in agreement with our intuition that a sequence of gambles biased towards winning will tend to continually increase the fortune as t increases, and vice verse for gambles biased towards losing. The variance function linearly increases as $t \to \infty$.

The R code below graphs a realization of $\mathcal{Y} = \{Y_t : t \in \mathbb{N}\}$ followed by a corresponding realization of $\mathcal{Z}$ for $\theta = 0.6$ for $t \leq 20$.

```
J = 1:20
X = rbinom(n = 20, size = 1, prob = 0.6)
Y = 2 * X - 1
qplot(x = J, y = Y, geom = "point", size = I(5),
      xlab = "$J$", ylab = "$\\mathcal{Y}(\\omega)$")
```

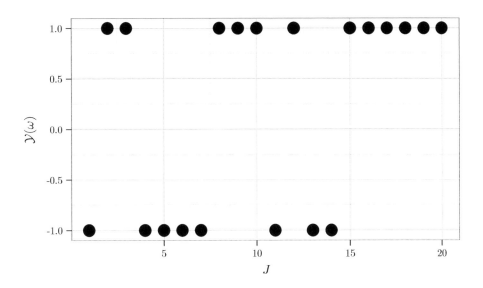

```
qplot(x = J, y = cumsum(Y), geom = "step", xlab = "$J$",
    ylab = "$\\mathcal{Z}(\\omega)$")
```

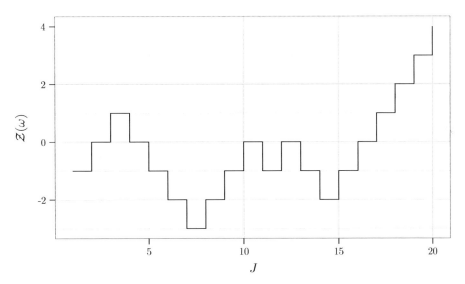

The R code below graphs multiple sample paths sampled from $\mathcal{Z}$, superimposed with the expectation function $m(t)$ for the case $\theta = 0.6$. The graph uses transparency to show high or low density of sample paths where multiple sample paths overlap. In this case, the gambles are biased towards winning, and the gambler's fortune will tend to increase in most cases (though a long unfortunate run is still possible). Note how the process tends to move linearly upwards as t increases, which is in agreement with the above derivation for $m(t)$. The in-

crease in the variability of the fortune as t increases is in agreement with the
linear increase of $v(t)$ derived above.

```
plot(c(1, 100), c(-30, 30), type = "n", xlab = "$J$",
    ylab = "$\\mathcal{Z}(\\omega)$")
Zn = vector()
for (i in 1:100) {
    X = rbinom(n = 100, size = 1, prob = 0.6)
    Z = cumsum(2 * X - 1)
    lines(1:100, c(Z), type = "s", col = rgb(0,
        0, 0, alpha = 0.2))
    Zn[i] = Z[100]
}
curve(x * (2 * 0.6 - 1), add = TRUE, col = rgb(1,
    0, 0), lwd = 5, lty = 2)
title("100 Sample Paths of Random Walk ($\\theta=0.6$)",
    cex.main = 1)
```

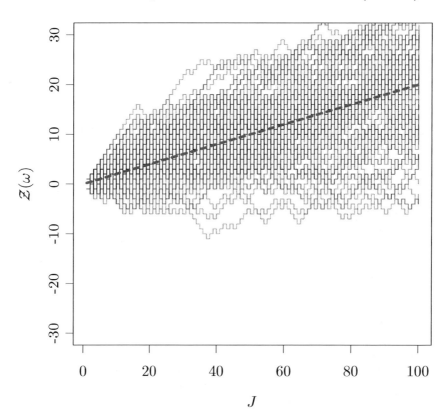

The R code below graphs the histogram of the random walk (or gambler's fortunes) at $t = 100$ based on the 100 sample paths computed above. This represents the distribution of Z_{100}. The average fortune seems to be around 20, which is precisely the value of $m(100)$ for $\theta = 0.6$. The shape of the histogram is in agreement with a sample from the binomial distribution of Z_n (see Exercise 3 at the end of this chapter).

```
qplot(Zn, ..density.., binwidth = 5, xlab = "$\\mathcal{Z}_{100}$",
    geom = "histogram")
```

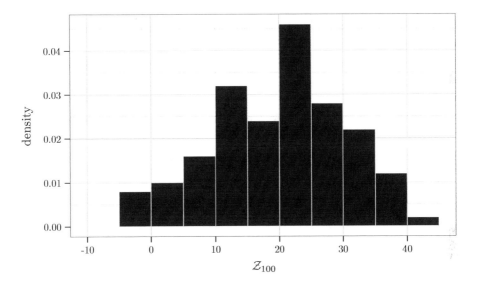

The R code below graphs the equal height contours of the correlation coefficient function, as a function of t_1, t_2, for the case of $\theta = 0.6$. The correlation coefficient $\rho(t_1, t_2) = 1$ for $t_1 = t_2$ indicating perfect correlation. As $|t_1 - t_2|$ increases, the correlation coefficient decays in agreement with our intuition that the random walk process at two disparate time points is not highly correlated.

```
x = seq(0, 10, length.out = 30)
R = expand.grid(x, x)
names(R) = c("t1", "t2")
R$covariance = pmin(R$t1, R$t2)/(sqrt(R$t1 * R$t2))
ggplot(R, aes(x = t1, y = t2, z = covariance)) +
    xlab("$t_1$") + ylab("$t_2$") + stat_contour() +
    opts(title = "Equal Height Contours of $\\rho(t_1,t_2)$")
```

Equal Height Contours of $\rho(t_1, t_2)$

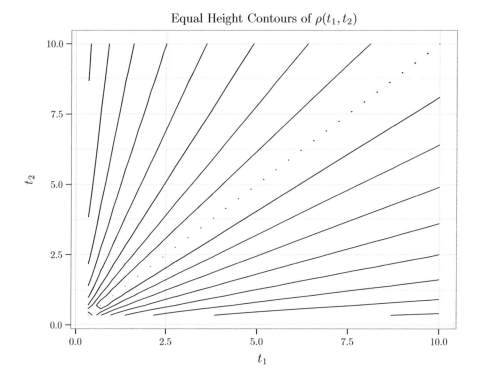

6.5   Random Processes and Measure Theory*

Consider a probability measure space $(\Omega, \mathcal{F}, \mathsf{P})$ and a random process $\mathcal{X} : \Omega \to \mathcal{H}$. In the case of discrete-time RPs, $\mathcal{H} = \mathbb{R}^\infty$ and we can consider the RP as a measurable function $\mathcal{X} : (\Omega, \mathcal{F}) \to (\mathbb{R}^\infty, \mathcal{B}(\mathbb{R}^\infty))$.

In some cases we are only interested in evaluating probabilities involving a subset of the random variables $\{X_t : t \in A \subset J\}$. In these cases, it is sometimes more convenient to work with a σ-algebra on Ω that is coarser than $\mathcal{F}$. For example, if we are only interested in probabilities involving X_k, we can replace $\mathcal{F}$ with a coarser σ-algebra $\mathcal{L}$ that makes $X_k : (\Omega, \mathcal{L}) \to (\mathbb{R}, \mathcal{B}(\mathbb{R}))$ a random variable. The coarsest such σ-algebra is denoted $\sigma(X_k)$. We define this concept below and extend it to a collection of random variables.

Definition 6.5.1. Given a function $X : \Omega \to \mathbb{R}$, we denote the smallest σ-algebra $\mathcal{L}$ on Ω that makes X a measurable function $X : (\Omega, \mathcal{L}) \to (\mathbb{R}, \mathcal{B}(\mathbb{R}))$ as $\sigma(X)$. In other words,

$$\sigma(X) = \{X^{-1}(B) : B \in \mathcal{B}(\mathbb{R})\}.$$

Definition 6.5.1 generalizes to collection of random variables as follows.

Definition 6.5.2. Given a collection of functions $X_\theta : \Omega \to \mathbb{R}$ indexed by $\theta \in \Theta$, we denote the smallest σ-algebra $\mathcal{L}$ on Ω that makes $X_\theta : (\Omega, \mathcal{L}) \to (\mathbb{R}, \mathcal{B}(\mathbb{R}))$,

$\theta \in \Theta$ measurable functions as $\sigma(\{X_\theta, \theta \in \Theta\})$. In other words, $\sigma(\{X_\theta, \theta \in \Theta\})$ is the intersection of all σ-algebras under which $X_\theta, \theta \in \Theta$ are measurable.

Next, we prove Kolmogorov's extension theorem (Proposition 6.2.1) for discrete-time random processes $(J = \mathbb{N})$.

Proof of Kolmogorov's Extension Theorem for Discrete-Time Processes. We separate the proof to two parts: uniqueness and existence.

In the uniqueness part, we assume two distributions P and P' on $(\Omega, \mathcal{F})$ that agree on all finite dimensional marginals and show that $\mathsf{P} = \mathsf{P}'$ almost everywhere. Note that the set of measurable cylinders is a π-system that generates the Borel σ-algebra of $\mathbb{R}^\infty$ (see Section F.5 and Example B.4.4), and that $\mathsf{P} = \mathsf{P}'$ on all finite dimensional marginals and therefore $\mathsf{P} = \mathsf{P}'$ on all measurable cylinders. It follows from Corollary E.3.1 that $\mathsf{P} = \mathsf{P}'$ almost everywhere establishing uniqueness.

To show existence, we assume that we have collection of finite dimensional marginal distributions over $X_n : n \in \mathbb{N}$ that do not contradict each other, and verify that $\mathcal{X} = (X_n : n \in \mathbb{N})$ is a random process. The agreement of the marginals of $\mathcal{X}$ and the given marginal distributions over $X_n, n \in \mathbb{N}$ holds by definition of $\mathcal{X}$. It remains to show that $\mathcal{X} : (\Omega, \mathcal{F}) \to (\mathbb{R}^\infty, \mathcal{B}(\mathbb{R}^\infty))$ is a measurable function. Recall that $\mathcal{B}(\mathbb{R}^\infty)$ is generated by open sets in $\mathbb{R}^\infty$, which are unions of finite intersections of measurable cylinders. Using the second countability of $\mathbb{R}^\infty$ we can ensure that the union contains at most a countably infinite number of terms. We thus have for an open set A in $\mathbb{R}^\infty$, the expression $A = \cup_{i \in \mathbb{N}} \cap_{j=1}^{k_i} A_{ij}$ where A_{ij} are measurable cylinders. It follows that

$$\mathcal{X}^{-1}(\cup_{i \in \mathbb{N}} \cap_{j=1}^{k_i} A_{ij}) = \cup_{i \in \mathbb{N}} \cap_{j=1}^{k_i} \mathcal{X}^{-1}(A_{ij})$$

which is a countable union of finite intersections of Borel sets in $\mathcal{F}$, and therefore $\mathcal{X}^{-1}(A) \in \mathcal{F}$. ∎

6.6 The Borel-Cantelli Lemmas and the Zero-One Law*

This section contains advanced material concerning probabilities of infinite sequence of events. The results rely on limits of sets, introduced in Section A.4. In particular, we use the concept of $\limsup A_n$ for a sequence of sets $A_n, n \in \mathbb{N}$ and its interpretation as A_n infinitely often (abbreviated i.o.).

Proposition 6.6.1 (The First Borel-Cantelli Lemma). *For any sequence of events A_n, $n \in \mathbb{N}$,*

$$\sum_{n \in \mathbb{N}} \mathsf{P}(A_n) < \infty \qquad \textit{implies} \qquad \mathsf{P}(A_n \ i.o.) = 0.$$

Proof. Note that $\limsup_n A_n \subset \cup_{k=m}^{\infty} A_k$ (see Section A.4), implying that

$$P(\limsup A_n) \leq P(\cup_{k \geq m} A_k) \leq \sum_{k \geq m} P(A_k).$$

Since $\sum_{n \in \mathbb{N}} P(A_n) < \infty$, it follows that the right hand side above converges to 0 as $m \to \infty$. Since the inequality above holds for all m, $P(\limsup A_n) = 0$. ∎

Lemma 6.6.1.

$$1 - x \leq \exp(-x), \qquad \forall x \in \mathbb{R}.$$

Proof. The first derivative of $1 - x$ is -1 and the second is 0. The first derivative of $\exp(-x)$ is $-\exp(-x)$ and the second is $\exp(-x)$. Both the two functions and their first derivatives agree at 0. Since $(\exp(-x))'' \geq (1 - x)''$ for all x, it follows that $(\exp(-x))'$ will be larger than $(1 - x)'$ for positive x and smaller for negative x. It follows that $\exp(-x)$ will decay more slowly as $x > 0$ increases and grow faster as $x < 0$ decreases. This establishes the desired inequality. ∎

The R code below graphs the two functions $1 - x$ (dashed) and $\exp(-x)$.

```
curve(exp(-x), from = -2, to = 3, xlab = "$x$",
    ylab = "$1-x$ (dashed) vs. $\\exp(-x)$ (solid)",
    lwd = 3)
curve(1 - x, add = TRUE, col = "red", lty = 2,
    lwd = 3)
```

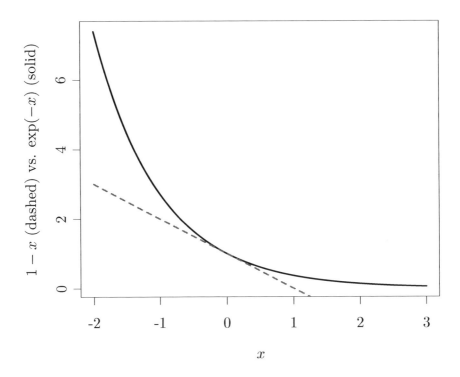

Proposition 6.6.2 (The Second Borel-Cantelli Lemma). *For a sequence of independent events $A_n, n \in \mathbb{N}$,*

$$\sum_{n \in \mathbb{N}} P(A_n) = +\infty \qquad implies \qquad P(A_n \ i.o.) = 1.$$

Proof. Since the events $A_n, n \in \mathbb{N}$ are independent, their complements $A_n^c, n \in \mathbb{N}$ are independent as well (see Proposition E.4.1). Using the bound in Lemma 6.6.1, we have

$$P\left(\bigcap_{k=n}^{n+r} A_n^c\right) = \prod_{k=n}^{n+r} (1 - P(A_k)) \leq \exp\left(-\sum_{k=n}^{n+r} P(A_k)\right). \qquad (6.1)$$

Since $\sum_{n \in \mathbb{N}} P(A_n) = +\infty$, the right hand side above tends 0 as $r \to \infty$. It follows that

$$P\left(\bigcap_{k=n}^{\infty} A_n^c\right) = 0$$

$$1 - P(\limsup A_n) = P((\limsup A_n)^c) = P\left(\bigcup_{n\in\mathbb{N}}\bigcap_{k\geq n} A_k^c\right) \leq P\left(\bigcap_{k\geq n} A_k^c\right) = 0.$$

∎

Definition 6.6.1. Let $A_n, n \in \mathbb{N}$ be a sequence of events. The tail σ-algebra is

$$\mathcal{T}(\{A_n : n \in \mathbb{N}\}) \stackrel{\text{def}}{=} \bigcap_{n\in\mathbb{N}} \sigma(\{A_m : m \geq n\}).$$

Intuitively, the tail σ-algebra contains events relating to the behavior of the tail of the sequence $A_n, n \in \mathbb{N}$.

Example 6.6.1. *The events A_k, A_k^c or $A_k \cap A_l$ are not members of the tail σ-algebra since there exists $\sigma(\{A_r, A_{r+1}, \ldots\})$ that does not contain these events.*

The event $\limsup A_n = \bigcap_{n\in\mathbb{N}} \bigcup_{k\geq n} A_k$ is an intersection of unions of tail events and therefore $\limsup A_n \subset \bigcup_{k\geq r} A_k$ for all r. As a result, $\limsup A_n \in \sigma(\{A_r, A_{r+1}, \ldots\})$ for all r, implying that $\limsup A_n$ is a member of the tail σ-algebra. This is in agreement with the interpretation of $\limsup A_n$ as A_n infinitely often, which depends only on the tail behavior and not on any particular A_k.

Proposition 6.6.3 (Kolmogorov's Zero-One Law). *Let $A_n, n \in \mathbb{N}$ be independent events and $A \in \mathcal{T}(\{A_n : n \in \mathbb{N}\})$. Then $P(A) = 0$ or $P(A) = 1$.*

Proof. By Corollary E.4.1, $\sigma(A_1), \ldots, \sigma(A_{n-1})$, $\sigma(\{A_n, A_{n+1}, \ldots\})$ are independent, and since $A \in \sigma(\{A_n, A_{n+1}, \ldots\})$, the sets $A, A_1, \ldots, A_{n-1}$ are independent. This holds for all n, implying that $A, A_1, A_2, \ldots$ are independent. Applying Corollary E.4.1 again, we have that $\sigma(A), \sigma(\{A_n : n \in \mathbb{N}\})$ are independent. Since $A \in \sigma(A)$ and $A \in \sigma(\{A_n : n \in \mathbb{N}\})$, A is independent of itself, leading to $P(A \cap A) = P(A)P(A) = (P(A))^2$, and implying that $P(A) = 0$ or $P(A) = 1$. ∎

6.7 Notes

More information on random processes is available in most rigorous probability textbooks, such as [17, 10, 5, 1, 33, 25]. Two random processes textbooks that avoid measure theory are [36, 26].

6.8   Exercises

1. Describe a collection of finite dimensional marginals that violates the consistency conditions.

2. Prove formally that the marginals defining the iid process satisfy the consistency conditions.

3. Prove that the random walk process $\mathcal{Z}$ has one dimensional binomial marginals.

4. Consider a discrete-time random walk model similar to the on in this chapter, with the following exception: the walker steps two steps forward with probability θ and one step backwards with probability $1 - \theta$. Characterize the process in terms of the RP categories (stationary, Markov, etc.), and derive its expectation, variance, and autocorrelation functions.

Chapter 7

Important Random Processes

The previous chapter introduced the basic concepts associated with random processes and explored a few basic examples. In this chapter we describe three popular random processes: the Markov chain process, the Poisson process, and the Gaussian process.

7.1 Markov Chains

7.1.1 Basic Definitions

Definition 7.1.1. A Markov chain is a discrete-time, discrete-state Markov process.

Since a Markov chain has Discrete-time, the index set J of the process $\mathcal{X} = \{X_t : t \in J\}$ is finite or countably infinite. We assume that $J = \{0, 1, \ldots, l\}$ in the finite case or $J = \{0, 1, 2, 3, \ldots\}$ in the infinite case. Alternative definitions, such as $J = \mathbb{Z}$, lead to similar results but are much less common.

Since a Markov chain has discrete-state, X_t are discrete RVs, or in other words $\mathsf{P}(X_t \in S_t) = 1$ for some finite or countably infinite set S_t. Taking $S = \cup_t S_t$, we can assume that the random variables X_t have a common discrete state space: $\mathsf{P}(X_t \in S) = 1$, for all $t \in J$. Note that S is finite or countably infinite set since it is a countable union of countably infinite sets (see Chapter A). With no loss of generality, we can identify S with $\{0, 1, \ldots, l\}$ (in the finite case) or $\{0, 1, 2, \ldots\}$ (in the countably infinite case). There is no loss of generality since we can simply relabel the elements of the set S with the elements of one of the sets above and adjust the probability function accordingly; for example, {red, black, green} becomes $\{0, 1, 2\}$.

The Markov assumption implies that for all k, t such that $1 \leq k < t \in \mathbb{N}$, we have

$$P(X_t = x_t | X_{t-1} = x_{t-1}, \ldots, X_{t-k} = x_{t-k}) = P(X_t = x_t | X_{t-1} = x_{t-1}). \quad (7.1)$$

Recalling Kolmogorov's extension theorem (Proposition 6.2.1), we can characterize an RP by specifying a consistent set of all finite-dimensional marginal distributions. Together with the Markov assumption (7.1), this implies that in order to characterize a Markov process $\mathcal{X}$, it is sufficient to specify

$$T_{ij}(t) \stackrel{\text{def}}{=} P(X_{t+1} = j | X_t = i), \quad \forall i, j \in S, \forall t \in J \quad \text{(transition probabilities)}$$

$$\rho_i \stackrel{\text{def}}{=} P(X_0 = i), \qquad\qquad\qquad \forall i \in S \quad \text{(initial probabilities)}.$$

The sufficiency follows from the fact that the probabilities above, together with the Markov assumption, are consistent and derive all other finite dimensional marginals. For example,

$$
\begin{aligned}
&P(X_5 = i, X_4 = j, X_2 = k) \\
&= P(X_5 = i | X_4 = j, X_2 = k)\, P(X_4 = i | X_2 = j)\, P(X_2 = k) \\
&= P(X_5 = i | X_4 = j)\, P(X_4 = j | X_2 = k)\, P(X_2 = k) \\
&= T_{ji}(4) \left(\sum_{l \in S} P(X_4 = j | X_3 = l)\, P(X_3 = l | X_2 = k) \right) \\
&\qquad\qquad \left(\sum_{r \in S} \sum_{s \in S} P(X_2 = k | X_1 = r)\, P(X_1 = r | X_0 = s)\, P(X_0 = s) \right) \\
&= T_{ji}(4) \left(\sum_{l \in S} T_{lj}(3) T_{kl}(2) \right) \left(\sum_{r \in S} \sum_{s \in S} T_{rk}(1) T_{sr}(0) \rho_s \right).
\end{aligned}
$$

We extend below the definition of the finite dimensional simplex $\mathbb{P}_n$ (Definition 5.1.2) to the countably infinite case below.

Definition 7.1.2. For a countably infinite set S, we define $\mathbb{P}_S$ as the set of all functions $f : S \to \mathbb{R}$ such that $f(s) \geq 0$ for all $s \in S$ and $\sum_{s \in S} f(s) = 1$.

Example 7.1.1. *The initial probabilities $\rho(s), s \in S$ are restricted to lie in $\mathbb{P}_S$.*

Definition 7.1.3. A homogenous Markov chain satisfies

$$T_{ij}(t) = T_{ij}(t'), \qquad \forall i, j \in S, \quad \forall t, t' \in J.$$

In this case, we refer to $T_{ij}(t)$ as simply T_{ij}.

We concentrate in the rest of this chapter on Homogenous Markov chains, as they are significantly simpler than the general case.

It is useful to consider the transition probabilities and initial probabilities using algebraic expression. If the state space S is finite, the transition probabilities

T form a matrix and the initial probabilities $\boldsymbol{\rho}$ form a vector. If the state space is infinite, the transition probabilities T form a matrix with an infinite number of rows and columns, and the initial probabilities $\boldsymbol{\rho}$ form a vector with an infinite number of components. It is convenient to use matrix algebra to define $TT = T^2$ and $\boldsymbol{\rho}^\top T$ (the notation $\boldsymbol{\rho}^\top$ transposes $\boldsymbol{\rho}$ from a column vector to a row vector) as follows:

$$[T^2]_{ij} = \sum_{k \in S} T_{ik} T_{kj} \tag{7.2}$$

$$[\boldsymbol{\rho}^\top T]_i = \sum_{k \in S} \rho_k T_{ki}. \tag{7.3}$$

When the state space S is finite, these algebraic operations reduce to the conventional product between matrices and between a vector and a matrix, as defined in Chapter C. We similarly define higher order powers, such as $T^k = T^{k-1} T$, and also denote T^0 as the matrix satisfying $[T^0]_{ij} = 1$ if $i = j$ and 0 otherwise (this matrix corresponds to the identity matrix). Note that in the general case $(T_{ij})^n \neq (T^n)_{ij}$. The first term is the n-power of a scalar and the second is an element of a matrix raised to the n-power. We use the notation T^n_{ij} for the latter:

$$T^n_{ij} \overset{\text{def}}{=} (T^n)_{ij}.$$

Proposition 7.1.1. *Let $\mathcal{X}$ be a homogenous Markov chain as described above. Then*

$$\mathsf{P}(X_{t+n} = j | X_t = i) = [T^n]_{ij}$$
$$p(X_n = i) = [\boldsymbol{\rho}^\top T^n]_i.$$

Proof. Since

$$[T^2]_{ij} = \sum_{k \in S} T_{ik} T_{kj} = \sum_{k \in S} \mathsf{P}(X_{t+1} = i | X_t = k) \, \mathsf{P}(X_t = k | X_{t-2} = j),$$

the matrix T^2 encodes the second order transition probabilities $[T^2]_{ij} = p(X_{t+2} = j | X_t = i)$. Using induction, we get that the matrix $T^n = T \cdot T \cdots T$ is the n-step transition matrix $\mathsf{P}(X_{t+n} = j | X_t = i) = [T^n]_{ij}$ which proves the first statement. Also,

$$[\rho^\top T]_i = \sum_{k \in S} \rho_k T_{ki} = \sum_{k \in S} \mathsf{P}(X_0 = k) \, \mathsf{P}(X_1 = i | X_0 = k) = \mathsf{P}(X_1 = k).$$

The second statement follows by induction. ∎

Note that the Markov property also implies that

$$\mathsf{P}(X_1 = x_1, \dots, X_n = x_n, X_{n+1} = x_{n+1}, \dots, X_{n+m} = x_{n+m} | X_0 = x_0)$$
$$= \mathsf{P}(X_1 = x_2, \dots, X_n = x_n | X_0 = x_0) \, \mathsf{P}(X_1 = x_{n+1}, \dots, X_m = x_{n+m} | X_0 = x_n). \tag{7.4}$$

7.1.2   Examples

Example 7.1.2. *In a Markov chain describing the precipitation on different days, we distinguish between three states: sunny, cloudy, and rainy denoted as consecutive non-negative integers. Assuming the following transition probabilities*

$$P(X_{t+1} = x_{t+1}|X_t = sunny) = \begin{cases} 0.5 & x_{t+1} = sunny \\ 0.4 & x_{t+1} = cloudy \\ 0.1 & x_{t+1} = rainy \end{cases}$$

$$P(X_{t+1} = x_{t+1}|X_t = cloudy) = \begin{cases} 0.4 & x_{t+1} = sunny \\ 0.5 & x_{t+1} = cloudy \\ 0.1 & x_{t+1} = rainy \end{cases}$$

$$P(X_{t+1} = x_{t+1}|X_t = rainy) = \begin{cases} 0.3 & x_{t+1} = sunny \\ 0.5 & x_{t+1} = cloudy \\ 0.2 & x_{t+1} = rainy \end{cases},$$

the transition matrix is

$$T = \begin{pmatrix} 0.5 & 0.4 & 0.1 \\ 0.4 & 0.5 & 0.1 \\ 0.3 & 0.5 & 0.2 \end{pmatrix}.$$

Assuming $P(X_0)$ *is uniformly distributed over the three states* $\rho = (1/3, 1/3, 1/3)$, *The following R code computes the probabilities of* $P(X_t)$ *for* $t = 0, 1, \ldots, 5$, *assuming* $P(X_0)$ *is uniformly distributed over the three states:* $\rho = (1/3, 1/3, 1/3)$.

```
s = factor(c("sunny", "cloudy", "rainy"), levels = c("rainy",
    "cloudy", "sunny"), ordered = TRUE)
P = data.frame(pmf = c(1/3, 1/3, 1/3), state = s,
    time = "$t=0$")
T = Tk = matrix(c(0.5, 0.4, 0.1, 0.4, 0.5, 0.1,
    0.3, 0.5, 0.2), byrow = T, nrow = 3)
for (i in 2:6) {
    P = rbind(P, data.frame(pmf = t(c(1/3, 1/3,
        1/3) %*% Tk), state = s, time = rep(paste("$t=",
        i - 1, "$", sep = ""), 3)))
    Tk = Tk %*% T
}
P


##         pmf   state   time
## 1  0.3333   sunny  $t=0$
## 2  0.3333  cloudy  $t=0$
## 3  0.3333   rainy  $t=0$
## 4  0.4000   sunny  $t=1$
## 5  0.4667  cloudy  $t=1$
## 6  0.1333   rainy  $t=1$
```

```
## 7   0.4267   sunny $t=2$
## 8   0.4600  cloudy $t=2$
## 9   0.1133   rainy $t=2$
## 10  0.4313   sunny $t=3$
## 11  0.4573  cloudy $t=3$
## 12  0.1113   rainy $t=3$
## 13  0.4320   sunny $t=4$
## 14  0.4569  cloudy $t=4$
## 15  0.1111   rainy $t=4$
## 16  0.4321   sunny $t=5$
## 17  0.4568  cloudy $t=5$
## 18  0.1111   rainy $t=5$
```

The numeric results above are displayed in a graph below.

```
qplot(x = state, ymin = 0, ymax = pmf, data = P,
    geom = "linerange", facets = "time~.", ylab = "$p_{X_t}$") +
    geom_point(aes(x = state, y = pmf)) + coord_flip()
```

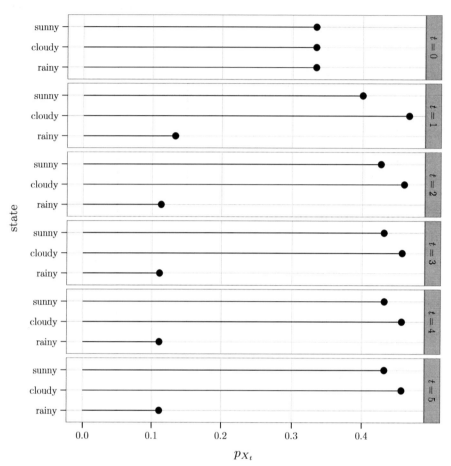

We observe that (i) even though the initial weather X_0 is uniformly distributed over the three states, the weather at time $t = 5$ is highly unlikely to be rainy, and (ii) it appears that

$$\lim_{t \to \infty} p_{X_t}(r) = \begin{cases} 0.4321 & r = sunny \\ 0.4558 & r = cloudy \\ 0.1111 & r = rainy \end{cases}.$$

We will confirm this experimental observation with theory in Section 7.1.6.

Example 7.1.3 (Random Walk in $\mathbb{Z}$). *We consider a Markov chain process* $\mathcal{X} = \{X_0, X_1, X_2, \ldots\}$ *with* $S = \mathbb{Z}$, $X_0 = 0$ *with probability one, and*

$$T_{ij} = \begin{cases} \theta & j = i + 1 \\ 1 - \theta & j = i - 1 \\ 0 & otherwise \end{cases}.$$

This Markov chain corresponds to the random walk model described in Section 6.4. The random variable X_t denotes the capital of a persistent gambler that wins \$1 with probability θ and loses \$1 with probability $1 - \theta$. It is assumed that the gambler starts with zero capital and is able to keep accumulating debt indefinitely if needed.

Example 7.1.4 (Gambler's Ruin). *We consider a variation of the Gambler Markov chain in which the gambler starts from \$k but is not able to accumulate any debt, effectively terminating further gambling if no more capital is available. Note that there is no limit on the amount of money the gambler wins, implying that the gambler plays against an infinitely wealthy casino. This situation reflects more accurately non-symmetric gambling between an individual gambler with a fixed initial capital and an extremely wealthy casino.*

The state space in this case is $S = \{0, 1, 2, \ldots\}$, the initial probabilities are $\rho_k = 1$ and $\rho_j = 0$ for $j \neq k$, and the transition probabilities are

$$
T_{ij} = \begin{cases}
1 & i = 0 = j \\
0 & i = 0 < j \\
\theta & j = i + 1 > 1 \\
1 - \theta & j = i - 1 > 0 \\
0 & \text{otherwise}
\end{cases}.
$$

The R code graphs 100 sample paths for $\theta = 0.6$. We use transparency (`alpha` parameter in the code below) in order to illustrate regions where many sample paths overlap. Such regions are shaded darker than regions with a single sample path or a few overlapping paths.

```
J = 1:100
plot(c(1, 100), c(-5, 30), type = "n", xlab = "$J$",
    ylab = "$\\mathcal{Z}(\\omega)$")
Zn100 <- Zn500 <- vector(len = 100)
for (i in 1:100) {
    X = rbinom(n = 500, size = 1, prob = 0.6)
    Z = cumsum(2 * X - 1) + 4
    Z[cumsum(Z == 0) > 0] = 0
    lines(J, c(Z[1:100]), type = "s", col = rgb(0,
        0, 0, alpha = 0.2))
    Zn100[i] = Z[100]
    Zn500[i] = Z[500]
}
title("100 Sample Paths ($\\theta=0.6, X_0=4$)",
    cex.main = 1)
```

100 Sample Paths $(\theta = 0.6, X_0 = 4)$

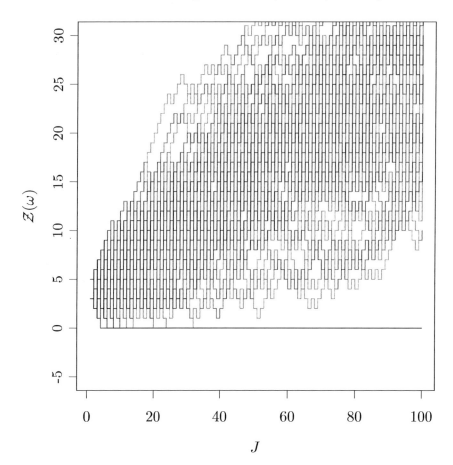

The R code below graphs a histogram of the capital of the random gambler at $t = 100$ over multiple samples. The shape of the histogram approximates p_{X_t}.

```
qplot(Zn100, main = "histogram (100 sample paths)",
    xlab = "$Z_{100}$")
```

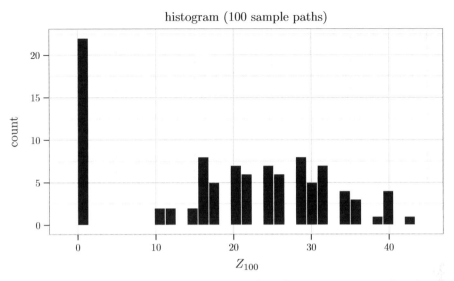

Comparing the graph above with the graph in Section 6.4 we see that the gambler (usually) does much worse in the graphs above than in the case of Section 6.4, even though the gambler starts with $4 instead of $0.

Repeating the previous graph for t = 500 below, we see that the two groups (ruined gamblers and newly rich gamblers) become more separated.

```
qplot(Zn500, main = "histogram (100 sample paths)",
    xlab = "$Z_{500}$")
```

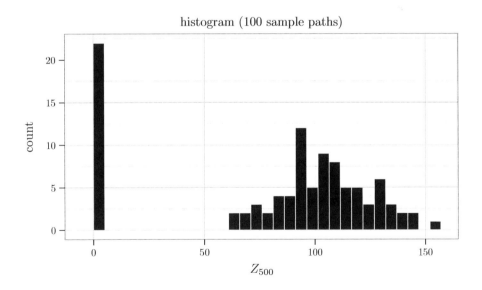

Example 7.1.5 (Symmetric Random Walk in $\mathbb{Z}^d$). *Consider a Markov chain process $\mathcal{X} = \{X_0, X_1, X_2, \ldots\}$ with $S = \mathbb{Z}^d$, and $X_0 = (0, \ldots, 0)$ with probability one. Each element of S has $2d$ neighbors obtained by increasing or decreasing by one a single component. The transition matrix T is defined to be $T_{ab} = 1/(2d)$ if $a, b \in \mathbb{Z}^d$ are neighbors and zero otherwise. If $d = 1$, this model reduces to the random walk model of Section 6.4 with $\theta = 1/2$.*

7.1.3 Transience, Persistence, and Irreducibility

Definition 7.1.4. For a Markov chain, we denote

$$f_{ij}^{(n)} = \mathsf{P}(X_1 \neq j, \ldots, X_{n-1} \neq j, X_n = j | X_0 = i)$$

$$f_{ij} = \sum_{n \in \mathbb{N}} f_{ij}^{(n)} = \mathsf{P}(\text{state } j \text{ is visited eventually} | X_0 = i).$$

A state i is persistent if $f_{ii} = 1$ and transient if $f_{ii} < 1$.

Note that the sequence of probabilities $f_{ij}^{(n)}$, $n \in \mathbb{N}$ correspond to a sequence of disjoint events, implying that the sum f_{ij} corresponds to the probability of their union.

Proposition 7.1.2. *For a Markov chain, we have*

$$\mathsf{P}\big(X_n = j \ i.o.|X_0 = i\big) = \begin{cases} 0 & f_{jj} < 1 \\ f_{ij} & f_{jj} = 1 \end{cases}$$

$$\mathsf{P}\big(X_n = i \ i.o.|X_0 = i\big) = \begin{cases} 0 & f_{jj} < 1 \\ 1 & f_{jj} = 1 \end{cases}.$$

The notation i.o. above corresponds to infinitely often (see Section A.4).

Proof. The probability that during the time interval $0, \ldots, t_k$, the Markov chain visits state j at times $t_1 < \cdots < t_k$ (but not in between) conditioned on $X_0 = 1$ is $f_{ij}^{(t_1)} f_{jj}^{(t_2 - t_1)} \cdots f_{jj}^{(t_k - t_{k-1})}$. In a similar way, the probability that the Markov chain visits state j at least k times conditioned on $X_0 = 1$ is $f_{ij} f_{jj}^{k-1}$. The event $X_n = i$ i.o. corresponds to letting $k \to \infty$, which yields the two results above. ∎

Proposition 7.1.3. *Consider a Markov chain and a state $s \in S$. (i) The state s is transient if and only if $\mathsf{P}(X_n = s \ i.o.|X_0 = s) = 0$ that in turn occurs if and only if $\sum_{n \in \mathbb{N}} T_{ii}^n < \infty$. (ii) The state s is persistent if and only if $\mathsf{P}(X_n = s \ i.o.|X_0 = s) = 1$ that in turn occurs if and only if $\sum_{n \in \mathbb{N}} T_{ii}^n = +\infty$.*

Proof. By the first Borel-Cantelli Lemma (Proposition 6.6.1), $\sum_{n \in \mathbb{N}} T_{ss}^n < \infty$ implies $\mathsf{P}(X_n = s \ \text{i.o.}|X_0 = s) = 0$ that in turn implies $f_{ss} < 1$ (see Proposition 7.1.2) and transience of s.

It remains to show that $f_{ss} < 1$ implies $\sum_n T_{ss}^n < \infty$. Consider the event of transitioning from state s to state s' after n steps. The *first* time the chain visits

s' may occur after 1 time step, after 2 time steps, and so on until n time steps. Using the law of total probability (Proposition 1.5.4) we have

$$T_{ss'}^n = \sum_{r=0}^{n-1} f_{ss'}^{(n-r)} T_{s's'}^r$$

$$\sum_{t=1}^{n} T_{ss}^t = \sum_{t=1}^{n} \sum_{r=0}^{t-1} f_{ss'}^{(n-r)} T_{s's'}^r$$

$$= \sum_{r=0}^{n-1} \sum_{t=r+1}^{n} f_{ss'}^{(n-r)} T_{s's'}^r$$

$$= \sum_{r=0}^{n-1} T_{s's'}^r \sum_{t=r+1}^{n} f_{ss'}^{(n-r)}$$

$$\leq f_{ss} \sum_{r=0}^{n} T_{ss}^r$$

$$(1 - f_{ss}) \sum_{t=1}^{n} T_{ss}^t \leq f_{ss}.$$

The third equality follows from the fact that the double summation ranges over the indices

$$\{t, r : 1 \leq t \leq n, 0 \leq r < t\} = \{t, r : r < t \leq n, 0 \leq r \leq n - 1\}.$$

The last inequality above follows from the fact that $T_{ss}^0 = 1$. This shows that $f_{ss} < 1$ implies $\sum_n T_{ss}^n < \infty$ and concludes the proof of (i).

The proof of (ii) follows from the proof of (i) and the fact that $\mathsf{P}(X_n = s \text{ i.o.}|X_0 = s) = 0$ is either 0 or 1, $\sum_{n \in \mathbb{N}} T_{ii}^n$ is either finite or infinite, and a state is either persistent or transient. ∎

Definition 7.1.5. A Markov chain is called irreducible if for any two states $i, j \in S$, we have $T_{ij}^n > 0$ for some n.

Proposition 7.1.4. *For an irreducible Markov chain, one of the following two alternatives hold:*
(i) All states are transient, $\mathsf{P}(\cup_{s \in S}\{X_n = s \text{ i.o } \}|X_0 = s') = 0$ *for all* $s' \in S$, *and* $\sum_{n \in \mathbb{N}} T_{ss'}^n < \infty$.
(ii) All states are persistent, $\mathsf{P}(\cap_{s \in S}\{X_n = s \text{ i.o } \}|X_0 = s') = 1$ *for all* $s' \in S$, *and* $\sum_{n \in \mathbb{N}} T_{ss'}^n = +\infty$.

Proof. By irreducibility, for each s, s' there exists t, t' such that $T_{ss'}^t > 0$ and $T_{s's}^{t'} > 0$. Since

$$T_{ss}^{t+t'+n} \geq T_{ss'}^{t+t'+n} T_{s's'}^n T_{s's}^{t'}$$

and $T_{ss'}^{t+t'+n}T_{s's}^{t'} > 0$, we have that $\sum_{n\in\mathbb{N}} T_{ss}^n < \infty$ implies $\sum_{n\in\mathbb{N}} T_{s's'}^n < \infty$. We thus have that if one state is transient then all states are transient. Similarly, if one state is persistent then all states are persistent.

If all states are transient, then by Proposition 7.1.3 $P(X_n = s \text{ i.o} \,|X_0 = s') = 0$ for all $s, s' \in S$ and $P(\cup_{s\in S}\{X_n = s \text{ i.o} \}|X_0 = s') = 0$ (as implied by Proposition 1.6.8, a countable union of sets with zero probability have zero probability). Denoting the time of first transition from s to s' by r, the law of total probability (Proposition 1.5.4) implies

$$\sum_{n=1}^{\infty} T_{ss'}^n = \sum_{n=1}^{\infty}\sum_{r=1}^{n} f_{ss'}^{(r)} T_{s's'}^{n-r}$$

$$= \sum_{r=1}^{\infty} f_{ss'}^{(r)} \sum_{m=1}^{\infty} T_{s's'}^m$$

$$\leq \sum_{m=1}^{\infty} T_{s's'}^m.$$

Since the states are transient, Proposition 7.1.4 implies that $\sum_{m=1}^{\infty} T_{s's'}^m < \infty$, which together with the inequality above implies $\sum_{n=1}^{\infty} T_{ss'}^n < \infty$.

If all states are persistent, Proposition 7.1.4 implies that $P(X_n = s \text{ i.o} \,|X_0 = s') = 0$ and

$$T_{ss'}^m = P(X_m = s'|X_0 = s)$$

$$= P(\{X_m = s'\} \cap \{X_n = s \text{ i.o} \}|X_0 = s)$$

$$\leq \sum_{n=m+1}^{\infty} P(X_m = s', X_{m+1} \neq s, \ldots, X_{n-1} \neq s, X_m = s|X_0 = s)$$

$$= \sum_{n=m+1}^{\infty} T_{ss'}^m f_{s's}^{(n-m)}$$

$$= f_{s's} \sum_{n=m+1}^{\infty} T_{ss'}^m.$$

By irreducibility, there exists m for which $T_{ss'}^m > 0$, implying that $f_{ss'} > 0$, which can only happen if $f_{ss'} = 1$. Since we did not constrain the choice of s, s', we have $f_{ss'} = 1$ for all $s, s' \in S$. This implies $P(X_n = s \text{ i.o} \,|X_0 = s') = 1$ for all s, s', and by de-Morgan's law

$$P(\cap_{s\in S}\{X_n = s \text{ i.o} \}|X_0 = s') = 1 - P(\cup_{s\in S}\{X_n = s \text{ i.o} \}^c|X_0 = s')$$

$$\geq 1 - \sum_{s\in S} P(\{X_n = s \text{ i.o} \}^c|X_0 = s'))$$

$$= 1 - 0.$$

Finally, $\sum_{n\in\mathbb{N}} T_{ss'}^n = +\infty$ since otherwise Proposition 6.6.1 would imply that $P(X_n = s \text{ i.o} \,|X_0 = s') = 0$, yielding a contradiction. ∎

Since for an irreducible Markov chain, all states are persistent or transient we refer to the Markov chain itself as persistent or transient.

Corollary 7.1.1. *An irreducible Markov chain with a finite state space S is persistent.*

Proof. Since $\sum_{s \in S} T^n_{s's} = 1$, it follows that

$$\sum_{n \in \mathbb{N}} \sum_{s \in S} T^n_{s's} = \infty$$

$$\sum_{s \in S} (\sum_{n \in \mathbb{N}} T^n_{s's}) = \infty.$$

Since S is finite we have $\sum_{n \in \mathbb{N}} T^n_{s's} = \infty$, which together with Proposition 7.1.4 implies persistence. ∎

7.1.4 Persistence and Transience of the Random Walk Process

The Markov chain corresponding to the symmetric random walk on $\mathbb{Z}^d$ (Example 7.1.5) is persistent when $d = 1$ or $d = 2$ and transient for $d \geq 3$. This is a remarkable result since it shows a qualitatively different behavior between the cases of one and two dimensions on one hand and three or more dimensions on the other hand.

By a symmetry argument, the probability T^n_{ss} is the same for all $s \in S$ and so we denote it as $q_n^{(d)}$.

If $d = 1$, a return to the starting point after n step is possible only if n is even. We compute below $q_{2n}^{(d)}$, $n \in \mathbb{N}$ ($q_{2n+1}^{(d)} = 0$ since $2n + 1$ is odd). We have

$$q_{2n}^{(d)} = \binom{2n}{n} 2^{-2n}$$

since there are $2n$-choose n ways to select n moves to the right and the remaining moves are automatically chosen as moves to the left, and each such sequence of moves occurs with probability $(1/2)^{2n}$. Using Stirling's Formula (Proposition 1.6.2), we have

$$q_n^{(1)} = \frac{(2n)!}{n!n!} 2^{-2n}$$

$$= \frac{(2n)!}{(2n)^{2n+1/2} e^{-2n}} \cdot \frac{n^{n+1/2} e^{-n}}{n!} \cdot \frac{(2n)^{2n+1/2} e^{-2n}}{n^{2n+1} e^{-2n}} 2^{-2n}$$

$$= \frac{(2n)!}{(2n)^{2n+1/2} e^{-2n}} \cdot \frac{n^{n+1/2} e^{-n}}{n!} \cdot 2^{1/2} n^{-1/2}$$

$$\sim (\pi n)^{-1/2}.$$

(See Section B.5 for an explanation of the notation $\sim$ above.) It follows from Proposition D.2.5 that $\sum_{n \in \mathbb{N}} q_n^{(1)} = \infty$, implying that the chain is persistent.

If $d = 2$, a return to the starting point after $2n + 1$, $n \in \mathbb{N}$ steps is impossible. A return to the starting point after $2n$, $n \in \mathbb{N}$ is possible only if the number of up and down steps are equal and the number of right and left steps are equal. The number of such sequences of length $2n$ multiplied by their probability $(1/4)^{-2n}$ gives

$$q_{2n}^{(2)} = \sum_{r=0}^{n} \frac{(2n)!}{r!r!(n-r)!(n-r)!} \frac{1}{4^{2n}}$$

$$= 4^{-2n} \binom{2n}{n} \sum_{r=0}^{n} \binom{n}{u} \binom{n}{n-u} \tag{7.5}$$

$$= 4^{-2n} \binom{2n}{n} \binom{2n}{n}. \tag{7.6}$$

The first equality above follows from Proposition 1.6.7, where r corresponds to the number of up moves and $n - r$ corresponds to the number of right moves. The last equality above holds since $\sum_{r=0}^{n} \binom{n}{u} \binom{n}{n-u}$ is the number of possible ways to select n out of $2n$ with no replacement and no order: first select u from the first half containing n items and then select $n - u$ items from the second half containing n items, for all possible values of u: $0 \le u \le n$. As in the case of $d = 1$, using Stirling's approximation on (7.6) yields $q_{2n}^{(d)} \sim (\pi n)^{-1}$. As before, Proposition D.2.5 leads then to $\sum_{n \in \mathbb{N}} q_n^{(2)} = \infty$ implying that the chain is persistent.

If $d = 3$ similar arguments lead to

$$q_{2n}^{(3)} = \sum \frac{(2n)!}{r!r!q!q!(n-r-q)!(n-r-q)!} \frac{1}{6^{2n}}$$

where the sum ranges over all non-negative r, q satisfying $r + q \le n$ (r is the number of steps up, q the number of steps right, and $n - r - q$ the number of steps in the third dimension). Conditioning on the number of moves in the third dimension, we can reduce the above equation to quantities containing $q_n^{(1)}$ and $q_n^{(2)}$:

$$q_{2n}^{(3)} = \sum_{l=0}^{n} \binom{2n}{2l} \left(\frac{1}{3}\right)^{2n-2l} \left(\frac{2}{3}\right)^{2l} q_{2n-2l}^{(1)} q_{2l}^{(2)}. \tag{7.7}$$

By bounding the above terms, it can be shown that $q_{2n}^{(3)} = O(n^{-3/2})$ (details are available at [5, Example 8.6]) which implies (see Proposition D.2.5) $\sum_{n \in \mathbb{N}} q_n^{(2)} < \infty$, showing that the Markov chain is transient. Similar recursive arguments for $d > 3$ lead to $q_n^d = O(n^{-k/2})$, implying (see Proposition D.2.5) $\sum_{n \in \mathbb{N}} q_n^{(d)} < \infty$, for $d \ge 3$, and consequentially showing the transience for dimensions higher than 3.

7.1.5 Periodicity in Markov Chains

Definition 7.1.6. The period of a state s, denoted by $d(s)$, is the greatest common divisor of all integers $n \in \mathbb{N}$ for which $T_{ss}^n > 0$. If the period of a state is 1, we say that the state is aperiodic.

Example 7.1.6. *In the symmetric random walk on $\mathbb{Z}^d$, $d = 1$ case (Example 7.1.4), every step has period 2. This follows from the fact that a return to each state is possible only after an even number of time periods: $2, 4, 6, 8, \ldots$. The greatest common divisor of all even numbers is 2. The same holds for the $d > 1$ case.*

Example 7.1.7. *In the case of the weather pattern Markov chain (Example 7.1.2), the period of all states is 1. This follows from the fact that it is possible to remain in each of the current states $\mathsf{P}(X_t = s | X_{t-1} = s) > 0$, implying that returns are feasible after $1, 2, 3, 4, 5, \ldots$ time periods.*

Proposition 7.1.5. *For an irreducible Markov chain, all states have the same period.*

Proof. Consider two states $s, s' \in S$ with periods a, b. By irreducibility, for each s, s' there exists m, n such that $T_{ss'}^m > 0$ and $T_{s's}^n > 0$. We also have

$$T_{ss}^{(m+n+c)} \geq T_{ss'}^{(m)} T_{s's'}^{(c)} T_{s's}^{(n)}, \qquad n \in \{0, 1, 2, \ldots\}. \tag{7.8}$$

In (7.8), if $c = 0$ the right hand side is positive, implying that the period a of s divides $m + n$. Repeating (7.8) for all c for which $T_{s's'}^c > 0$, we have that a divides $m + c + n$ and therefore it also divides c. It follows that a divides all all c for which $T_{s's'}^c > 0$, leading to $a \leq b$. Repeating the argument with s, s' flipped yields $b \leq a$ indicating that the periods of every two states are equal in an irreducible Markov chain. ∎

As a result of the above proposition, we refer to the common period as the period of the Markov chain. If it is 1, we say that the Markov chain is aperiodic.

Proposition 7.1.6. *Consider an irreducible aperiodic Markov chain. Then*

$$\forall s, s' \in S \quad \exists N \in \mathbb{N}, \text{ such that } n > N \implies T_{ss'}^n > 0.$$

Proof. We first prove the following fact: a set of positive integers A with a greatest common divisor 1 and closed under addition ($a, b \in A$ implies $a + b \in A$) contains all integers greater than a certain N'.

We define $A' = \{a - b : a, b \in A\} \cup A \cup \{-a : a \in A\}$ and refer to the smallest positive element of A' as d. For $x \in A'$, we have $x = qd + r$, $0 \leq q < d$. Since A' is closed under addition and subtraction $r = x - qd \in A'$ and since d is the smallest positive element in A', we have $r = 0$. This implies that A' contains all multiples of d. Consequentially, d divides all elements in A' and therefore all element in A, implying that $d = 1$ (since A is a set with a greatest common divisor of 1). We thus have $1 \in A'$ expressed as $1 = x - y$ for $x, y \in A$. We show next that A contains all elements greater than $N' = (x + y)^2$. For $a > N'$ we write $a = q(x+y)+r$ for some q and $0 \leq r < x+y$. Since $a > N' \geq (r+1)(x+y)$, we have $q = (a - r)/(x + y) > r$ implying that $q + r \geq q - r > 0$. This leads to

$$a = q(x+y)+r = q(x+y)+r(x-y) = (q+r)x+(q-r)y, \qquad q+r > 0, \quad q-r > 0,$$

which together with the fact that A is closed under addition implies that $a \in A$. This concludes the proof that A contains all integers greater than N'.

We denote the set of $n \in \mathbb{N}$ for which $T^n_{s's'} > 0$ by A. Since $T^{a+b}_{s's'} \geq T^a_{s's'} T^b_{s's'}$, it follows that A is closed under addition. Since the chain is aperiodic, it follows that A has a greatest common divisor of 1. Using the result above, we have that A contains all integers greater than N'. Setting $N = N' + r$, where $T^r_{ss'} > 0$ we have than whenever $n > N$, $T^n_{ss'} \geq T^r_{ss'} T^{n-r}_{s's'} > 0$. ∎

7.1.6   The Stationary Distribution

Definition 7.1.7. Consider a Markov chain with transition probabilities $T_{ss'} = P(X_t = s'|X_{t-1} = s)$, $s, s' \in S$. A function $q : S \to \mathbb{R}$ is a stationary distribution if the following expressions hold

$$\sum_{s \in S} q(s)T_{ss'} = q(s'), \qquad \forall s' \in S \tag{7.9}$$

$$q(s) \geq 0, \qquad \forall s \in S \tag{7.10}$$

$$\sum_{s \in S} q(s) = 1. \tag{7.11}$$

Using the algebraic notation in (7.2)-(7.3), we may treat T as a (potentially infinite) matrix and $\boldsymbol{q} = (q(s) : s \in S)$ as a (potentially infinite) column vector. The first condition above then becomes

$$\boldsymbol{q}^\top T = \boldsymbol{q}^\top \qquad \text{or} \qquad T^\top \boldsymbol{q} = \boldsymbol{q}$$

implying that $\boldsymbol{q}$ is an eigenvector of the matrix $T^\top$ with eigenvalue 1. The second and third conditions in the definition above ensures that $\boldsymbol{q}$ represents a distribution over S. Note that if $\boldsymbol{q}$ corresponds to the stationary distribution of a Markov chain, then we have

$$(T^k)^\top \boldsymbol{q} = \boldsymbol{q}, \qquad \forall q \in \mathbb{N}.$$

Proposition 7.1.7 (Equilibrium Property). *If an irreducible aperiodic Markov chain has a stationary distribution $q(s), s \in S$, then the Markov chain is persistent, and the stationary distribution is unique, strictly positive, and*

$$\lim_{n \to \infty} T^n_{ss'} = q(s'), \qquad \forall s, s' \in S.$$

Proof. If the Markov chain is transient, then $\sum_{n \in \mathbb{N}} T^n_{ss'}$ for all $s, s' \in S$, implying that $T^n_{ss'} \to \infty$ for all $s, s' \in S$. This, together with the equation $\sum_{s \in S} q(s) T^n_{ss'} = q(s')$ implies that $q(s) = 0$ for all s, a contradiction to the requirement $\sum_s q(s) = 1$. This shows that the Markov chain is persistent.

We construct a new Markov chain process with a state space $S \times S$ and transition probabilities $\tilde{T}_{sr,s'r'} = T_{ss'} T_{rr'}$. Note that $\tilde{T}$ indeed represents transition probabilities since it is non-negative and

$$\sum_{s'} \sum_{r'} \tilde{T}_{sr,s'r'} = \sum_{s'} T_{ss'} \sum_{r'} T_{rr'} = 1.$$

The new Markov chain represents two independent Markov chains processes progressing in parallel, and so $\tilde{T}^n_{sr,s'r'} = T^n_{ss'} T^n_{rr'}$.

Since the two independent Markov chains are irreducible, there exists N such that $n > N$ implies $T^n_{ss'} T^n_{rr'}$ is positive (Proposition 7.1.6), implying that $\tilde{T}^n_{sr,s'r'}$ for $n > N$, making the new coupled Markov chain irreducible. The aperiodicity of the original Markov chain implies that the coupled Markov chain is aperiodic as well. The function $q(s, r) = q(s) q(r)$ on $S \times S$ is a stationary distribution of the new Markov chain:

$$\sum_{s,r} q(s, r) \tilde{T}_{sr,s'r'} = \sum_{s,r} q(s) q(r) T_{ss'} T_{rr'}$$

$$= \sum_s q(s) T_{ss'} \sum_r q(r) T_{rr'}$$

$$= q(s') q(r')$$

$$= q(s', r')$$

$$\sum_{s,r} q(s, r) = \sum_r q(s) \sum_s q(s) = 1 \cdot 1$$

$$q(s, r) = q(s) q(r) \geq 0.$$

Since the coupled Markov chain is aperiodic, irreducible and has a stationary distribution, it is also persistent.

Denoting the two component Markov chains composing the coupled chain by X_n and Y_n, and $\tau = \min\{n : X_n = Y_n = c\}$ (note that since the coupled chain is persistent, it visits (c, c) i.o. and therefore τ is finite), we have

$$\mathsf{P}(X_n = r, Y_n = r', \tau = m | X_0 = s, Y_0 = s')$$
$$= \mathsf{P}(\tau = m | X_0 = s, Y_0 = s') T^{n-m}_{cr} T^{n-m}_{cr'}, \qquad m \leq n.$$

Summing over all possible values of r' in the equation above gives $\mathsf{P}(X_n = r, \tau = m | X_0 = s, Y_0 = s') = \mathsf{P}(\tau = m | X_0 = s, Y_0 = s') T^{n-m}_{cr}$ and summing over all possible values of r in the equation above gives $\mathsf{P}(Y_n = r', \tau = m | X_0 = s, Y_0 = s') = \mathsf{P}(\tau = m | X_0 = s, Y_0 = s') T^{n-m}_{cr'}$. Setting $r = r'$ and summing over $m = 1, 2, \ldots, n$ we get

$$\mathsf{P}(Y_n = r, \tau \leq m | X_0 = s, Y_0 = s') = \mathsf{P}(X_n = r, \tau \leq m | X_0 = s, Y_0 = s'). \quad (7.12)$$

Thus, even though the two chains started from different states s and s', conditioned on the chains meeting at τ, their distributions for times $n \geq \tau$ are identical.

From (7.13) we have

$$
\begin{aligned}
&\mathsf{P}(X_n = r | X_0 = s, Y_0 = s') \\
&= \mathsf{P}(X_n = r, \tau \leq n | X_0 = s, Y_0 = s') + \mathsf{P}(X_n = r, \tau > n | X_0 = s, Y_0 = s') \\
&\leq \mathsf{P}(X_n = r, \tau \leq n | X_0 = s, Y_0 = s') + \mathsf{P}(\tau > n | X_0 = s, Y_0 = s') \\
&= \mathsf{P}(Y_n = r, \tau \leq n | X_0 = s, Y_0 = s') + \mathsf{P}(\tau > n | X_0 = s, Y_0 = s') \\
&\leq \mathsf{P}(Y_n = r | X_0 = s, Y_0 = s') + \mathsf{P}(\tau > n | X_0 = s, Y_0 = s').
\end{aligned}
$$

Repeating the equation above with X_n and Y_n interchanged yields

$$
|T^n_{sr} - T^n_{s'r}| = |\mathsf{P}(X_n = r | X_0 = s, Y_0 = s') - \mathsf{P}(Y_n = r | X_0 = s, Y_0 = s')|
$$
$$
\leq \mathsf{P}(\tau > n | X_0 = s, Y_0 = s').
$$

Since τ is finite, the right hand side above converges to 0 (as $n \to \infty$), implying that the left hand side also converges to 0.

$$
\lim_{n \to \infty} |T^n_{sr} - T^n_{s'r}| = |\mathsf{P}(X_n = r | X_0 = s, Y_0 = s') - \mathsf{P}(Y_n = r | X_0 = s, Y_0 = s')| = 0.
$$
(7.13)

Again, this implies that the two chains "mix" or behave identically for large n even though they started from different initial states.

Using the fact that q is the stationary distribution of the chain we have

$$
\rho(r) - T^n_{s'r} = \sum_{s \in S} \rho(s) T^n_{sr} - 1 \cdot T^n_{s'r} = \sum_{s \in S} \rho(s)(T^n_{sr} - T^n_{s'r}),
$$

which converges to 0 by (7.13). To show that q is positive, we let $r \to \infty$ in the following inequality

$$
T^{(n+m+r)}_{ss} \geq T^n_{ss'} T^r_{s's'} T^m_{s's},
$$

yielding

$$
q(s) \geq T^n_{ss'} q(s') T^m_{s's}.
$$

Since the chain is persistent we can choose n and m such that $T^n_{ss'} > 0$ and $T^m_{s's} > 0$. It thus follows that if $q(s')$ is positive then $q(s)$ is positive. Since $\sum q(s) = 1$, we have that $q(s) > 0$ for all $s \in S$. ∎

The stationary distribution is extremely important in understanding the behavior of Markov chains (it if exists). As the following corollary shows, the distribution of X_n for large n will converge to the stationary distribution. In particular, this occurs regardless of the initial probabilities $\mathsf{P}(X_0)$. In other words, for large n, the distribution of X_n is approximately independent of n and is approximately independent of the initial distribution ρ or the initial state X_0. Algebraically, this implies that the rows of T^n converge to the vector q and thus

$$
\lim_{n \to \infty} \rho^\top T^n = q, \qquad \forall \rho \in \mathbb{P}_S.
$$

Corollary 7.1.2. *If an irreducible aperiodic Markov chain has a stationary distribution $q(s), s \in S$, then*

$$\lim_{n \to \infty} p_{X_n}(s) = q(s), \qquad \forall s \in S.$$

Proof. Denoting the initial probabilities as $\rho(s) = \mathsf{P}(X_0 = s), s \in S$, we have

$$\lim_{n \to \infty} p(X_n = s) = \lim_{n \to \infty} \sum_{r \in S} \rho(r) T_{rs}^n = \sum_{r \in S} \rho(r) \lim_{n \to \infty} T_{rs}^n$$
$$= 1 \cdot \lim_{n \to \infty} T_{rs}^n = q(s).$$

■

Example 7.1.8. *Recall Example 7.1.2, where we observed that for a uniform initial distribution X_0, the distribution of X_n converged to*

$$\lim_{t \to \infty} p_{X_t}(r) = \begin{cases} 0.4321 & r = sunny \\ 0.4558 & r = cloudy \\ 0.1111 & r = rainy \end{cases}.$$

We verify below that this is indeed a stationary distribution of the Markov chain, that the rows of T^n converges to this distribution, and that $\lim_n p(X_n)$ equals this distribution for a variety of initial probabilities.

```
print(T <- matrix(c(0.5, 0.4, 0.1, 0.4, 0.5, 0.1,
     0.3, 0.5, 0.2), byrow = T, nrow = 3))
```

```
##      [,1] [,2] [,3]
## [1,] 0.5  0.4  0.1
## [2,] 0.4  0.5  0.1
## [3,] 0.3  0.5  0.2
```

```
print(q <- c(0.4321, 0.4558, 0.1111))
```

```
## [1] 0.4321 0.4558 0.1111
```

```
# verify that q is a stationary distribution
sum(q)
```

```
## [1] 0.999
```

```
c(0.4321, 0.4558, 0.1111) %*% T
```

```
##        [,1]   [,2]   [,3]
## [1,] 0.4317 0.4563 0.111
```

```
# raise transition matrix to the 10 power
```

```
print(T10 <- T %^% 10)

##            [,1]    [,2]    [,3]
## [1,] 0.4321 0.4568 0.1111
## [2,] 0.4321 0.4568 0.1111
## [3,] 0.4321 0.4568 0.1111

# distribution of chain at t=10 for 3
# different initial distributions
c(1, 0, 0) %*% T10

##            [,1]    [,2]    [,3]
## [1,] 0.4321 0.4568 0.1111

c(0, 1, 0) %*% T10

##            [,1]    [,2]    [,3]
## [1,] 0.4321 0.4568 0.1111

c(0, 0, 1) %*% T10

##            [,1]    [,2]    [,3]
## [1,] 0.4321 0.4568 0.1111
```

7.2 Poisson Processes

The Poisson process is a continuous-time discrete-state Markov process $\{X_t : t \geq 0\}$ where $X_t \in \mathbb{N}$ counts the number of independent arrivals or events occurring in the time interval $[0, t]$. The Poisson process additionally assumes the following:

- The probability of getting a single arrival or event in a small time interval $[t, t + \Delta]$ is $\Delta \cdot \lambda$, where $\lambda > 0$ is commonly referred to as the intensity of the process.

- The probability of getting more than a single event in a small time interval is negligible.

The Poisson process is often used to model the count of independent arrivals. Examples include arrivals of cars in an intersection and arrival of phone calls at a telephone switchboard. The Poisson process marginals X_t are random variables counting the number of such arrivals in the time interval $[0, t]$.

We proceed with a more formal definition of the Poisson process, and follow with some properties.

7.2.1 Postulates and Differential Equation

The Poisson process is motivated by the following basic postulates:

1. The process $\{X_t : t \geq 0\}$ is a continuous time discrete state Markov process for which

$$P_{ij}(t) \stackrel{\text{def}}{=} \mathsf{P}(X_{t+u} = j | X_u = i).$$

Note specifically that $P_{ij}(t)$ does not depend on u.

2.

$$\lim_{h \to 0} \mathsf{P}(X_{t+h} - X_t = 1 | X_t = x)/h = \lambda,$$

or, alternatively using the growth notation (see Section B.5):

$$\mathsf{P}(X_{t+h} - X_t = 1 | X_t = x) = \lambda h + o(h).$$

3. $\mathsf{P}(X_0 = 0) = 1$

4. $\mathsf{P}(X_{t+h} - X_t = 0 | X_t = x) = 1 - \lambda h + o(h).$

Postulates 2 and 4 imply that the probability of multiple events occurring in a short time interval is negligible (formally, it decays to zero faster than the interval size). Thus, the distribution of the number of arrivals in a short time interval $[t, t + h]$ is approximately a Bernoulli distribution with parameter λh.

Denoting $P_m(t) = \mathsf{P}(X_t = m)$, the above postulates imply

$$P_0(t + h) = P_0(t)P_0(h) = P_0(t)(1 - \lambda h + o(h)),$$

which in turn imply

$$(P_0(t + h) - P_0(t))/h = -P_0(t)(\lambda h + o(h))/h = -P_0(t)(\lambda + o(1)).$$

This leads to the differential equation:

$$P_0'(t) = -\lambda P_0(t). \tag{7.14}$$

Subject to the initial condition $X_0 = 0$ (postulate 3), the solution of the above equation is

$$P_0(t) = e^{-\lambda t}, \tag{7.15}$$

as can be verified by substituting (7.15) in (7.14). Using the law of total probability (Proposition 1.5.4), we get that, for some constant C,

$$P_m(t + h) - P_m(t) = P_m(t)P_0(h) + P_{m-1}(t)P_1(h) + \sum_{j=2}^{m} P_{m-j}(t)P_j(h) - P_m(t)$$

$$= P_m(t)(1 - \lambda h + o(h)) + P_{m-1}(t)(\lambda h + o(h))) + o(h)C - P_m(t)$$

$$= -\lambda P_m(t)h + \lambda P_{m-1}(t)h + o(h),$$

which leads to the following differential equation:

$$P_m'(t) = \lim_{h \to 0} \frac{P_m(t + h) - P_m(t)}{h} = -\lambda P_m(t) + \lambda P_{m-1}(t). \tag{7.16}$$

The Poisson process is closely related to a number of important random variables, including the Poisson RV, the exponential RV, the binomial RV, and the uniform RV. We describe these relationships below.

7.2.2   Relationship to Poisson Distribution

Solving the differential equation (7.16) recursively for $P_m(t)$ using $P_0(t) = e^{-\lambda t}$ and $P_m(0) = 0$ yields the solution

$$P_m(t) = \frac{\lambda^m t^m}{m!} \exp(-\lambda t), \tag{7.17}$$

implying that

$$X_t \sim \text{Poisson}(\lambda t), \qquad t \geq 0.$$

Alternatively, it is possible to verify that (7.17) is the solution of (7.16) by substituting (7.17) in (7.16) and verifying that equality holds.

Since a sum of independent Poisson RVs is a Poisson RV (see Section 3.6) the distribution of the number of events in the interval $[s, t]$ is

$$X_t - X_s \sim \text{Poisson}(\lambda(t - s)).$$

Using the moments of the Poisson distribution (see Section 3.6), we get the following expressions for the mean and auto-covariance functions:

$$m(t) = \lambda t$$
$$C(t, s) = \mathsf{E}((X_s - \lambda s)(X_t - \lambda t)) = \mathsf{E}((X_s - \lambda s)((X_t - X_s - \lambda t + \lambda s) + (X_s - \lambda s)))$$
$$= \mathsf{Var}(X_s) + \mathsf{E}((X_s - \lambda s)(X_t - X_s - \lambda t + \lambda s))$$
$$= \mathsf{Var}(X_s) + \mathsf{E}(X_s - \lambda s)\,\mathsf{E}(X_t - X_s - \lambda t + \lambda s)$$
$$= \mathsf{Var}(X_s) = \lambda s = \lambda \min(s, t).$$

The equation above assumes $s \leq t$. A similar derivation holds in the case of $t \leq s$.

7.2.3   Relationship to Exponential Distribution

Proposition 7.2.1. *The time to first arrival and the times between subsequent arrivals in the Poisson process are independent $exp(\lambda)$ RVs.*

Proof. We have

$$\mathsf{P}(\text{time of first arrival} \leq z) = 1 - \mathsf{P}(X_z = 0) = 1 - e^{-\lambda z},$$

which coincides with the exponential distribution cdf. In other words, the time to first arrival has an exponential distribution with parameter λ.

Using the memoryless property of the exponential distribution (see Chapter 3) and postulate 1, we can get a similar result, assuming that we start measuring time from $t \neq 0$. ∎

The code below shows a sample path of the Poisson process with $\lambda = 1$ overlaid with the expectation function $m(t) = 1 \cdot t$. The generation of the sample path uses the proposition above by sampling the wait times and translating these values to the corresponding arrival times. Note how the arrivals represented by discontinuous jumps below occur at arbitrary t values, rather than integers, demonstrating the continuous-time nature of the process.

```
Z = rexp(30, 1)
t = seq(0, sum(Z), length.out = 200)
X = t * 0
for (i in 1:30) {
    Zc = cumsum(Z)
    X[t >= Zc[i]] = i
}
plot(c(0, sum(Z)), c(0, 30), type = "n", xlab = "$t$",
    ylab = "")
curve(x * 1, add = TRUE, col = rgb(1, 0, 0), lwd = 2,
    lty = 2, xlab = "$t$", ylab = "$X_t$")
lines(t, X, lwd = 3, type = "s")
title("A Poisson process sample path")
```

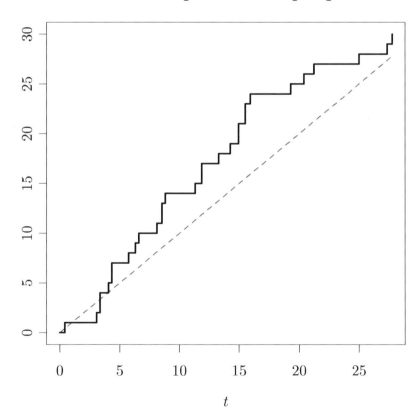

7.2.4   Relationship to Binomial Distribution

Recalling postulates 2 and 4 of the Poisson process, which imply that an arrival in $[h, h + \Delta]$ is approximately a Bernoulli RV (for small Δ) with parameter λh, we have that X_t is approximately the sum of t/Δ independent Bernoulli RVs, and therefore is approximately a binomial RV. See also Section 3.6, which asserts that the binomial distribution with $n \to \infty$, $\theta \to 0$ and $n\theta \to \lambda$ is approximately a Poisson distribution.

Another relationship between the Poisson process and the binomial distribution is that for $u < t$ and $k < n$

$$
\begin{aligned}
P(X_u = k | X_t = n) &= \frac{P(X_u = k)\,P(X_t - X_u = n - k)}{P(X_t = n)} \\
&= \frac{\text{Pois}(k\,;\lambda u)\text{Pois}(n - k\,;\lambda(t - u))}{\text{Pois}(n,\,;\lambda t)} \\
&= \binom{n}{k}(u/t)^k(1 - u/t)^{n-k} = \text{Bin}(k\,;n, u/t).
\end{aligned}
$$

In other words, given n arrivals by time t, the distribution of k arrivals by time u is binomial with parameters n and u/t.

7.2.5   Relationship to Uniform Distribution

We have

$$
\begin{aligned}
P(X_r = 1 | X_t = 1) &= \frac{P(X_r = 1 \cap X_t = 1)}{P(X_t = 1)} = \frac{P(X_r = 1 \cap X_t - X_r = 0)}{P(X_t = 1)} \\
&= \frac{P(X_r = 1)\,P(X_t - X_r = 0)}{P(X_t = 1)} \\
&= \frac{(\lambda x)e^{-\lambda x}}{1}\frac{(\lambda(t - r))^0 e^{-\lambda(t-r)}}{1}\frac{1}{(\lambda t)^1 e^{-\lambda t}} = \frac{r}{t},
\end{aligned}
$$

which is the cdf of the uniform distribution on the interval $[0, t]$. Thus, given that one arrival has occurred in the interval $[0, t]$, the precise time of that arrival is uniformly distributed in that interval.

7.3   Gaussian Processes

Definition 7.3.1. A Gaussian process is a continuous-time continuous-state RP $\{X_t : t \in J \subset \mathbb{R}^d\}$ in which all finite-dimensional marginal distributions follow a multivariate Gaussian distribution

$$
\boldsymbol{X} = (X_{\boldsymbol{t}_1}, \ldots, X_{\boldsymbol{t}_k}) \sim N((m(\boldsymbol{t}_1), \ldots, m(\boldsymbol{t}_k)), C(\boldsymbol{t}_1, \ldots, \boldsymbol{t}_k)), \qquad \boldsymbol{t}_1, \ldots, \boldsymbol{t}_k \in \mathbb{R}^d.
$$

In order to invoke Kolmogorov's extension theorem and ensure the above definition is rigorous, we need the covariance function C to (i) produce legitimate covariance matrices (symmetric positive definite matrices), and (ii) produce consistent finite dimensional marginals, in the sense of Definition 6.2.1. The second requirement of consistency of the finite dimensional marginals can be ensured by defining the covariance function to be

$$[C(\boldsymbol{t}_1, \ldots, \boldsymbol{t}_k))]_{ij} = C'(\boldsymbol{t}_i, \boldsymbol{t}_j)$$

for some function $C' : \mathbb{R}^d \times \mathbb{R}^d \to \mathbb{R}$.

Example 7.3.1. *Denoting by T the matrix whose rows are formed by $\boldsymbol{t}_i, i = 1, \ldots, k$, we consider a Gaussian process defined by the following covariance function*

$$[C(\boldsymbol{t}_1, \ldots, \boldsymbol{t}_k)]_{ij} = \alpha\langle \boldsymbol{t}_i, \boldsymbol{t}_j \rangle + \beta\delta_{ij},$$

where $\alpha, \beta > 0$, and $\delta_{ij} = 1$ if $i = j$ and 0 otherwise, or in vector notation

$$C(\boldsymbol{t}_1, \ldots, \boldsymbol{t}_k) = \alpha TT^\top + \beta I, \beta\delta_{ij}.$$

The matrix $C(\boldsymbol{t}_1, \ldots, \boldsymbol{t}_k))$ is symmetric, and is also positive definite: for all $\boldsymbol{v} \neq \boldsymbol{0}$:

$$\boldsymbol{v}^\top C(\boldsymbol{t}_1, \ldots, \boldsymbol{t}_k)\boldsymbol{v} = \alpha\boldsymbol{v}^\top TT^\top\boldsymbol{v} + \boldsymbol{v}^\top \beta I\boldsymbol{v} = \alpha\|\boldsymbol{w}\|^2 + \beta\|\boldsymbol{v}\|_2^2 > 0.$$

which makes it a suitable covariance function.

As α, β increase the variance increases making the process more likely to vary further from the expectation function m. If $\beta \ll \alpha$, the first term in the covariance function becomes dominant, implying that the variance grows similarly to $\langle \boldsymbol{t}_1, \boldsymbol{t}_2 \langle$. If $d = 1$, this means that X_t, X_s are positively correlated if t, s have the same sign and negatively correlated if t, s have opposing signs. Furthermore, the degree of correlation in absolute value increases as $|t|, |s|$ increase. If $\alpha \ll \beta$ the second term in the covariance function becomes dominant, implying that as the process values at two distinct times $\boldsymbol{t}_1, \boldsymbol{t}_2$ is uncorrelated and therefore independent. The resulting process has marginals $X_t, t \in \mathbb{R}^k$ that are independent $N(m(\boldsymbol{t}), \beta)$ random variables.

The R code below graphs samples from this random process in four cases: first term dominant with low variance, second term dominant with low variance, first term dominant with high variance, and second term dominant with high variance. In all cases the expectation function is a sinusoidal curve $m(t) = \sin(t)$. Note that in general, the sample paths are not smooth curves.

```
X = seq(-3, 3, length.out = 100)
m = sin(X)
n = 30
I = diag(1, nrow = 100, ncol = 100)
Y1 = rmvnorm(n, m, X %o% X/100 + I/1000)
plot(X, Y1[1, ], type = "n", xlab = "$t$", ylab = "$X_t$",
    main = "first term dominant with low variance")
for (s in 1:n) lines(X, Y1[s, ], col = rgb(0,
    0, 0, alpha = 0.5))
```

first term dominant with low variance

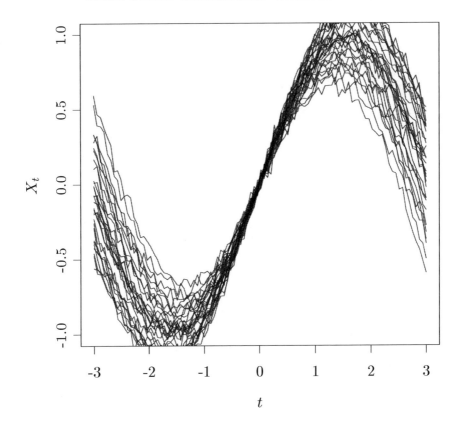

```
Y2 = rmvnorm(n, m, X %o% X/1000 + I/100)
plot(X, Y2[1, ], type = "n", xlab = "$t$", ylab = "$X_t$",
    main = "second term dominant with low variance")
for (s in 1:n) lines(X, Y2[s, ], col = rgb(0,
    0, 0, alpha = 0.5))
```

second term dominant with low variance

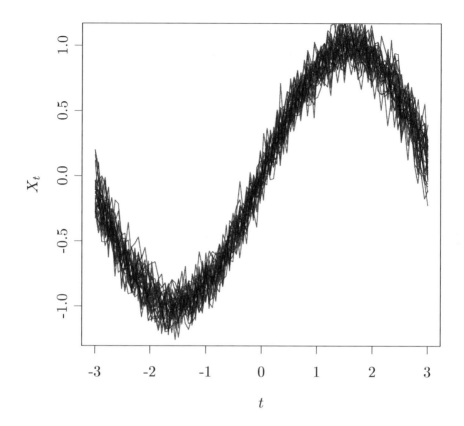

```
Y3 = rmvnorm(n, m, X %o% X + I/10)
plot(X, Y3[1, ], type = "n", xlab = "$t$", ylab = "$X_t$",
    main = "first term dominant with high variance")
for (s in 1:n) lines(X, Y3[s, ], col = rgb(0,
    0, 0, alpha = 0.5))
```

first term dominant with high variance

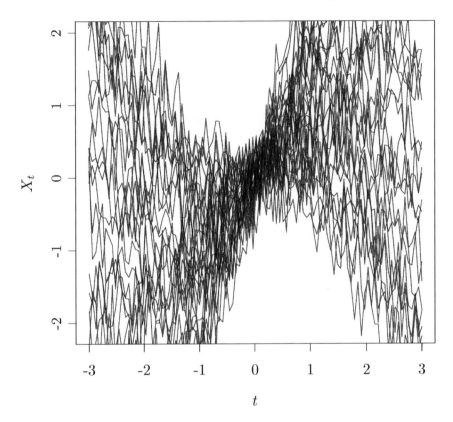

```
Y4 = rmvnorm(n, m, X %o% X + I * 10)
plot(X, Y4[1, ], type = "n", xlab = "$t$", ylab = "$X_t$",
    main = "second term dominant with high variance")
for (s in 1:n) lines(X, Y4[s, ], col = rgb(0,
    0, 0, alpha = 0.5))
```

second term dominant with high variance

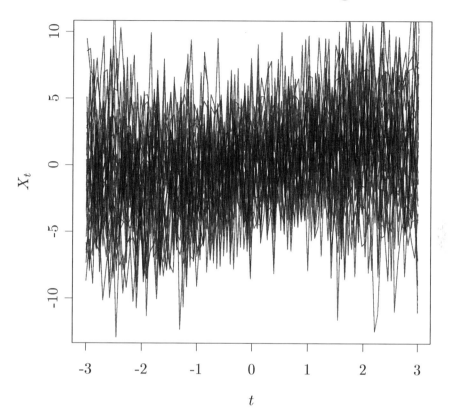

7.3.1 The Wiener Process

An interesting special case of the Gaussian process is the Wiener process.

Definition 7.3.2. The Wiener process $\mathcal{Z} = \{Z_t, t \geq 0\}$ is a Gaussian process with

$$
\begin{aligned}
Z_0 &= 0, && \text{with probability 1,} \\
m_{\mathcal{Z}}(t) &= 0, && \forall t \geq 0, \\
C_{\mathcal{Z}}(t,s) &= \alpha \min(t,s), && \alpha > 0, \quad \forall s, t \geq 0.
\end{aligned}
$$

We motivate the Wiener process by deriving it informally as the limit of a discrete-time discrete-valued random walk process with step size h.

Let $Z_t = \sum_{i=1}^{n} hY_n$, for some $h > 0$, where $Y_n = 2X_n - 1$ and $X_n \stackrel{iid}{\sim} \text{Ber}(1/2), n \in \mathbb{N}$. Since $\mathsf{E}(Y_n) = \mathsf{E}(2X_n - 1) = 2 \cdot 1/2 - 1 = 0$ and $\mathsf{Var}(Y_n) = \mathsf{Var}(2X_n - 1) = 4\,\mathsf{Var}(X_n) = 4(1/2)(1 - 1/2) = 1$, we have $\mathsf{E}(hY_n) = 0$ and $\mathsf{Var}(hY_n) = h^2$.

We set $h = \sqrt{\alpha}\delta$ and let $\delta \to 0, h \to 0$, or alternatively $h = \sqrt{\alpha}\sqrt{t/n}$ and let $n \to \infty$. This yields

$$Z_t = \lim_{n \to \infty} \sqrt{\alpha}\sqrt{t/n} \sum_{i=1}^{n} Y_n = \sqrt{\alpha t} \lim_{n \to \infty} \frac{\sum_{i=1}^{n} Y_n}{\sqrt{n}}.$$

Intuitively, the segment $[0, t]$ is divided to n steps of size $\delta = t/n$ each, and Z_t measures the position of the symmetric random walk at time t or equivalently after n steps of size δ. By the central limit theorem (see Section 8.9) $\lim_{n \to \infty} \sum_{i=1}^{n} Y_n/\sqrt{n}$ approaches a Gaussian RV with mean zero and variance one. Since Z_t approaches $\sqrt{\alpha t}$ times a $N(0, 1)$ RV, we have as $n \to \infty$

$$Z_t \sim N(0, \alpha t).$$

To show that the limit of the random walk above corresponds to Definition 7.3.2, it remains to show that (i) $\mathcal{Z}$ is a Gaussian process and (ii) it has a zero expectation function and its auto-covariance function is $C_{\mathcal{Z}}(t, s) = \alpha \min(t, s)$.

The increments of $\mathcal{Z}$ are independent and have the same distribution. For example $Z_5 - Z_3$ is independent of $Z_2 - Z_0 = Z_2$ and have the same distribution. Since

$$f_{Z_{t_1}, \ldots, Z_{t_k}}(z) = f_{Z_{t_1}}(z_1) f_{Z_{t_2} - Z_{t_1}}(z_2 - z_1) \cdots f_{Z_{t_k} - Z_{t_{k-1}}}(z_k - z_{k-1}),$$

the pdf of a finite dimensional marginal $f_{Z_{t_1}, \ldots, Z_{t_k}}$ is a product of independent univariate Gaussians, which has a multivariate Gaussian distribution.

The mean function $m_{\mathcal{Z}}(t) = 0$ as it is a sum of zero mean random variables. The auto-covariance function is

$$C_{\mathcal{Z}}(t, s) = \mathsf{E}\left(\left(\lim \sum_{i=1}^{t/\delta} hY_i\right)\left(\lim \sum_{i=1}^{s/\delta} hY_i\right)\right) = \lim h^2 \mathsf{E}\left(\sum_{i=1}^{t/\delta} \sum_{j=1}^{s/\delta} Y_iY_j\right)$$

$$= \lim h^2 \sum_{i=1}^{t/\delta} \sum_{j=1}^{s/\delta} \mathsf{E}(Y_iY_j) = \lim h^2 \sum_{i=1}^{\min(s,t)/\delta} \mathsf{E}(Y_i^2)$$

$$= \lim h^2 \min(s, t) \, \mathsf{Var}(Y)/\delta = \lim \alpha\delta \min(s, t)/\delta$$

$$= \alpha \min(s, t),$$

where in the third equation above we used the fact that $\mathsf{E}(Y_iY_j) = 0$ for $i \neq j$ (since Y_i, Y_j are independent RVs with mean 0).

7.4   Notes

More information on Markov chains, the Poisson process, and Gaussian process is available in [36, 26, 5], listed in increasing level of mathematical rigor. In particular, our description of Markov chains is similar to the description in [5]. Some information on Gaussian processes applications in data mining appears in [32].

7.5 Exercises

1. Prove formally that all first and second-dimensional marginals of a Markov chain are specified by the transition probabilities T and the initial probabilities ρ.

2. Prove formally that (7.4) follows from the Markov property of a Markov chain.

3. Show that a non-symmetric random walk on $\mathbb{Z}$ (Example 7.1.4 with $d = 1$ and probability of moving forward being different from the probability of moving backward) is transient.

Chapter 8

Limit Theorems

We consider in this chapter several important limit theorems. We start by exploring different types of convergences, and then move on to the law of large numbers and the central limit theorem. We emphasize the multivariate case of random vectors with $d > 1$, but for the sake of intuition it is useful to keep the univariate case in mind.

8.1 Modes of Stochastic Convergence

We list below the three major types or modes of convergences associated with random vectors.

Definition 8.1.1. Let $\boldsymbol{X}^{(n)}, n \in \mathbb{N}$ be a sequence of random vectors and $\boldsymbol{X}$ be a random vector.

- $\boldsymbol{X}^{(n)}$ converges in probability to $\boldsymbol{X}$, denoted by $\boldsymbol{X}^{(n)} \xrightarrow{\text{P}} \boldsymbol{X}$, if

$$\lim_{n \to \infty} \mathsf{P}(\|\boldsymbol{X}^{(n)} - \boldsymbol{X}\| \geq \epsilon) = 0, \qquad \forall \epsilon > 0.$$

- $\boldsymbol{X}^{(n)}$ converges with probability 1 to $\boldsymbol{X}$, denoted by $\boldsymbol{X}^{(n)} \xrightarrow{\text{as}} \boldsymbol{X}$, if

$$\mathsf{P}\left(\lim_{n \to \infty} \|\boldsymbol{X}^{(n)} - \boldsymbol{X}\| = 0\right) = 1.$$

  Note that $\lim_{n \to \infty} \|\boldsymbol{X}^{(n)} - \boldsymbol{X}\| = 0$ represent the event

$$\left\{\omega : \lim_{n \to \infty} \|\boldsymbol{X}^{(n)}(\omega) - \boldsymbol{X}(\omega)\| = 0\right\} \subset \Omega.$$

- $\boldsymbol{X}^{(n)}$ converges in distribution to $\boldsymbol{X}$, denoted by $\boldsymbol{X}^{(n)} \rightsquigarrow \boldsymbol{X}$, if

$$\lim_{n \to \infty} F_{\boldsymbol{X}^{(n)}}(\boldsymbol{x}) = F_{\boldsymbol{X}}(\boldsymbol{x}) \quad \text{for all } \boldsymbol{x} \text{ at which } F_{\boldsymbol{X}}(\boldsymbol{x}) \text{ is continuous.}$$

We make the following comments.

- In the definitions above, the limit RV X may be deterministic, in other words $X = c \in \mathbb{R}^d$ with probability 1. In this case we use notations such as $X^{(n)} \xrightarrow{\text{as}} c$ in the one dimensional case or $X^{(n)} \xrightarrow{\text{as}} c$ in higher dimensions.

- There is a fundamental difference between convergence in distribution and the other two types of convergence. Convergence in distribution merely implies that the distribution of $X^{(n)}$ is similar to that of X for large n. Specifically, it does not say anything about $X^{(n)}$ and X taking on similar values with high probability. Convergence in probability and convergence with probability 1 imply that for large n, the values of $X^{(n)}$ and X are similar (see the following example).

- The following section shows that convergence with probability one implies convergence in probability, which in turn implies convergence in distribution. The converse is not true in general.

Example 8.1.1. *If X and $X^{(n)}, n \in \mathbb{N}$ are independent uniform RVs in $[a, b]$, we have $X^{(n)} \rightsquigarrow X$ since the distribution of all RVs is identical. But we certainly do not have convergence in probability or with probability 1 since the RVs are independent and typically take on substantially different values.*

8.2   Relationships Between the Modes of Convergences

Proposition 8.2.1.

$$X^{(n)} \xrightarrow{\text{as}} X \quad \textit{if and only if} \quad \mathsf{P}(\|X^{(n)} - X\| \geq \epsilon \text{ i.o.}) = 0, \quad \forall \epsilon > 0.$$

Proof. The event $(X^{(n)} \xrightarrow{\text{as}} X)^c$ is equivalent to the event $\cup_{\epsilon > 0}\{\|X^{(n)} - X\| \geq \epsilon \text{ i.o.}\}$. It follows that the event $X^{(n)} \xrightarrow{\text{as}} X$ is equivalent to $\mathsf{P}(\|X^{(n)} - X\| \geq \epsilon \text{ i.o.}) = 0$ for all $\epsilon > 0$.   ∎

See Section A.4 for the definition of i.o. or infinitely often.

Proposition 8.2.2.

$$X^{(n)} \xrightarrow{\text{as}} X \quad \textit{implies} \quad X^{(n)} \xrightarrow{\text{p}} X$$
$$X^{(n)} \xrightarrow{\text{p}} X \quad \textit{implies} \quad X^{(n)} \rightsquigarrow X.$$

Proof. We first show that convergence with probability 1 implies convergence in probability. Since

$$\{\|X^{(n)} - X\| \geq \epsilon \text{ i.o.}\} = \limsup_n\{\|X^{(n)} - X\| \geq \epsilon\},$$

the event $\boldsymbol{X}^{(n)} \xrightarrow{\text{as}} \boldsymbol{X}$ implies

$$\lim_n \mathsf{P}(\|\boldsymbol{X}^{(n)} - \boldsymbol{X}\| \geq \epsilon) \leq \limsup_n \mathsf{P}(\|\boldsymbol{X}^{(n)} - \boldsymbol{X}\| \geq \epsilon)$$

$$\leq \mathsf{P}\left(\limsup_n \|\boldsymbol{X}^{(n)} - \boldsymbol{X}\| \geq \epsilon\right) = 0.$$

The inequality $\limsup \mathsf{P}(A_n) \leq \mathsf{P}(\limsup A_n)$ follows from Fatou's lemma (see Chapter F) applied to the sequence of indicator functions $f_n = I_{A_n}$ and the measure $\mu = P$. The last equality follows from the previous proposition.

We next show that convergence in probability implies convergence in distribution. Denoting $\mathbf{1} = (1, \ldots, 1)$, we have that if $\boldsymbol{X}^{(n)} \leq \boldsymbol{x}$ then either $\boldsymbol{X} \leq \boldsymbol{x} + \epsilon\mathbf{1}$, or $\|\boldsymbol{X} - \boldsymbol{X}^{(n)}\| > \epsilon$, or both (we interpret inequality between two vectors as a sequence of inequalities for the corresponding components: $\boldsymbol{u} \leq \boldsymbol{v}$ implies $u_i \leq v_i$, $i = 1, \ldots, d$). Similarly, if $\boldsymbol{X} \leq \boldsymbol{x} - \epsilon\mathbf{1}$ then either $\boldsymbol{X}^{(n)} \leq \boldsymbol{x}$ or $\|\boldsymbol{X} - \boldsymbol{X}^{(n)}\| > \epsilon$, or both. This implies that for all n,

$$F_{\boldsymbol{X}}^{(n)}(\boldsymbol{x}) \leq \mathsf{P}(\boldsymbol{X} \leq \boldsymbol{x} + \epsilon\mathbf{1}) + \mathsf{P}(\|\boldsymbol{X} - \boldsymbol{X}^{(n)}\| > \epsilon)$$

$$= F_{\boldsymbol{X}}(\boldsymbol{x} + \epsilon\mathbf{1}) + \mathsf{P}(\|\boldsymbol{X} - \boldsymbol{X}^{(n)}\| > \epsilon)$$

$$F_{\boldsymbol{X}}(\boldsymbol{x} - \epsilon\mathbf{1}) \leq \mathsf{P}(\boldsymbol{X}^{(n)} \leq \boldsymbol{x}) + \mathsf{P}(\|\boldsymbol{X} - \boldsymbol{X}^{(n)}\| > \epsilon)$$

$$= F_{\boldsymbol{X}^{(n)}}(\boldsymbol{x}) + \mathsf{P}(\|\boldsymbol{X} - \boldsymbol{X}^{(n)}\| > \epsilon).$$

Since $\boldsymbol{X}^{(n)} \xrightarrow{\text{P}} \boldsymbol{X}$, we have $\mathsf{P}(\|\boldsymbol{X} - \boldsymbol{X}^{(n)}\| > \epsilon) \to 0$ and letting $n \to \infty$ in the two inequalities above, we get

$$F_{\boldsymbol{X}}(\boldsymbol{x} - \epsilon\mathbf{1}) \leq \liminf F_{\boldsymbol{X}^{(n)}}(\boldsymbol{x}) \leq \limsup F_{\boldsymbol{X}^{(n)}}(\boldsymbol{x}) \leq F_{\boldsymbol{X}}(\boldsymbol{x} + \epsilon\mathbf{1}).$$

The left hand side and the right hand side converge to $F_{\boldsymbol{X}}(\boldsymbol{x})$ as $\epsilon \to 0$ at points $\boldsymbol{x}$ where $F_{\boldsymbol{X}}$ is continuous, implying that $F_{\boldsymbol{X}^{(n)}}(\boldsymbol{x}) \to F_{\boldsymbol{X}}(\boldsymbol{x})$. ∎

The following example shows that convergence in probability may occur even if convergence with probability one does not occur.

Example 8.2.1. *Consider $\Omega = [0, 1]$ with P being the uniform distribution over Ω. The sequence of random variables $X^{(1)} = I_{(0,1/2]}$, $X^{(2)} = I_{(1/2,1]}$, $X^{(3)} = I_{(0,1/4]}$, $X^{(5)} = I_{(1/4,1/2]}$, $X^{(6)} = I_{(1/2,3/4]}$, and so on, does not converge with probability one to any limit (for all ω, $X^{(n)}(\omega)$ is a divergent sequence). On the other hand, $X^{(n)} \xrightarrow{\text{P}} 0$ since $\mathsf{P}(|X^{(n)}| \geq \epsilon) \to 0$ for all $\epsilon > 0$.*

Proposition 8.2.3. *If $\boldsymbol{c} \in \mathbb{R}^d$ then*

$$\boldsymbol{X}^{(n)} \rightsquigarrow \boldsymbol{c} \quad \text{if and only if} \quad \boldsymbol{X}^{(n)} \xrightarrow{\text{P}} \boldsymbol{c}.$$

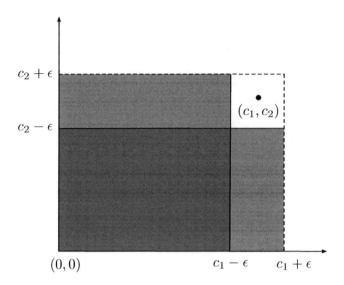

Figure 8.1: This figure illustrates the proof of Proposition 8.2.3. The white square region may be expressed as the region contained within the dashed rectangle minus the two shaded rectangles plus the intersection of the two shaded rectangles (since it was subtracted twice).

Proof. It suffices to prove that convergence in distribution to a constant vector implies probability in probability (convergence in probability always implies convergence in distribution). We prove the result below for two dimensions $d = 2$. The cases of other dimensions are similar.

We have (see Figure 8.1 for an illustration)

$$\mathsf{P}(\|\boldsymbol{X}^{(n)} - \boldsymbol{c}\| \le \sqrt{2}\epsilon) = \mathsf{P}(\|\boldsymbol{X}^{(n)} - \boldsymbol{c}\|^2 \le 2\epsilon^2)$$
$$\ge \mathsf{P}(\boldsymbol{c} - \epsilon(1,1) < \boldsymbol{X}^{(n)} \le \boldsymbol{c} + \epsilon(1,1))$$
$$= \mathsf{P}(\boldsymbol{X}^{(n)} \le \boldsymbol{c} + \epsilon(1,1)) - \mathsf{P}(\boldsymbol{X}^{(n)} \le \boldsymbol{c} + \epsilon(1,-1))$$
$$- \mathsf{P}(\boldsymbol{X}^{(n)} \le \boldsymbol{c} + \epsilon(-1,1)) + \mathsf{P}(\boldsymbol{X}^{(n)} \le \boldsymbol{c} + \epsilon(1,1)).$$

In the first inequality above, we used the fact that if $|a_1 - b_1| \le \epsilon$ and $|a_2 - b_2| \le \epsilon$ then $\|\boldsymbol{a} - \boldsymbol{b}\|^2 \le 2\epsilon^2$. In the second equality we used the principle of inclusion-exclusion (see Figure 8.1 for an illustration). If we have convergence in distribution $\boldsymbol{X}^{(n)} \rightsquigarrow \boldsymbol{c}$, then the last term in the inequality above converges to $0 + 0 + 0 + 1$, which implies convergence in probability $\boldsymbol{X}^{(n)} \xrightarrow{\text{P}} \boldsymbol{c}$.   ∎

Proposition 8.2.4. *The convergence* $\boldsymbol{X}^{(n)} \xrightarrow{\text{P}} \boldsymbol{X}$ *occurs if and only if every sequence of natural numbers* $n_1, n_2, \ldots \in \mathbb{N}$ *has a subsequence* $r_1, r_2, \ldots \in \{n_1, n_2, \ldots\}$ *such that* $\boldsymbol{X}^{(r_k)} \xrightarrow{\text{as}} \boldsymbol{X}$ *as* $k \to \infty$.

Proof. We assume that $\boldsymbol{X}^{(n)} \overset{\mathrm{P}}{\to} \boldsymbol{X}$ and consider a sequence of positive numbers ϵ_i such that $\sum_i \epsilon_i < \infty$. For each ϵ_i, we can find a natural number n'_i such that $\mathsf{P}(\|\boldsymbol{X}^{(n)} - \boldsymbol{X}\| \geq \epsilon_i) < \epsilon_i$ for all $n > n'_i$. We can assume without loss of generality that $n'_1 < n'_2 < n'_3 < \cdots$ (otherwise replace n'_i with $\max(n'_1, n'_2, \ldots, n'_i)$).

Defining A_i to be the event $\{\|\boldsymbol{X}^{(n'_i)} - \boldsymbol{X}\| \geq \epsilon_i\}$, we have $\sum_i \mathsf{P}(A_i) \leq \sum_i \epsilon_i < \infty$ and by the first Borell-Cantelli Lemma (Proposition 6.6.1) we have $\mathsf{P}(A_i \text{ i.o.}) = 0$. Since $\lim_{k \to \infty} \epsilon_k = 0$, this implies that for all $\epsilon > 0$, $\mathsf{P}(\{\|\boldsymbol{X}^{(n'_i)} - \boldsymbol{X}\| \geq \epsilon\} \text{ i.o.}) = 0$, which by Proposition 8.2.1 implies that $\boldsymbol{X}^{(n'_i)} \overset{\mathrm{as}}{\to} \boldsymbol{X}$ as $i \to \infty$. We have thus shown that there exists a subsequence $n'_1, n'_2, \ldots$ of $1, 2, \ldots$ along which convergence with probability 1 occurs.

Considering now an arbitrary sequence $n_1, n_2, \ldots$ of natural numbers, we have $\boldsymbol{X}^{(n_i)} \overset{\mathrm{P}}{\to} \boldsymbol{X}$ as $i \to \infty$, and repeating the above argument with $n_1, n_2, n_3, \ldots$ replacing $1, 2, 3, \ldots$ we can find a subsequence $r_1, r_2, r_3, \ldots$ of $n_1, n_2, n_3, \ldots$ along which $\boldsymbol{X}^{(r_i)} \overset{\mathrm{as}}{\to} \boldsymbol{X}$ as $i \to \infty$.

To show that converse, we assume that $\boldsymbol{X}^{(n)} \overset{\mathrm{P}}{\not\to} \boldsymbol{X}$. Then there exists $\epsilon > 0$ and $\delta > 0$ such that $\mathsf{P}(\|\boldsymbol{X}^{(k)} - \boldsymbol{X}\| \geq \epsilon) > \delta$ for infinitely many k, which we denote $n_1, n_2, \ldots$. This implies that there exists no subsequence of $n_1, n_2, \ldots$ along which $\boldsymbol{X}^{(n_i)} \overset{\mathrm{as}}{\to} \boldsymbol{X}$. ∎

8.3 Dominated Convergence Theorem for Random Vectors*

We have the following extensions of the Dominated Convergence Theorem (see Chapter F).

Proposition 8.3.1 (Dominated Convergence Theorem for Random Variables). *Consider a sequence of RVs $X^{(n)}$ for which $|X^{(n)}| \leq Y$ for some RV Y with $\mathsf{E}|Y| < \infty$. Then*

$$X^{(n)} \overset{\mathrm{as}}{\to} X \qquad implies \qquad \mathsf{E}(X^{(n)}) \to \mathsf{E}(X).$$

Proof. We construct a discrete-time random process $\mathcal{X} = (X^{(n)}, n \in \mathbb{N})$ and consider the Dominated Convergence Theorem with μ being the distribution of the random process $\mathcal{X}$, $f_n = X^{(n)}$, and $f = X$ (see Proposition F.3.9). In this case $\int f_n \, d\mu$ becomes $\mathsf{E}(X^{(n)})$ and $\int f \, d\mu$ becomes $\mathsf{E}(X)$. ∎

Proposition 8.3.2 (Dominated Convergence Theorem for Random Vectors). *If $\boldsymbol{X}^{(n)} \overset{\mathrm{as}}{\to} \boldsymbol{X}$ and $\|\boldsymbol{X}\|^r \leq Y$ for some RV Y with $\mathsf{E}Y < \infty$ and some $r > 0$, then*

$$\mathsf{E}(\|\boldsymbol{X}^{(n)} - \boldsymbol{X}\|^r) \to 0.$$

Proof. $\boldsymbol{X}^{(n)} \overset{\mathrm{as}}{\to} \boldsymbol{X}$ and $\|\boldsymbol{X}\|^r \leq Y$ implies $\|\boldsymbol{X}^{(n)}\| \leq Y$ with probability 1, which implies

$$\|\boldsymbol{X}^{(n)} - \boldsymbol{X}\|^r \leq (\|\boldsymbol{X}^{(n)}\| + \|\boldsymbol{X}\|)^r \leq (Y^{1/r} + Y^{1/r})^r = 2^r Y$$

with probability 1. The result then follows from applying the dominated convergence theorem for random variables to the sequence of RVs $\|\boldsymbol{X}^{(n)} - \boldsymbol{X}\|^r, n \in \mathbb{N}$ (that sequence converges with probability 1 to 0 since $\boldsymbol{X}^{(n)} \overset{as}{\to} X$). ∎

The name of the proposition above is motivated by the fact that convergence $\mathsf{E}(\|\boldsymbol{X}^{(n)} - \boldsymbol{X}\|^r) \to 0$ is sometimes called convergence in the r-mean and denoted $\boldsymbol{X}^{(n)} \to^r \boldsymbol{X}$. The dominated convergence theorem for random vectors implies convergence in r-mean of the vector $\boldsymbol{X}^{(n)}$ to the vector $\boldsymbol{X}$, which is similar to the dominated convergence theorem for random variables for $r = 1$.

8.4   Scheffe's Theorem

Proposition 8.4.1 (Scheffe's Theorem). *The following two statements hold.*
(a) If $0 \le \boldsymbol{X}^{(n)} \overset{as}{\to} \boldsymbol{X}$ and $\mathsf{E}(\boldsymbol{X}^{(n)}) \to \mathsf{E}(\boldsymbol{X}) < \infty$, then

$$\mathsf{E}(\|\boldsymbol{X}^{(n)} - \boldsymbol{X}\|) \to 0.$$

(b) If $f_{\boldsymbol{X}^{(n)}}(\boldsymbol{x}) \to f_{\boldsymbol{X}}(\boldsymbol{x})$ for all $\boldsymbol{x}$, then

$$\int |f_{\boldsymbol{X}^{(n)}}(\boldsymbol{x}) - f_{\boldsymbol{X}}(\boldsymbol{x})|\, d\boldsymbol{x} \to 0.$$

Proof. We start with proving part (a) in the case of one dimension. Note that for any real number c we have $|c| = c + 2\max(-c, 0)$ (this can be verified independently for a positive and a negative c). It follows that

$$\mathsf{E}(|X^{(n)} - X|) = \mathsf{E}(X^{(n)} - X) + 2\,\mathsf{E}(\max(X - X^{(n)}, 0)).$$

The first term converges to 0 since $X^{(n)} \overset{as}{\to} X$. The second term converges to 0 by the dominated convergence theorem for random variables applied to the sequence of RVs $\max(X^{(n)} - X)$ (note that $0 \le \max(X - X^{(n)}, 0) \le \max(X, 0)$ and $\mathsf{E}(\max(X, 0)) < \infty$). The proof for the multivariate case follows by applying the triangle inequality $\|\boldsymbol{X}^{(n)} - \boldsymbol{X}\| \le \sum_{i=1}^{d} |X_i^{(n)} - X_i|$ and then applying the one dimensional case for each term separately.

The proof of part (b) is similar. We have

$$|f_{\boldsymbol{X}^{(n)}}(\boldsymbol{x}) - f_{\boldsymbol{X}}(\boldsymbol{x})| = f_{\boldsymbol{X}^{(n)}}(\boldsymbol{x}) - f_{\boldsymbol{X}}(\boldsymbol{x}) + 2\max(f_{\boldsymbol{X}}(\boldsymbol{x}) - f_{\boldsymbol{X}^{(n)}}(\boldsymbol{x}), 0)$$

$$\int |f_{\boldsymbol{X}^{(n)}}(\boldsymbol{x}) - f_{\boldsymbol{X}}(\boldsymbol{x})|\, d\boldsymbol{x} = \int f_{\boldsymbol{X}^{(n)}}(\boldsymbol{x}) - f_{\boldsymbol{X}}(\boldsymbol{x})\, d\boldsymbol{x}$$

$$+ 2\int \max(f_{\boldsymbol{X}}(\boldsymbol{x}) - f_{\boldsymbol{X}^{(n)}}(\boldsymbol{x}), 0)\, d\boldsymbol{x}.$$

The first integral in the right hand side above is 0 since both densities integrate to 1. The second integral converges to 0 by the dominated convergence theorem applied to $\max(f_{\boldsymbol{X}}(\boldsymbol{x}) - f_{\boldsymbol{X}^{(n)}}(\boldsymbol{x}), 0)$ (note that $\max(f_{\boldsymbol{X}}(\boldsymbol{x}) - f_{\boldsymbol{X}^{(n)}}(\boldsymbol{x}), 0) \le \max(f_{\boldsymbol{X}}(\boldsymbol{x}), 0)$) and the Lebesgue measure. ∎

Corollary 8.4.1. *If* $\lim_{n\to\infty} f_{\boldsymbol{X}^{(n)}}(\boldsymbol{x}) = f_{\boldsymbol{X}}(\boldsymbol{x})$ *for all* $\boldsymbol{x}$, *then*

$$\lim_{n\to\infty} \sup_A |\mathsf{P}(\boldsymbol{X}^{(n)} \in A) - \mathsf{P}(\boldsymbol{X} \in A)| = 0$$

where the supremum ranges over all measurable sets.

Proof.

$$\sup_A |\mathsf{P}(\boldsymbol{X}^{(n)} \in A) - \mathsf{P}(\boldsymbol{X} \in A)| = \sup_A \left| \int_A (f_{\boldsymbol{X}^{(n)}}(\boldsymbol{x}) - f_{\boldsymbol{X}}(\boldsymbol{x}))\, d\boldsymbol{x} \right|$$

$$\leq \int_A |f_{\boldsymbol{X}^{(n)}}(\boldsymbol{x}) - f_{\boldsymbol{X}}(\boldsymbol{x})|\, d\boldsymbol{x}$$

$$= \int |f_{\boldsymbol{X}^{(n)}}(\boldsymbol{x}) - f_{\boldsymbol{X}}(\boldsymbol{x})|\, d\boldsymbol{x} \to 0,$$

where the convergence follows from the Scheffe's Theorem. ∎

The above corollary shows that pointwise convergence of the pdfs is stronger than convergence in distribution: the former implies $\sup_A |\mathsf{P}(\boldsymbol{X}^{(n)} \in A) - \mathsf{P}(\boldsymbol{X} \in A)| \to 0$, while the latter corresponds to convergence $\mathsf{P}(\boldsymbol{X}^{(n)} \in A) \to \mathsf{P}(\boldsymbol{X} \in A)$ only for some sets of the form $A = (-\infty, a]$ (for some values a).

8.5 The Portmanteau Theorem

Proposition 8.5.1 (The Portmanteau Theorem). *The following statements are equivalent.*

1. $\boldsymbol{X}^{(n)} \rightsquigarrow \boldsymbol{X}$.

2. $\mathsf{E}(h(\boldsymbol{X}^{(n)})) \to \mathsf{E}(h(\boldsymbol{X}))$ *for all continuous functions* $h : \mathbb{R}^d \to \mathbb{R}$ *that are non-zero only on a closed and bounded set.*

3. $\mathsf{E}(h(\boldsymbol{X}^{(n)})) \to \mathsf{E}(h(\boldsymbol{X}))$ *for all bounded continuous functions* $h : \mathbb{R}^d \to \mathbb{R}$.

4. $\mathsf{E}(h(\boldsymbol{X}^{(n)})) \to \mathsf{E}(h(\boldsymbol{X}))$ *for all bounded measurable functions* $h : \mathbb{R}^d \to \mathbb{R}$ *for which* $\mathsf{P}(\boldsymbol{X} \in \{\boldsymbol{x} : h \text{ is continuous at } \boldsymbol{x}\}) = 1$.

We use the following lemma in the proof below.

Lemma 8.5.1. *Let* $h : \mathbb{R}^d \to \mathbb{R}$ *be a bounded measurable function for which*

$$\mathsf{P}(\boldsymbol{X} \in \{\boldsymbol{x} : h \text{ is continuous at } \boldsymbol{x}\}) = 1. \qquad (*)$$

Then for every $\epsilon > 0$, *there exists bounded continuous functions* m, M *such that* $m \leq h \leq M$ *and* $\mathsf{E}(M(\boldsymbol{X}) - m(\boldsymbol{X})) < \epsilon$.

Proof*. We define the following sequences of functions

$$m_k(\boldsymbol{x}) = \inf\{h(\boldsymbol{y}) + k\|\boldsymbol{x} - \boldsymbol{y}\| : \boldsymbol{y} \in \mathbb{R}^d\}, \quad k \in \mathbb{N}$$
$$M_k(\boldsymbol{x}) = \sup\{h(\boldsymbol{y}) - k\|\boldsymbol{x} - \boldsymbol{y}\| : \boldsymbol{y} \in \mathbb{R}^d\}, \quad k \in \mathbb{N}.$$

Since (i) $m_k(\boldsymbol{x}) \le m_{k+1}(\boldsymbol{x})$, (ii) $h_{k+1}(\boldsymbol{x}) \le h_k(\boldsymbol{x})$, (iii) $m_k(\boldsymbol{x}) \le h(\boldsymbol{x})$ ($h(\boldsymbol{x}) = \{h(\boldsymbol{y}) + k\|\boldsymbol{x} - \boldsymbol{y}\| : \boldsymbol{y} \in \mathbb{R}^d\}$ for $\boldsymbol{y} = \boldsymbol{x}$), and similarly (iv) $h \le M_k$, we have

$$m_1(\boldsymbol{x}) \le m_2(\boldsymbol{x}) \le \cdots \le h(\boldsymbol{x}) \le \cdots \le M_2(\boldsymbol{x}) \le M_1(\boldsymbol{x}).$$

We note the following.

1. Since monotonic limits of bounded sequences necessarily converge, the limits $\lim_{k\to\infty} m_k(\boldsymbol{x})$ and $\lim_{k\to\infty} M_k(\boldsymbol{x})$ are well defined and

$$\lim_{k\to\infty} m_k(\boldsymbol{x}) \le h(\boldsymbol{x}) \le \lim_{k\to\infty} M(\boldsymbol{x}).$$

2. The functions m_k, M_k are continuous since

$$
\begin{aligned}
m_k(\boldsymbol{x}) &= \inf\{h(\boldsymbol{y}) + k\|\boldsymbol{x} - \boldsymbol{y}\| : \boldsymbol{y} \in \mathbb{R}^d\} \\
&\le \inf\{h(\boldsymbol{y}) + k\|\boldsymbol{z} - \boldsymbol{y}\| : \boldsymbol{y} \in \mathbb{R}^d\} + \|\boldsymbol{y} - \boldsymbol{z}\| \\
&= m_k(\boldsymbol{z}) + k\|\boldsymbol{y} - \boldsymbol{z}\|,
\end{aligned}
$$

implying that

$$|m_k(\boldsymbol{x}) - m_k(\boldsymbol{z})| \le k\|\boldsymbol{y} - \boldsymbol{z}\|$$

(and similarly in the case of M_k).

3. Since h is bounded, m_k and M_k are bounded as well (m_k is an infimum of a bounded function plus a non-negative term and similarly for M_k).

4. If $\boldsymbol{x}$ is a continuity point of h, then

$$\lim_{k\to\infty} m_k(\boldsymbol{x}) = h(\boldsymbol{x}) = \lim_{k\to\infty} M(\boldsymbol{x}).$$

This follows from the following argument. For $\epsilon > 0$, select δ such that $\|\boldsymbol{x} - \boldsymbol{y}\| < \delta$ implies $|h(\boldsymbol{x}) - h(\boldsymbol{y})| < \epsilon$. Then if $r > (h(\boldsymbol{x}) - \inf_{\boldsymbol{w}} h(\boldsymbol{w}))/\delta$, we have for an arbitrary $\epsilon > 0$

$$\lim_k m_k(\boldsymbol{x}) \ge m_r(\boldsymbol{x})$$

$$= \min\left\{\inf_{\boldsymbol{y}:\|\boldsymbol{x}-\boldsymbol{y}\|<\delta} h(\boldsymbol{y}) - r\|\boldsymbol{x} - \boldsymbol{y}\|, \inf_{\boldsymbol{y}:\|\boldsymbol{x}-\boldsymbol{y}\|\ge\delta} h(\boldsymbol{y}) - r\|\boldsymbol{x} - \boldsymbol{y}\|\right\}$$

$$\ge \min\left\{h(\boldsymbol{x}) - \epsilon, \inf_{\boldsymbol{w}} h(\boldsymbol{w}) + (h(\boldsymbol{x}) - \inf_{\boldsymbol{w}} h(\boldsymbol{w}))\frac{\delta}{\delta}\right\} = h(\boldsymbol{x}) - \epsilon,$$

and similarly in the case of M_k.

5. Observation 4 above, together with Equation (*) implies

$$\mathsf{E}\lim_k m_k(\boldsymbol{X}) = \mathsf{E}(h(\boldsymbol{X})) = \mathsf{E}\lim_k M_k(\boldsymbol{X}).$$

Observation 5 above, together with monotone convergence theorem (Proposition F.3.4) give

$$\mathsf{E}\, m_k(\boldsymbol{X}) \nearrow \mathsf{E}\lim_k m_k(\boldsymbol{X}) = \mathsf{E}(h(\boldsymbol{X})),$$

$$\mathsf{E}\, M_k(\boldsymbol{X}) \searrow \mathsf{E}\lim_k M_k(\boldsymbol{X}) = \mathsf{E}(h(\boldsymbol{X})),$$

implying that for all $\epsilon > 0$, there exists r such that whenever $k > r$, we have $\mathsf{E}(m_k(\boldsymbol{X}) - M_k(\boldsymbol{X})) < \epsilon$.

∎

Proof of The Portmanteau Theorem*. Statement 4 implies statement 3 since continuous functions are measurable. Statement 3 implies statement 2 since continuous function on a compact set (in $\mathbb{R}^d$ a compact set is a closed and bounded set) are bounded. It remains to show that $(4) \Rightarrow (1) \Rightarrow (2) \Rightarrow (3) \Rightarrow (4)$.

$(4) \Rightarrow (1)$: For every $\boldsymbol{z}$ that is a continuity point of $F_{\boldsymbol{X}}$ we construct a function $h_{\boldsymbol{z}}$ such that $\mathsf{E}(h_{\boldsymbol{z}}(\boldsymbol{X}^{(n)})) \to \mathsf{E}(h_{\boldsymbol{z}}(\boldsymbol{X}))$ implies $F_{\boldsymbol{X}^{(n)}}(\boldsymbol{z}) \to F_{\boldsymbol{X}}(\boldsymbol{z})$. This will prove the claim that $\boldsymbol{X}^{(n)} \rightsquigarrow \boldsymbol{X}$. Specifically, we define $h_{\boldsymbol{z}}(\boldsymbol{x}) = 1$ if $\boldsymbol{x} \le \boldsymbol{z}$ (the inequality holds for all components of the two vectors) and $h_{\boldsymbol{z}}(\boldsymbol{x}) = 0$ otherwise. Since $\mathsf{E}\, h_{\boldsymbol{z}}(\boldsymbol{x}) = F_{\boldsymbol{x}}(\boldsymbol{z})$, the convergence $\mathsf{E}(h_{\boldsymbol{z}}(\boldsymbol{X}^{(n)})) \to \mathsf{E}(h_{\boldsymbol{z}}(\boldsymbol{X}))$ implies $F_{\boldsymbol{X}^{(n)}}(\boldsymbol{z}) \to F_{\boldsymbol{X}}(\boldsymbol{z})$. The functions $h_{\boldsymbol{z}}(\boldsymbol{x})$ satisfy the conditions of (4): they are bounded and measurable, and since

$$\lim_{\epsilon \to 0} F_{\boldsymbol{X}}(\boldsymbol{z} + \epsilon \cdot (1, \cdots, 1)) - F_{\boldsymbol{X}}(\boldsymbol{z} - \epsilon \cdot (1, \cdots, 1)) = 0$$

(recall that $\boldsymbol{z}$ is a continuity point of $F_{\boldsymbol{X}}$), the set of points at which $h_{\boldsymbol{z}}$ is discontinuous has probability 0 (under the distribution of $\boldsymbol{X}$).

$(1) \Rightarrow (2)$: We assume $\boldsymbol{X}^{(n)} \rightsquigarrow \boldsymbol{X}$ and consider a continuous function h that is zero outside a bounded and closed set C. Since C is compact, and continuous functions on a compact set are uniformly continuous (Proposition B.3.5), h is uniformly continuous. Given $\epsilon > 0$, we can find $\delta > 0$ such that $\|\boldsymbol{x} - \boldsymbol{y}\| < \delta$ implies $\|h(\boldsymbol{x}) - h(\boldsymbol{y})\| < \epsilon$. Since C is bounded we can obtain a finite partition Q of C into rectangular cells of the form $\{\boldsymbol{x} : \boldsymbol{a} < \boldsymbol{x} \le \boldsymbol{b}\}$ such that the distance between points within each cell is less than δ. We can also assume, without loss of generality, that the boundaries between the cells have probability zero under $\boldsymbol{X}$ (since there can only be a countable number of regions with positive probability).

Based on this partition, we can define a simple function (function that takes on finitely many values) whose values are constant on the partition cells

$$h'(\boldsymbol{x}) = \sum_{i=1}^{r} a_i I_{A_i}(\boldsymbol{x}) \qquad \text{where} \qquad a_i = h\left(\max_{\boldsymbol{x} \in A_i}(x_1), \dots, \max_{\boldsymbol{x} \in A_i}(x_d)\right).$$

In fact, we can write the function h' in the following form

$$h'(\boldsymbol{x}) = \sum_{i=1}^{m} \beta_i I_{(-\infty, b_i]}(\boldsymbol{x})$$

using the method described in the proof of Proposition 8.2.3 (see also Figure 8.1). Note that by construction of the partition, each point $\boldsymbol{b}_i$ is a continuity point of $F_{\boldsymbol{X}}$ and therefore $\boldsymbol{X}^{(n)} \rightsquigarrow \boldsymbol{X}$ implies

$$\lim_{n \to \infty} \mathsf{E}(h'(\boldsymbol{X}^{(n)})) = \lim_{n \to \infty} \sum_{i=1}^{m} \beta_i F_{\boldsymbol{X}^{(n)}}(\boldsymbol{x}) = \sum_{i=1}^{m} \beta_i F_{\boldsymbol{X}}(\boldsymbol{x}) = \mathsf{E}(h'(\boldsymbol{X})). \qquad (8.1)$$

We finally note that

$$|\mathsf{E}(h(\boldsymbol{X}^{(n)})) - \mathsf{E}(h(\boldsymbol{X}))|$$
$$= |\mathsf{E}(h(\boldsymbol{X}^{(n)})) + \mathsf{E}(h'(\boldsymbol{X}^{(n)})) - \mathsf{E}(h'(\boldsymbol{X}^{(n)})) + \mathsf{E}(h'(\boldsymbol{X})) - \mathsf{E}(h'(\boldsymbol{X})) - \mathsf{E}(h(\boldsymbol{X}))|$$
$$\leq |\mathsf{E}(h(\boldsymbol{X}^{(n)})) - \mathsf{E}(h'(\boldsymbol{X}^{(n)}))| + |\mathsf{E}(h'(\boldsymbol{X}^{(n)})) - \mathsf{E}(h'(\boldsymbol{X}))| + |\mathsf{E}(h'(\boldsymbol{X})) - \mathsf{E}(h(\boldsymbol{X}))|$$
$$\leq \epsilon + |\mathsf{E}(h'(\boldsymbol{X}^{(n)})) - \mathsf{E}(h'(\boldsymbol{X}))| + \epsilon$$
$$\to 2\epsilon \qquad \text{as } n \to \infty.$$

Above, the first inequality follows from the triangle inequality, the second inequality follows from the uniform convergence of h and the construction of the partition Q and h', and the convergence follows from (8.1). Since this holds for all $\epsilon > 0$, we have $|\mathsf{E}(h(\boldsymbol{X}^{(n)})) - \mathsf{E}(h(\boldsymbol{X}))| \to 0$ or $\mathsf{E}(h(\boldsymbol{X}^{(n)})) \to \mathsf{E}(h(\boldsymbol{X}))$.

$(2) \Rightarrow (3)$: For an arbitrary continuous function h such that $\|h(\boldsymbol{x})\| \leq M$ and an arbitrary $\epsilon > 0$ we will show that $|\mathsf{E}(h(\boldsymbol{X}^{(n)})) - \mathsf{E}(h(\boldsymbol{X}))| \to \epsilon$. Since this holds for all $\epsilon > 0$, this implies $\mathsf{E}(h(\boldsymbol{X}^{(n)})) \to \mathsf{E}(h(\boldsymbol{X}))$.

It is possible to find α such that $\mathsf{P}(\|\boldsymbol{X}\| \geq \alpha) < \epsilon/(2M)$, and a continuous function $0 \leq h'(\boldsymbol{x}) \leq 1$ such that $h'(\boldsymbol{x}) = 0$ if $\|\boldsymbol{x}\| \geq \alpha + 1$ and $h'(\boldsymbol{x}) = 1$ if $\|\boldsymbol{x}\| \leq \alpha$. It follows that $\mathsf{E}(h'(\boldsymbol{X})) \geq 1 - \epsilon/(2M)$ and

$$|\mathsf{E}(h(\boldsymbol{X}^{(n)})) - \mathsf{E}(h(\boldsymbol{X}))|$$
$$= |\mathsf{E}(h(\boldsymbol{X}^{(n)})) - \mathsf{E}(h(\boldsymbol{X})) + \mathsf{E}(h(\boldsymbol{X}^{(n)})h'(\boldsymbol{X}^{(n)}))$$
$$\quad - \mathsf{E}(h(\boldsymbol{X}^{(n)})h'(\boldsymbol{X}^{(n)})) + \mathsf{E}(h(\boldsymbol{X})h'(\boldsymbol{X})) - \mathsf{E}(h(\boldsymbol{X})h'(\boldsymbol{X}))|$$
$$\leq |\mathsf{E}(h(\boldsymbol{X}^{(n)})) - \mathsf{E}(h(\boldsymbol{X}^{(n)})h'(\boldsymbol{X}^{(n)}))| + |\mathsf{E}(h(\boldsymbol{X}^{(n)})h'(\boldsymbol{X}^{(n)})) - \mathsf{E}(h(\boldsymbol{X})h'(\boldsymbol{X}))|$$
$$\quad + |\mathsf{E}(h(\boldsymbol{X})h'(\boldsymbol{X})) - \mathsf{E}(h(\boldsymbol{X}))|$$
$$\to |\mathsf{E}(h(\boldsymbol{X}^{(n)})) - \mathsf{E}(h(\boldsymbol{X}^{(n)})h'(\boldsymbol{X}^{(n)}))| + 0 + |\mathsf{E}(h(\boldsymbol{X})h'(\boldsymbol{X})) - \mathsf{E}(h(\boldsymbol{X}))|$$
$$\to \epsilon/2 + \epsilon/2 = \epsilon.$$

The first inequality above follows from the triangle inequality. The first convergence follows from statement (2) and the fact that hh' is a continuous function that vanishes outside a bounded and closed set. The second convergence follows from

$$|\mathsf{E}(h(\boldsymbol{X}^{(n)})) - \mathsf{E}(h(\boldsymbol{X}^{(n)})h'(\boldsymbol{X}^{(n)}))| \le \mathsf{E}(|h(\boldsymbol{X}^{(n)})| \cdot |1 - h'(\boldsymbol{X}^{(n)})|)$$
$$\le M\,\mathsf{E}(|1 - h'(\boldsymbol{X}^{(n)})|)$$
$$= M\,\mathsf{E}(1 - h'(\boldsymbol{X}^{(n)}))$$
$$= M(1 - \mathsf{E}\,h'(\boldsymbol{X}^{(n)}))$$
$$\to M(1 - \mathsf{E}\,h'(\boldsymbol{X}))$$
$$\le M\epsilon/(2M) = \epsilon/2$$

(with a similar bound applying to the term $|\mathsf{E}(h(\boldsymbol{X})) - \mathsf{E}(h(\boldsymbol{X})h'(\boldsymbol{X}))|$).

(3) $\Rightarrow$ (4): Let h be a bounded measurable function, for which $\mathsf{P}(\boldsymbol{X} \in \{\boldsymbol{x} : h \text{ is continuous at } \boldsymbol{x}\}) = 1$. Given $\epsilon > 0$, we use the previous lemma to obtain continuous and bounded functions m, M such that $m \le h \le M$ and

$$\mathsf{E}(h(\boldsymbol{X})) - \epsilon \le \mathsf{E}\,m(\boldsymbol{X})$$
$$= \lim \mathsf{E}\,m(\boldsymbol{X}^{(n)})$$
$$\le \liminf \mathsf{E}(h(\boldsymbol{X}^{(n)}))$$
$$\le \limsup \mathsf{E}(h(\boldsymbol{X}^{(n)}))$$
$$\le \lim \mathsf{E}\,M(\boldsymbol{X}^{(n)})$$
$$= \mathsf{E}(M(\boldsymbol{X}))$$
$$\le \mathsf{E}(h(\boldsymbol{X})) + \epsilon.$$

Since ϵ is arbitrary, $\liminf \mathsf{E}(h(\boldsymbol{X}^{(n)})) = \limsup \mathsf{E}(h(\boldsymbol{X}^{(n)}))$, and $\mathsf{E}(h(\boldsymbol{X}^{(n)})) \to \mathsf{E}(h(\boldsymbol{X}))$. ∎

8.6 The Law of Large Numbers

Proposition 8.6.1 (Markov Inequality).

$$\mathsf{P}(|X| \ge \alpha) \le \frac{\mathsf{E}(|X|^k)}{\alpha^k}, \qquad \forall \alpha > 0, \quad \forall k \in \mathbb{N}.$$

Proof. If X is continuous,

$$\mathsf{E}(|X|^k) = \int_{-\infty}^{\infty} |x|^k f_X(x)dx = \int_{0}^{\infty} |x|^k f_X(x)dx$$
$$\ge \int_{\alpha}^{\infty} |x|^k f_X(x)dx \ge \int_{a}^{\infty} a^k f_X(x)dx = \alpha^k \,\mathsf{P}(|X| \ge a).$$

If X is discrete, replace the integrals above with sums. The general case follows from Lebesgue integration theory. ∎

Corollary 8.6.1 (Chebyshev Inequality).

$$P(|X - E(X)| \geq \alpha) \leq \frac{\mathrm{Var}(X)}{\alpha^2}, \qquad \forall \alpha > 0.$$

Proof. Apply Markov inequality $P(Z^2 \geq \alpha^2) \leq E(Z^2)/\alpha^2$ to the RV Z^2 where $Z = |X - E(X)|$. This gives

$$P(|X - E(X)| \geq \alpha) = P(Z \geq \alpha) = P(Z^2 \geq \alpha^2) \leq \frac{\mathrm{Var}(X)}{\alpha^2}.$$

∎

Proposition 8.6.2 (The Strong Law of Large Numbers). *For a sequence of iid random vectors $\boldsymbol{X}^{(n)}, n \in \mathbb{N}$ with finite fourth moments $(E(\boldsymbol{X}^{(1)^4}) < \infty)$,*

$$\frac{1}{n} \sum_{i=1}^{n} \boldsymbol{X}^{(i)} \overset{\mathrm{as}}{\to} E(\boldsymbol{X}^{(1)}).$$

Proof. Convergence with probability 1 of the d-dimensional sequence of RV occurs if and only if each of the components converge to the corresponding limit. It follows that it is sufficient to prove the above theorem in the scalar case.

Applying Cauchy-Schwartz inequality (Proposition B.4.1) to the inner product $g(X^2, 1) = E(X^2 \cdot 1)$ we have that $E(X^2)^2 \leq E(X^4)$. Similarly, we have $E(X)^2 \leq E(X^2)$. This shows that $E(X^r), 1 \leq r \leq 4$ is finite if $E(X^4) < \infty$.

We assume below that $E(X^{(1)}) = 0$. We can then prove the general case by applying the obtained result to the sequence $Y^{(n)} = X^{(n)} - E(X^{(1)})$: $n^{-1} \sum Y^{(n)} \overset{\mathrm{as}}{\to} 0$, which implies $n^{-1} \sum X^{(n)} \overset{\mathrm{as}}{\to} E(X^{(1)})$.

We have

$$E\left((\sum_{i=1}^{n} X^{(i)})^4 \right) = E\left(\sum_{i=1}^{n} (X^{(i)})^4 + 6 \sum_{1 \leq i < j \leq n} (X^{(i)})^2 (X^{(i)})^2 \right).$$

The missing terms in the sum above are either $E(X^{(i)} X^{(j)} X^{(k)} X^{(l)})$ for distinct i, j, k, l or $E((X^{(i)})^3 X^{(j)})$ for distinct i, j. In either case, these terms are zero since expectation of a product of independent RVs is a product of the corresponding expectation, and since $E(X^{(i)}) = 0$. The reason for the coefficient 6 is that if we fix the index of the first term in $X^{(a)} X^{(b)} X^{(c)} X^{(d)}$ to be $a = i$, there are three remaining variables that can receive the same index i (the remaining indices get the value j). Alternatively, the index of the first variable may be j, yielding 3 more terms with the required multiplicity of indices.

This leads to the bound

$$E\left(\left(\sum_{i=1}^{n} X^{(i)} \right)^4 \right) \leq nM + 6 \binom{n}{2} M' = nM + 3n(n-1)M' \leq n^2 M''$$

for some finite $M, M', M'' < \infty$. Applying Markov's inequality with $k = 4$, we get

$$\mathsf{P}\left(\left|n^{-1}\sum_{i=1}^{n}X^{(i)}\right| \geq \epsilon\right) = \mathsf{P}\left(\left|\sum_{i=1}^{n}X^{(i)}\right| \geq n\epsilon\right) \leq Mn^{-2}\epsilon^{-4},$$

which combined with the first Borell-Cantelli lemma (Proposition 6.6.1) shows that

$$\mathsf{P}\left(\left|n^{-1}\sum_{i=1}^{n}X^{(i)}\right| \geq \epsilon \text{ i.o.}\right) = 0.$$

Proposition 8.2.1 completes the proof. ∎

The proposition above remains true under the weaker condition of finite first moment $\mathsf{E}(X^{(1)}) < \infty$. The proof, however, is considerably more complicated in this case [5].

Corollary 8.6.2 (The Weak Law of Large Numbers). *For a sequence of iid random vectors $X^{(n)}, n \in \mathbb{N}$ with finite fourth moment $\mathsf{E}(X^{(1)^4}) < \infty$,*

$$\frac{1}{n}\sum_{i=1}^{n}X^{(i)} \overset{\mathsf{P}}{\to} \mathsf{E}(X^{(1)}).$$

Proof. This follows from the strong law of large numbers and Corollary 8.2.1. ∎

The weak law of large numbers (WLLN) shows that the sequence of random vectors $Y^{(n)} = \frac{1}{n}\sum_{i=1}^{n}X^{(i)}$, $n = 1, 2, \ldots$ is increasingly similar in values to the expectation vector μ. Thus for large n, $Y^{(n)}$ may be intuitively considered as a deterministic RV whose value is μ with high probability.

An important special case occurs when $X^{(i)}$ are iid indicator variables $X^{(i)}(\omega) = I_A(\omega)$ with respect to some set $A \subset \mathbb{R}$. In this case, the average $\frac{1}{n}\sum_{i=1}^{n}X^{(i)} = \frac{1}{n}\sum_{i=1}^{n}I_A$ becomes the relative frequency of occurrence in the set A and the expectation $\mathsf{E}(X^{(i)})$ becomes $\mathsf{E}(I_A) = \mathsf{P}(A)$. We thus have a convergence of relative frequencies as $n \to \infty$ to their expectation.

We demonstrate the WLLN below by drawing samples $X^{(n)}$, $i = 1, \ldots, n$ iid from a uniform distribution over $\{1, \ldots, 6\}$ (fair die) and displaying $X^{(n)}$ together with the averages $Y^{(n)} = \sum_{i=1}^{n}X^{(n)}/n$ for different n values.

```
# draw 10 samples from U({1,...,6})
S = sample(1:6, 10, replace = TRUE)
S
```

```
##  [1] 3 4 3 2 2 2 6 5 3 3
```

```
# compute running averages of 1,...,10
# samples
```

```
cumsum(S)/1:10
```

```
##   [1] 3.000 3.500 3.333 3.000 2.800 2.667
##   [7] 3.143 3.375 3.333 3.300
```

The example above shows that as n increases, the averages get closer and closer to the expectation 3.5. Below is a graph for larger n values. Note how the variations around 3.5 diminish for large n.

```
S = seq(from = 1, to = 6, length = 200)
plot(1:200, S, type = "n", xlab = "$n$", ylab = "$(X_1+\\cdots+X_n)/n$")
for (s in 1:100) lines(1:200, cumsum(sample(1:6,
    200, replace = TRUE))/1:200, col = rgb(0,
    0, 0, alpha = 0.3))
```

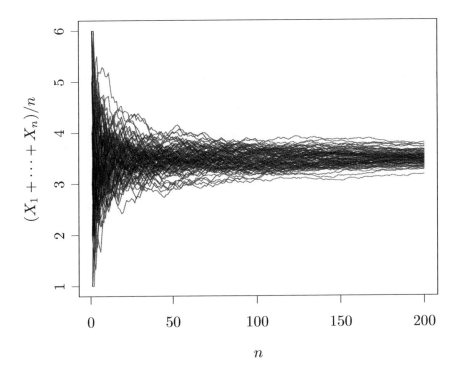

This is one instantiation of the sequence of die rolls. Other instantiations will produce other graphs. The WLLN states that all such graphs will become arbitrarily close to 3.5 as n increases.

8.7 The Characteristic Function*

We use in this section the following standard results associated with complex numbers. The notation i denotes the complex number $i = \sqrt{-1}$ and z refers to the complex number $z = a + ib, a, b \in \mathbb{R}$.

1. $\exp(ix) = \cos x + i \sin x$

2. $|a + bi| = \sqrt{a^2 + b^2}$

3. $\overline{a + bi} = a - bi$

The notation $\mathbb{C}$ corresponds to the set of all complex numbers.

Definition 8.7.1. The characteristic function associated with the random vector $\boldsymbol{X}$ is the following function

$$\phi : \mathbb{R}^d \to \mathbb{C}, \qquad \phi_{\boldsymbol{X}}(\boldsymbol{t}) = \mathsf{E}(\exp(i\boldsymbol{t}^\top \boldsymbol{X}))$$

where the expectation is taken with respect to the distribution of $\boldsymbol{X}$. We sometimes omit the index $\boldsymbol{X}$ and simply denote the characteristic function by $\phi(\boldsymbol{t})$.

Proposition 8.7.1. *Let ϕ be the characteristic function of a random vector $\boldsymbol{X}$. Then,*

1. *$\phi(\boldsymbol{0}) = 1$.*

2. *$|\phi(\boldsymbol{t})| \leq 1$.*

3. *$\phi(\boldsymbol{t})$ is a continuous function.*

4. *$\phi_{-\boldsymbol{X}}(\boldsymbol{t}) = \overline{\phi_{\boldsymbol{X}}(\boldsymbol{t})}$.*

5. *$\phi_{a\boldsymbol{X}+\boldsymbol{b}}(\boldsymbol{t}) = e^{i\boldsymbol{t}^\top \boldsymbol{b}} \phi_{\boldsymbol{X}}(a\boldsymbol{t})$.*

6. *If $\boldsymbol{X}^{(n)}, n = 1, \ldots, N$ are independent RVs, then $\phi_{\sum_{n=1}^N \boldsymbol{X}^{(n)}}(\boldsymbol{t}) = \prod_{n=1}^N \phi_{\boldsymbol{X}^{(n)}}(\boldsymbol{t})$.*

7. *If $\mathsf{E}\|\boldsymbol{X}\| < \infty$ then $\nabla\phi = i\,\mathsf{E}(\boldsymbol{X})$.*

8. *If $\mathsf{E}\|\boldsymbol{X}\|^2 < \infty$ then $\nabla^2\phi = -\mathsf{E}(\boldsymbol{X}\boldsymbol{X}^\top)$.*

Proof. Statement 1 follows from $\phi(\boldsymbol{0}) = \mathsf{E}(1) = 1$. Statement 2 follows from

$$|\exp(ix)| = |\cos x + i\sin x| = \cos^2 x + i^2 \sin^2(x) = \cos^2 x - \sin^2(x)$$
$$\leq \cos^2 x + \sin^2 x = 1$$

and $|\mathsf{E}(X)| \leq \sup|X|\,\mathsf{E}(1) = \sup|X|$. To prove statement 3, note that if $\boldsymbol{t} \to \boldsymbol{r}$ then $\exp(i\boldsymbol{t}^\top \boldsymbol{x}) \to \exp(i\boldsymbol{r}^\top \boldsymbol{x})$, which implies by Proposition 8.3.1 that

$$\phi(\boldsymbol{t}) - \phi(\boldsymbol{r}) = \mathsf{E}(\exp(i\boldsymbol{t}^\top X) - \exp(i\boldsymbol{r}^\top \boldsymbol{X}))$$
$$\leq \mathsf{E}(\|\exp(i\boldsymbol{t}^\top X) - \exp(i\boldsymbol{r}^\top \boldsymbol{X})\|) \to 0.$$

Statement 4 follows from the following change of integration measure (see Proposition F.3.12):

$$\phi_{-\boldsymbol{X}}(\boldsymbol{t}) = \int \exp(i\boldsymbol{t}^\top \boldsymbol{x})\, dF_{-\boldsymbol{X}}(\boldsymbol{x}) = \int \exp(-i\boldsymbol{t}^\top \boldsymbol{x})\, dF_{\boldsymbol{X}}(\boldsymbol{x}).$$

Statement 5 follows from the change of integration measure (see Proposition F.3.12):

$$\phi_{a\boldsymbol{X}+b}(\boldsymbol{t}) = \int \exp(i\boldsymbol{t}^\top \boldsymbol{x})\, dF_{a\boldsymbol{X}+b}(\boldsymbol{x}) = \int \exp(i\boldsymbol{t}^\top (a\boldsymbol{X} + b))\, dF_{\boldsymbol{X}}(\boldsymbol{X})$$
$$= \exp(i\boldsymbol{t}^\top b) \int \exp(ia\boldsymbol{t}^\top \boldsymbol{X})\, dF_{\boldsymbol{X}}(\boldsymbol{X}).$$

The proof of statement 5 is similar to the proof of Proposition 4.8.1. The last two statement can be proven by expanding the exponential in ϕ using a Taylor series expansion, differentiating the Taylor series term by term at $t = 0$, and noting that only the leading term remains.   ∎

Note that part 2 of the proposition above implies that the characteristic function always exists.

Proposition 8.7.2. *The characteristic function of a $N(\boldsymbol{\mu}, \Sigma)$ random vector is*

$$\phi(\boldsymbol{t}) = \exp(i\boldsymbol{t}^\top \boldsymbol{\mu} - \boldsymbol{t}^\top \Sigma \boldsymbol{t}/2).$$

Proof. We start by showing that in the univariate case $\phi(t) = \exp(-t^2/2)$.

$$\phi(t) = \mathsf{E}\exp(itx) = \mathsf{E}(\cos tx + i\sin tx) = \mathsf{E}(\cos tx) + 0$$
$$= \int \cos(tx)\exp(-x^2)/\sqrt{2\pi}\, dx.$$

where the third equality above holds since the $\sin(tx)\exp(-tx^2)$ is an even function ($f(x) = -f(-x)$) whose integral is zero. Differentiating ϕ with respect to t, we have

$$\phi'(t) = -\int x\sin(tx)\exp(-x^2)/\sqrt{2\pi}\, dx$$
$$= \int \sin(tx)\frac{d\exp(-x^2)/\sqrt{2\pi}}{dx}\, dx$$
$$= \sin(tx)\exp(-x^2)/\sqrt{2\pi}\Big|_{-\infty}^{\infty} - \int t\cos(tx)\exp(-x^2)/\sqrt{2\pi}\, dx$$
$$= 0 - \int t\cos(tx)\exp(-x^2)/\sqrt{2\pi}\, dx$$
$$= -t\phi(t).$$

We have thus obtained a differential equation $\phi'(t) = -t\phi(t)$ (subject to the initial condition $\phi(0) = \mathsf{E}(1) = 1$) whose only solution is $\phi(t) = \exp(-t^2/2)$.

An alternative proof uses the completion of the square method as in Proposition 3.9.2.

Part 6 of the proposition above shows that

$$\text{if} \quad \boldsymbol{X} \sim N(\boldsymbol{0}, I) \quad \text{then} \quad \phi_{\boldsymbol{X}}(\boldsymbol{t}) = \exp(-\boldsymbol{t}^{\top}\boldsymbol{t}/2).$$

Given an arbitrary $\boldsymbol{X} \sim N(\boldsymbol{\mu}, \Sigma)$ we can use the transformation $\boldsymbol{X} = \Sigma^{1/2}\boldsymbol{Y} + \boldsymbol{\mu}$ where $\boldsymbol{Y} \sim N(\boldsymbol{0}, I)$ (as in Proposition 5.2.3). This yields

$$
\begin{aligned}
\mathsf{E}(\exp(i\boldsymbol{t}^{\top}\boldsymbol{X})) &= \mathsf{E}\left(\exp\left(i\boldsymbol{t}^{\top}(\Sigma^{1/2}\boldsymbol{Y} + \boldsymbol{\mu})\right)\right) \\
&= \exp(i\boldsymbol{t}^{\top}\boldsymbol{\mu})\,\mathsf{E}\left(\exp(i(\Sigma^{1/2}\boldsymbol{t})^{\top}\boldsymbol{Y})\right) \\
&= \exp(i\boldsymbol{t}^{\top}\boldsymbol{\mu})\,\mathsf{E}\left(\exp(i\boldsymbol{u}^{\top}\boldsymbol{Y})\right) \\
&= \exp(i\boldsymbol{t}^{\top}\boldsymbol{\mu})\exp\left(-\boldsymbol{u}^{\top}\boldsymbol{u}/2\right) \\
&= \exp(i\boldsymbol{t}^{\top}\boldsymbol{\mu})\exp\left(-(\Sigma^{1/2}\boldsymbol{t})^{\top}(\Sigma^{1/2}\boldsymbol{t})/2\right) \\
&= \exp(i\boldsymbol{t}^{\top}\boldsymbol{\mu} - \boldsymbol{t}^{\top}\Sigma\boldsymbol{t}/2).
\end{aligned}
$$

∎

Lemma 8.7.1. *Assuming that $a > 0$,*

$$\int \exp(-ax^2 + bx)\,dx = \sqrt{\frac{\pi}{a}}\exp(b^2/(4a)).$$

Proof. Writing $-ax^2 + bx = -a(x - b/(2a))^2 + b^2/(4a)$, we have

$$\int e^{-ax^2+bx}\,dx = e^{b^2/(4a)}\int e^{-a(x-b/(2a))^2}\,dx = e^{b^2/(4a)}\sqrt{\pi/a},$$

where the last equality follows from the fact that the Gaussian pdf integrates to one. ∎

8.8 Levy's Continuity Theorem and the Cramer-Wold Device

Proposition 8.8.1 (Levy's Continuity Theorem).

$$\boldsymbol{X}^{(n)} \rightsquigarrow \boldsymbol{X} \qquad \text{if and only if} \qquad \phi_{\boldsymbol{X}^{(n)}}(\boldsymbol{t}) \to \phi_{\boldsymbol{X}}(\boldsymbol{t}), \forall \boldsymbol{t} \in \mathbb{R}^d.$$

Proof. We assume that $\boldsymbol{X}^{(n)} \rightsquigarrow \boldsymbol{X}$. Since $\exp(i\boldsymbol{t}^{\top}\boldsymbol{X}) = \cos\boldsymbol{t}^{\top}\boldsymbol{X} + i\sin\boldsymbol{t}^{\top}\boldsymbol{X}$ we have that ϕ is continuous and bounded as a function of $\boldsymbol{X}$, which together with implication $1 \Rightarrow 3$ implies the pointwise convergence of the characteristic function.

Conversely, we assume that $\forall \boldsymbol{t} \in \mathbb{R}^d$, $\phi_{\boldsymbol{X}^{(n)}}(\boldsymbol{t}) \to \phi_{\boldsymbol{X}}(\boldsymbol{t})$ and show that for any continuous function g that is zero outside a bounded and closed set, we have $E(g(\boldsymbol{X}^{(n)})) \to E(g(\boldsymbol{X}))$. Using the Portmanteau theorem, this implies that $\boldsymbol{X}^{(n)} \rightsquigarrow \boldsymbol{X}$. Since g is continuous on a compact set, it is uniformly continuous and we can select for all $\epsilon > 0$ a $\delta > 0$ such that $\|\boldsymbol{x} - \boldsymbol{y}\| < \delta$ implies $|g(\boldsymbol{x}) - g(\boldsymbol{y})| < \epsilon$.

Denoting by $\boldsymbol{Z}$ a $N(\boldsymbol{0}, \sigma^2 I)$ random vector that is independent of $\boldsymbol{X}$ and the sequence $\boldsymbol{X}^{(n)}$, we have

$$
\begin{aligned}
& |E(g(\boldsymbol{X}^{(n)})) - E(g(\boldsymbol{X}))| \\
&\quad = |E(g(\boldsymbol{X}^{(n)})) - E(g(\boldsymbol{X})) + E(g(\boldsymbol{X}^{(n)} + \boldsymbol{Z})) \\
&\qquad - E(g(\boldsymbol{X}^{(n)} + \boldsymbol{Z})) + E(g(\boldsymbol{X} + \boldsymbol{Z})) - E(g(\boldsymbol{X} + \boldsymbol{Z}))| \\
&\quad \leq |E(g(\boldsymbol{X}^{(n)})) - E(g(\boldsymbol{X}^{(n)} + \boldsymbol{Z}))| + |E(g(\boldsymbol{X}^{(n)} + \boldsymbol{Z})) - E(g(\boldsymbol{X} + \boldsymbol{Z}))| \\
&\qquad + |E(g(\boldsymbol{X} + \boldsymbol{Z})) - E(g(\boldsymbol{X}))|.
\end{aligned}
$$

The first term above is bounded by 2ϵ since for σ sufficiently small

$$
\begin{aligned}
|E(g(\boldsymbol{X}^{(n)})) - E(g(\boldsymbol{X}^{(n)} + \boldsymbol{Z}))| &\leq E(|g(\boldsymbol{X}^{(n)})) - E(g(\boldsymbol{X}^{(n)} + \boldsymbol{Z}))|I(\|\boldsymbol{Z}\| \leq \delta) \\
&\quad + E(|g(\boldsymbol{X}^{(n)})) - E(g(\boldsymbol{X}^{(n)} + \boldsymbol{Z}))|I(\|\boldsymbol{Z}\| > \delta) \\
&\leq E(\epsilon) + 2\left(\sup_{\boldsymbol{w}} |g(\boldsymbol{w})|\right) P(\|\boldsymbol{Z}\| > \delta) \\
&\leq 2\epsilon.
\end{aligned}
$$

The third term above is also bounded by 2ϵ due to a similar argument. It remains to show that the second term converges to zero: $E(g(\boldsymbol{X}^{(n)} + \boldsymbol{Z})) \to E(g(\boldsymbol{X} + \boldsymbol{Z}))$. We will then have that $|E(g(\boldsymbol{X}^{(n)})) - E(g(\boldsymbol{X}))| \to 0$, implying that $E(g(\boldsymbol{X}^{(n)})) \to E(g(\boldsymbol{X}))$ (for all continuous functions g that are zero outside a bounded and closed set), which together with the Portmanteau theorem implies $\boldsymbol{X}^{(n)} \rightsquigarrow \boldsymbol{X}$.

We show below that $E(g(\boldsymbol{X}^{(n)} + \boldsymbol{Z})) \to E(g(\boldsymbol{X} + \boldsymbol{Z}))$. We have

$$
\begin{aligned}
E(g(\boldsymbol{X}^{(n)} + \boldsymbol{Z})) &= \frac{1}{(\sqrt{2\pi}\sigma)^d} \iint g(\boldsymbol{x} + \boldsymbol{z}) \exp(-\boldsymbol{z}^\top \boldsymbol{z}/(2\sigma^2)) \, d\boldsymbol{z} dF_{\boldsymbol{X}^{(n)}} \qquad (8.2) \\
&= \frac{1}{(\sqrt{2\pi}\sigma)^d} \iint g(\boldsymbol{u}) \exp(-(\boldsymbol{u} - \boldsymbol{x})^\top (\boldsymbol{u} - \boldsymbol{x})/(2\sigma^2)) \, d\boldsymbol{u} dF_{\boldsymbol{X}^{(n)}} \\
&= \frac{1}{(\sqrt{2\pi}\sigma)^d} \iint g(\boldsymbol{u}) \prod_{j=1}^{d} \exp\left(-\frac{(u_j - x_j)^2}{2\sigma^2}\right) d\boldsymbol{u} dF_{\boldsymbol{X}^{(n)}} \\
&= \frac{1}{(\sqrt{2\pi}\sigma)^d} \iint g(\boldsymbol{u}) \prod_{j=1}^{d} \frac{\sigma}{\sqrt{2\pi}} \int \exp\left(it_j(u_j - x_j) - \sigma^2 t_j^2/2\right) \\
&\qquad\qquad\qquad\qquad\qquad\qquad dt_j d\boldsymbol{u} dF_{\boldsymbol{X}^{(n)}} \\
&= \frac{1}{(2\pi)^d} \iiint g(\boldsymbol{u}) \exp\left(it^\top(\boldsymbol{u} - \boldsymbol{x}) - \sigma^2 t^\top t/2\right) dt d\boldsymbol{u} dF_{\boldsymbol{X}^{(n)}} \\
&= \frac{1}{(2\pi)^d} \iint g(\boldsymbol{u}) \exp\left(it^\top \boldsymbol{u} - \sigma^2 t^\top t/2\right) \phi_{\boldsymbol{X}^{(n)}}(-t) \, dt d\boldsymbol{u},
\end{aligned}
$$

with (8.3) labelling the fourth displayed line.

where $u = x + z$. Note that we used Lemma 8.7.1 in the fourth equality and Proposition 8.7.2 in the last equality.

Since g is continuous and non-zero only on a closed and bounded set, $g(u)$ may be made into a distribution by adding a constant to it and dividing by a constant. This implies that $E(g(X^{(n)} + Z))$ may be considered as an expectation over a two random vectors U having density $c(g(u) + b)$ and T have a Gaussian density. The argument of that expectation is the bounded function $\exp(it^\top u) \phi_{X^{(n)}}(-t)$, and so by the dominated convergence theorem for random variables (Proposition 8.3.1)

$$\frac{1}{(2\pi)^d} \iint g(u) \exp\left(it^\top u - \sigma^2 t^\top t / 2\right) \phi_{X^{(n)}}(-t) \, dt du$$

$$\to \frac{1}{(2\pi)^d} \iint g(u) \exp\left(it^\top u - \sigma^2 t^\top t / 2\right) \phi_X(-t) \, dt du.$$

Repeating the derivation in Equation (8.2) with X substituting $X^{(n)}$ we see that

$$E(g(X^{(n)} + Z)) = \frac{1}{(2\pi)^d} \iint g(u) \exp\left(it^\top u - \sigma^2 t^\top t / 2\right) \phi_X(-t) \, dt du,$$

implying that $E(g(X^{(n)} + Z)) \to E(g(X + Z))$. ∎

Note that Levy's continuity theorem above is similar to Proposition 2.4.2. The former equates convergence in distribution to convergence of characteristic functions. The latter equates convergence in distribution to convergence of the moment generating functions. An advantage of Levy's theorem is that in many cases the moment generating function does not exist, while the characteristic function always exist.

The following result shows a way to prove multivariate convergence in distribution using a variety of univariate convergence results.

Corollary 8.8.1 (Cramer-Wold Device). *If $t^\top X^{(n)} \rightsquigarrow t^\top X$ for all vectors $t \in \mathbb{R}^d$, then $X^{(n)} \rightsquigarrow X$.*

Proof. Using the continuity theorem, convergence in distribution occurs if the characteristic functions converge. This occurs since for all $t \in \mathbb{R}^d$,

$$\phi_{X^{(n)}}(t) = E(\exp(it^\top X^{(n)})) = \phi_{t^\top X^{(n)}}(1) \to \phi_{t^\top X}(1) = \phi_X(t).$$

∎

8.9 The Central Limit Theorem

The central limit theorem shows that the distribution of $n^{-1} \sum_{i=1}^n X^{(i)}$ for large n is Gaussian centered around $\mu = E(X)$ with variance Σ/n. Thus, not only can we say that $n^{-1} \sum_{i=1}^n X^{(i)}$ is close to μ (as the law of large numbers implies), we can say that its distribution is a bell shaped curve centered at μ whose variance decays linearly with n.

Proposition 8.9.1 (Central Limit Theorem). *For a sequence of iid d-dimensional random vectors $\boldsymbol{X}^{(n)}, n \in \mathbb{N}$ with finite expectation vector $\boldsymbol{\mu}$ and covariance matrix Σ, we have*

$$\sqrt{n}\left(\frac{1}{n}\sum_{i=1}^{n}\boldsymbol{X}^{(i)} - \boldsymbol{\mu}\right) \rightsquigarrow N(\boldsymbol{0}, \Sigma).$$

Proof. We prove the statement using Levy's continuity theorem and show that the characteristic function of $\sqrt{n}\left(\frac{1}{n}\sum_{i=1}^{n}\boldsymbol{X}^{(n)} - \boldsymbol{\mu}\right)$ converges (as $n \to \infty$) to the characteristic function of a $N(\boldsymbol{0}, \Sigma)$ random vector.

We denote by $\tilde{\phi}$ the characteristic functions of $\boldsymbol{X}^{(k)} - \mathsf{E}(\boldsymbol{X}^{(k)})$, $k \in \mathbb{N}$ ($\boldsymbol{X}^{(k)}$ are iid and so they have the same characteristic function). Since $\tilde{\phi}(\boldsymbol{0}) = 1$, $\nabla\tilde{\phi}(\boldsymbol{0}) = \boldsymbol{0}$, and $\lim_{\epsilon \to \boldsymbol{0}} \nabla^2\tilde{\phi}(\boldsymbol{\epsilon}) = \Sigma$, we have

$$\lim_{n\to\infty} \phi_{\sqrt{n}(\sum_{k=1}^{n}\boldsymbol{X}^{(k)}/n - \boldsymbol{\mu})}(\boldsymbol{t}) = \lim_{n\to\infty} \phi_{(\sum_{k=1}^{n}\boldsymbol{X}^{(k)}/n - \boldsymbol{\mu})}(\boldsymbol{t}/\sqrt{n})$$

$$= (\tilde{\phi}(\boldsymbol{t}/\sqrt{n}))^n$$

$$= \lim_{n\to\infty}\left(1 + \frac{1}{n}\boldsymbol{t}^\top\int_0^1\int_0^1 z\nabla^2\phi(zw\boldsymbol{t}/\sqrt{n})\,dz\,dw\,\boldsymbol{t}\right)^n$$

$$= \left(\lim_{n\to\infty}\frac{1}{n}\boldsymbol{t}^\top\int_0^1\int_0^1 z\nabla^2\phi(zw\boldsymbol{t}/\sqrt{n})\,dz\,dw\,\boldsymbol{t}\right)^n$$

$$= \exp(-\boldsymbol{t}^\top\Sigma\boldsymbol{t}/2)$$

$$= \phi_{N(\boldsymbol{0},\Sigma)}.$$

The second to last equality follows from Proposition D.2.2, and the last equality follows from Proposition 8.7.2. ∎

Example 8.9.1. *In 1-d, for any sequence of continuous iid RVs $X^{(1)}, X^{(2)}, \ldots$ with expectation μ and variance σ^2 ($d = 1$), we have*

$$\lim_{n\to\infty} \mathsf{P}\left(\sqrt{n}\left(\frac{1}{n}\sum_{i=1}^{n}X^{(i)} - \mu\right) \in (a, b)\right) = \int_a^b \frac{1}{\sqrt{2\pi\sigma^2}}e^{-z^2/(2\sigma^2)}\,dz.$$

We make the following comments concerning the central limit theorem (CLT):

- Statisticians generally agree as a rule of thumb that the approximation in the CLT is accurate when $n > 30$ and $d = 1$. However, the CLT approximation can also be effectively used in many cases where $n < 30$ or $d > 1$.

- Since a linear combination of a multivariate Gaussian random vector is a multivariate Gaussian random vector, we have the following implications:

$$\sqrt{n}\left(\frac{1}{n}\sum_{i=1}^{n}\boldsymbol{X}^{(i)} - \boldsymbol{\mu}\right) \sim N(\boldsymbol{0}, \Sigma) \quad \Rightarrow \quad \frac{1}{n}\sum_{i=1}^{n}\boldsymbol{X}^{(i)} - \boldsymbol{\mu} \sim N(\boldsymbol{0}, n^{-1}\Sigma)$$

$$\Rightarrow \quad \frac{1}{n}\sum_{i=1}^{n}\boldsymbol{X}^{(i)} \sim N(\boldsymbol{\mu}, n^{-1}\Sigma)$$

$$\Rightarrow \quad \sum_{i=1}^{n}\boldsymbol{X}^{(i)} \sim N(n\boldsymbol{\mu}, n\Sigma).$$

As a result, we can say that intuitively, $\frac{1}{n}\sum_{i=1}^{n}\boldsymbol{X}^{(i)}$ has a distribution that is approximately $N(\boldsymbol{\mu}, n^{-1}\Sigma)$ for large n i.e., it is approximately a Gaussian centered at the expectation and whose variance is decaying linearly with n. Similarly, intuitively, the distribution of $\sum_{i=1}^{n}\boldsymbol{X}^{(i)}$ for large n is approximately $N(n\boldsymbol{\mu}, n^2\Sigma/n) = N(n\boldsymbol{\mu}, n\Sigma)$.

- The CLT is very surprising. It states that if we are looking for a distribution of a random vector $\boldsymbol{Y}$ that is a sum or average of many other iid random vectors $\boldsymbol{X}^{(i)}$, $i = 1, \ldots, n$, the distribution of $\boldsymbol{Y}$ will be close to normal regardless of the distribution of the original $\boldsymbol{X}^{(i)}$. This explains the fact that many quantities in the real world appear to have a Gaussian distribution or close to it. Examples include physical measurements like height and weight (they may be the result of a sum of many independent RVs expressing genetic makeup), atmospheric interference (sum of many small interference on a molecular scale), and movement of securities prices in finance (sum of many buy or sell orders).

- The condition of iid in the CLT above may be relaxed. More general results state that a sum of many independent (but not identically distributed) RVs is Gaussian under relatively weak conditions. Even the independence assumption may be relaxed as long as it is not the case that every example depends on most of the other examples.

The code below displays the histograms of averages of $n = 1, 3, 30, 90$ iid samples from a Exp(λ) distribution (with $\mu = 1$). Note how, as n increases, the averages tend to concentrate increasingly around the average 1, which reflects the weak law of large numbers. The CLT is evident in that the histogram for large n resembles a bell-shaped curve centered at the expectation. Note also how the histogram width decreases from 30 to 90 as n increases according to Σ/n, in agreement with the decay of the asymptotic variance.

```
iter = 2000
avg0 <- avg1 <- avg2 <- avg3 <- rep(0, iter)
for (i in 1:iter) {
    S = rexp(90)   # sample from exp(1) distribution
    avg0[i]  = S[1]
```

```
    avg1[i] = mean(S[1:3])
    avg2[i] = mean(S[1:30])
    avg3[i] = mean(S[1:90])
}
SR = stack(list(`$n=1$` = avg0, `$n=3$` = avg1,
    `$n=30$` = avg2, `$n=90$` = avg3))
names(SR) = c("averages", "n")
ggplot(SR, aes(x = averages, y = ..density..)) +
    scale_x_continuous(limits = c(0, 3)) + facet_grid(n ~
    .) + geom_histogram()
```

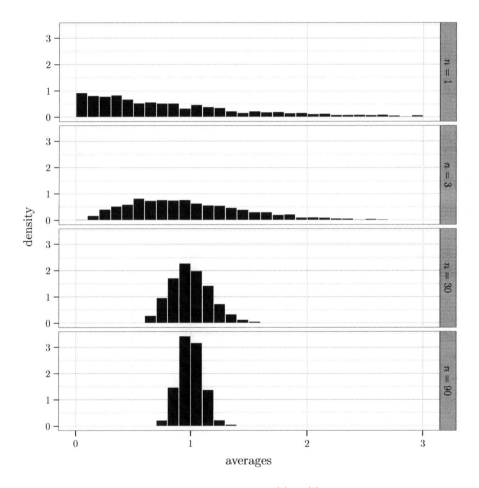

Example 8.9.2. *In a restaurant, orders $X^{(1)}, X^{(2)}, \ldots$ are received in an iid fashion with expectation $20 and standard deviation $10 (note that the distribution of the orders is unknown). Assuming that there are 100 orders a day, the expected daily revenue is $2000. The owner is interested in computing the probability that the daily revenue is less than or equal to half the expected daily revenue*

i.e., $1000. Note that we cannot compute the probability precisely since we don't know the precise distribution of the orders, only the expectation and variance. To approximate that quantity using the CLT, we need to transform it to a form that enables us to invoke the CLT

$$\mathsf{P}\left(\sum_{i=1}^{100} X^{(i)} \le 1000\right) = \mathsf{P}\left(10\left(\frac{\sum_{i=1}^{100} X^{(i)}}{100} - 20\right) \le 10\frac{1000}{100} - 200\right)$$

$$\approx \int_{-\infty}^{-100} \frac{1}{\sqrt{2\pi \cdot 100}} e^{-x^2/(2 \cdot 100)} \, dx \approx 7.6e^{-24}.$$

8.10 Continuous Mapping Theorem

A well known property of continuous functions is that they preserve limits. In other words, if f is continuous and $a_n \to a$ then $f(a_n) \to f(a)$. The following proposition shows that a similar result applies to the three different modes of stochastic convergence.

Proposition 8.10.1 (Continuous Mapping Theorem). *Let $f : \mathbb{R}^d \to \mathbb{R}^m$ be a function for which $\mathsf{P}(f(X) \text{ is continuous}) = 1$. Then,*

$$X^{(n)} \xrightarrow{\text{P}} X \qquad \text{implies} \qquad f(X^{(n)}) \xrightarrow{\text{P}} f(X)$$
$$X^{(n)} \rightsquigarrow X \qquad \text{implies} \qquad f(X^{(n)}) \rightsquigarrow f(X)$$
$$X^{(n)} \xrightarrow{\text{as}} X \qquad \text{implies} \qquad f(X^{(n)}) \xrightarrow{\text{as}} f(X).$$

Proof. We prove the first statement using Proposition 8.2.4. It is sufficient to show that for every sequence $n_1, n_2, \ldots$ we have a subsequence $m_1, m_2, \ldots$ along which $f(X^{(m_i)}) \xrightarrow{\text{P}} f(X)$. A second use of Proposition 8.2.4, shows that we can find a subsequence $m_1, m_2, \ldots$ of $n_1, n_2, \ldots$ along which $X^{(n)}$ converges to X with probability 1. Since continuous functions preserve limits this implies that $f(X^{(n)})$ converges to $f(X)$ along that subsequence with probability 1, and the first statement follows.

We prove the second statement using the portmanteau theorem. It is sufficient to show that for a bounded and continuous function h, we have $\mathsf{E}(h(f(X^{(n)}))) \to \mathsf{E}(h(f(X)))$. Since f, h are continuous with probability 1 and h is bounded, the function $g = h \circ f$ is also continuous with probability 1 and bounded. It follows from the portmanteau theorem that $\mathsf{E}(g(X^{(n)})) \to \mathsf{E}(g(X))$, proving the second statement.

To prove the third statement, note that we have with probability 1 a continuous function of a convergent sequence. Using the fact that continuous functions preserve limits, we have convergence to the required limit with probability 1. ∎

Example 8.10.1. *If $X^{(n)} \xrightarrow{\text{P}} X$ and $Y^{(n)} \xrightarrow{\text{P}} Y$, then*

1. $X^{(n)} + Y^{(n)} \xrightarrow{\text{P}} X + Y$,

2. $X^{(n)}Y_1^{(n)} \xrightarrow{P} XY_1$,

3. $X^{(n)}/Y_1^{(n)} \xrightarrow{P} X/Y_1$ provided that the denominators are not zero.

8.11   Slustky's Theorems

Proposition 8.11.1 (Slutsky's Theorem).

$$X^{(n)} \rightsquigarrow X \quad and \quad (X^{(n)} - Y^{(n)}) \xrightarrow{P} 0 \quad implies \quad Y^{(n)} \rightsquigarrow X,$$

$$X^{(n)} \xrightarrow{P} X \quad and \quad (X^{(n)} - Y^{(n)}) \xrightarrow{P} 0 \quad implies \quad Y^{(n)} \xrightarrow{P} X,$$

$$X^{(n)} \xrightarrow{as} X \quad and \quad (X^{(n)} - Y^{(n)}) \xrightarrow{as} 0 \quad implies \quad Y^{(n)} \xrightarrow{as} X.$$

Proof. To prove the first statement, it is sufficient to show that for an arbitrary continuous function h that is zero outside a closed and bounded set, $E(h(Y^{(n)})) \to E(h(X))$ (using the portmanteau theorem). Since a continuous function on a closed and bounded set in $\mathbb{R}^d$ (compact) is uniformly continuous, for all $\epsilon > 0$ we can find $\delta > 0$ such that $\|x - y\| < \delta$ implies $\|h(x) - h(y)\| < \epsilon$. Also, as a continuous function on a closed and bounded set in $\mathbb{R}^d$, h is also bounded, say by M, and therefore

$$
\begin{aligned}
|E(h(Y^{(n)})) - E(h(X))| &\leq |E(h(Y^{(n)})) - E(h(X^{(n)}))| + |E(h(X^{(n)})) - E(h(X))| \\
&\leq |E(h(Y^{(n)})) - E(h(X^{(n)}))|I(\|X^{(n)} - Y^{(n)}\| \leq \delta) \\
&\quad + |E(h(Y^{(n)})) - E(h(X^{(n)}))|I(\|X^{(n)} - Y^{(n)}\| > \delta) \\
&\quad + |E(h(X^{(n)})) - E(h(X))| \\
&\leq \epsilon + 2M\,P(\|X^{(n)} - Y^{(n)}\| > \delta) + |E(h(X^{(n)})) - E(h(X))|.
\end{aligned}
$$

The second term in the equation above converges to 0 as $n \to \infty$ since $(X^{(n)} - Y^{(n)}) \xrightarrow{P} 0$. The third term also converges to zero since $X^{(n)} \rightsquigarrow x$. Since ϵ was arbitrarily chosen, it follows that $|E(h(Y^{(n)})) - E(h(X))|$, or $E(h(Y^{(n)})) \to E(h(X))$.

To prove the second statement, we note that if $\|Y^{(n)} - X\| > \epsilon$ occurs, then by the triangle inequality $\|Y^{(n)} - X^{(n)}\| + \|X^{(n)} - X\| > \epsilon$ occurs, implying that

$$\{\|Y^{(n)} - X\| > \epsilon\} \subset \{\|Y^{(n)} - X^{(n)}\| > \epsilon/2\} \cup \{\|X^{(n)} - X\| > \epsilon/2\}.$$

This implies that

$$P(\|Y^{(n)} - X\| > \epsilon) \leq P(\|Y^{(n)} - X^{(n)}\| > \epsilon/2) + P(\|X^{(n)} - X\| > \epsilon/2)$$

which converges to 0 as $n \to \infty$ since $X^{(n)} \xrightarrow{P} x$ and $(X^{(n)} - Y^{(n)}) \xrightarrow{P} 0$.

The third statement follows from arithmetic of deterministic limits, which apply since we have convergence with probability 1. ∎

Corollary 8.11.1.

$$\boldsymbol{X}^{(n)} \rightsquigarrow \boldsymbol{x} \quad and \quad \boldsymbol{Y}^{(n)} \rightsquigarrow \boldsymbol{c} \quad implies \quad (\boldsymbol{X}^{(n)}, \boldsymbol{Y}^{(n)}) \rightsquigarrow (\boldsymbol{X}, \boldsymbol{c}),$$

$$\boldsymbol{X}^{(n)} \xrightarrow{\text{P}} \boldsymbol{x} \quad and \quad \boldsymbol{Y}^{(n)} \xrightarrow{\text{P}} \boldsymbol{c} \quad implies \quad (\boldsymbol{X}^{(n)}, \boldsymbol{Y}^{(n)}) \xrightarrow{\text{P}} (\boldsymbol{X}, \boldsymbol{c}),$$

$$\boldsymbol{X}^{(n)} \xrightarrow{\text{as}} \boldsymbol{x} \quad and \quad \boldsymbol{Y}^{(n)} \xrightarrow{\text{as}} \boldsymbol{c} \quad implies \quad (\boldsymbol{X}^{(n)}, \boldsymbol{Y}^{(n)}) \xrightarrow{\text{as}} (\boldsymbol{X}, \boldsymbol{c}).$$

Proof. Since $\boldsymbol{Y}^{(n)} \rightsquigarrow \boldsymbol{c}$, we also have $\boldsymbol{Y}^{(n)} \xrightarrow{\text{P}} \boldsymbol{c}$, implying that

$$\mathsf{P}(\|(\boldsymbol{X}^{(n)}, \boldsymbol{Y}^{(n)}) - (\boldsymbol{X}^{(n)}, \boldsymbol{c})\| > \epsilon) = \mathsf{P}(\|\boldsymbol{Y}^{(n)} - \boldsymbol{c}\| > \epsilon) \to 0.$$

Let h be an arbitrary continuous and bounded function. The statement $\mathsf{E}(h(\boldsymbol{X}^{(n)}, \boldsymbol{c})) \to \mathsf{E}(h(\boldsymbol{X}, \boldsymbol{c}))$ follows from the fact that $\boldsymbol{X}^{(n)} \rightsquigarrow \boldsymbol{X}$ and the portmanteau theorem. Combining this with Slutsky's theorem shows that $(\boldsymbol{X}^{(n)}, \boldsymbol{Y}^{(n)}) \rightsquigarrow (\boldsymbol{X}, \boldsymbol{c})$, which proves the first statement.

To prove the second statement, we note that

$$\mathsf{P}(\|(\boldsymbol{X}^{(n)}, \boldsymbol{Y}^{(n)}) - (\boldsymbol{X}, \boldsymbol{Y})\| > \epsilon) \leq \mathsf{P}(\|\boldsymbol{X}^{(n)} - \boldsymbol{X}\| > \epsilon/\sqrt{2})$$
$$+ \mathsf{P}(\|\boldsymbol{Y}^{(n)} - \boldsymbol{Y}\| > \epsilon/\sqrt{2}),$$

and that the two terms in the right hand side above converge to 0 as $n \to \infty$.

The third statement follows from arithmetic of deterministic limits, which apply since we have convergence with probability 1. ∎

Corollary 8.11.2. *If $\boldsymbol{f}$ is a continuous function, then*

$$\boldsymbol{X}^{(n)} \rightsquigarrow \boldsymbol{X} \quad and \quad \boldsymbol{Y}^{(n)} \rightsquigarrow \boldsymbol{c} \quad imply \quad \boldsymbol{f}(\boldsymbol{X}^{(n)}, \boldsymbol{Y}^{(n)}) \rightsquigarrow \boldsymbol{f}(\boldsymbol{X}^{(n)}, \boldsymbol{c}).$$

Proof. This follows from previous corollary and the continuous mapping theorem. ∎

Example 8.11.1. *If $X^{(n)} \rightsquigarrow X$ and $Y^{(n)} \xrightarrow{\text{P}} c$ for some constant c, then*

1. $X^{(n)} Y^{(n)} \rightsquigarrow Xc$

2. $X^{(n)} + Y^{(n)} \rightsquigarrow X + c$

3. $X^{(n)} / Y^{(n)} \rightsquigarrow X/c$,

where we assume in the last expression that $c \neq 0$.

8.12 Notes

More information on the topics in this chapter is available in [17, 10, 5, 1, 33, 25]. Our description follows most closely [5] and the first chapter of [19]. Applications of limit theorems in statistics are described in [40, 19, 47, 13] and applications in information theory are described in [41, 12].

The theory described in this chapter has been complemented by additional important results in limit theory. The uniform strong law of large numbers [19] extends the strong law of large numbers to apply uniformly over a set of parameters. Berry-Essen theory explore the accuracy of the Gaussian approximation in the central limit theorem [4]. Multiple extensions of the iid central limit theorem are available, for example in the research papers [22, 3, 29, 34] or the monograph [15]. A converse to Shceffe's theorem is described in [9, 44]. The law of the iterated logarithm complements the law of large numbers and the central limit theorem in describing the behavior of a sum of iid RVs [5].

8.13 Exercises

1. Let $X^{(n)} \xrightarrow{P} X$ and $Y^{(n)} \xrightarrow{P} Y$. Prove that $X^{(n)} - Y^{(n)} \xrightarrow{P} X - Y$.

2. Comment on whether the WLLN implies the CLT and on whether the CLT implies the WLLN. Motivate your answer.

Appendix:
Mathematical Prerequisites

Appendix A

Set Theory

This chapter describes set theory, a mathematical theory that underlies all of modern mathematics.

A.1   Basic Definitions

Definition A.1.1. A set is an unordered collection of elements.

Sets may be described by listing their elements between curly braces, for example $\{1, 2, 3\}$ is the set containing the elements 1, 2, and 3. Alternatively, we an describe a set by specifying a certain condition whose elements satisfy, for example $\{x : x^2 = 1\}$ is the set containing the elements 1 and -1 (assuming x is a real number).

We make the following observations.

- There is no importance to the order in which the elements of a set appear. Thus $\{1, 2, 3\}$, is the same set as $\{3, 2, 1\}$.

- An element may either appear in a set or not, but it may not appear more than one time.

- Sets are typically denoted by an uppercase letter, for example A or B.

- It is possible that the elements of a set are sets themselves, for example $\{1, 2, \{3, 4\}\}$ is a set containing three elements (two scalars and one set). We typically denote such sets with calligraphic notation, for example $\mathcal{U}$.

Definition A.1.2. If a is an element in a set A, we write $a \in A$. If a is not an element of A, we write $a \notin A$. The empty set, denoted by $\emptyset$ or $\{\}$, does not contain any element.

Definition A.1.3. A set A with a finite number of elements is called a finite set and its size (number of elements) is denoted by $|A|$. A set with an infinite number of elements is called an infinite set.

Definition A.1.4. We denote $A \subset B$ if all elements in A are also in B. We denote $A = B$ if $A \subset B$ and $B \subset A$, implying that the two sets are identical. The difference between two sets $A \setminus B$ is the set of elements in A but not in B. The complement of a set A with respect to a set Ω is $A^c = \Omega \setminus A$ (we may omit the set Ω if it is obvious from context). The symmetric difference between two sets A, B is

$$A \triangle B = \{x : x \in A \setminus B \text{ or } x \in B \setminus A\}.$$

Example A.1.1. *We have $\{1, 2, 3\} \setminus \{3, 4\} = \{1, 2\}$ and $\{1, 2, 3\} \triangle \{3, 4\} = \{1, 2, 4\}$. Assuming $\Omega = \{1, 2, 3, 4, 5\}$, we have $\{1, 2, 3\}^c = \{4, 5\}$.*

In many cases we consider multiple sets indexed by a finite or infinite set. For example $U_\alpha, \alpha \in A$ represents multiple sets, one set for each element of A.

Example A.1.2. *Below are three examples of multiple sets, $U_\alpha, \alpha \in A$. The first example shows two sets: $\{1\}$ and $\{2\}$. The second example shows multiple sets: $\{1, -1\}$, $\{2, -2\}$, $\{3, -3\}$, and so on. The third example shows multiple sets, each containing all real numbers between two consecutive natural numbers.*

$$U_i = \{i\}, \qquad\qquad i \in A = \{1, 2\},$$
$$U_i = \{i, -i\}, \qquad\qquad i \in A = \mathbb{N} = \{1, 2, 3, \ldots\},$$
$$U_\alpha = \{\alpha + r : 0 \le r \le 1\}, \qquad\qquad \alpha \in A = \mathbb{N} = \{1, 2, 3, \ldots\}.$$

Definition A.1.5. For multiple sets $U_\alpha, \alpha \in A$ we define the union and intersection operations as follows:

$$\bigcup_{\alpha \in A} U_\alpha = \{u : u \in U_\alpha \text{ for one or more } \alpha \in A\}$$

$$\bigcap_{\alpha \in A} U_\alpha = \{u : u \in U_\alpha \text{ for all } \alpha \in A\}.$$

Figure A.1 illustrates these concepts in the case of two sets A, B with a non-empty intersection.

Definition A.1.6. The sets $U_\alpha, \alpha \in A$ are disjoint or mutually disjoint if $\cap_{\alpha \in A} U_\alpha = \emptyset$ and are pairwise disjoint if $\alpha \neq \beta$ implies $U_\alpha \cap U_\beta = \emptyset$. A union of pairwise disjoint sets $U_\alpha, \alpha \in A$ is denoted by $\uplus_{\alpha \in A} U_\alpha$.

Example A.1.3.

$$\{a, b, c\} \cap \{c, d, e\} = \{c\}$$
$$\{a, b, c\} \cup \{c, d, e\} = \{a, b, c, d, e\}$$
$$\{a, b, c\} \setminus \{c, d, e\} = \{a, b\}.$$

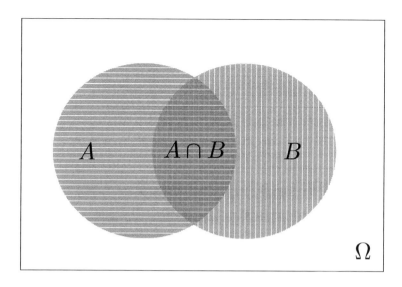

Figure A.1: Two circular sets A, B, their intersection $A \cap B$ (gray area with horizontal and vertical lines), and their union $A \cup B$ (gray area with either horizontal or vertical lines or both). The set $\Omega \setminus (A \cup B) = (A \cup B)^c = A^c \cap B^c$ is represented by white color.

Example A.1.4. *If $A_1 = \{1\}, A_2 = \{1, 2\}, A_3 = \{1, 2, 3\}$ we have*

$$\{A_i : i \in \{1, 2, 3\}\} = \{\{1\}, \{1, 2\}, \{1, 2, 3\}\}$$

$$\bigcup_{i \in \{1,2,3\}} A_i = \{1, 2, 3\}$$

$$\bigcap_{i \in \{1,2,3\}} A_i = \{1\}.$$

The properties below are direct consequences of the definitions above.

Proposition A.1.1. *For all sets $A, B, C \subset \Omega$,*

1. *Union and intersection are commutative and distributive:*

$$A \cup B = B \cup A, \qquad (A \cup B) \cup C = A \cup (B \cup C)$$
$$A \cap B = B \cap A, \qquad (A \cap B) \cap C = A \cap (B \cap C)$$

2. *$(A^c)^c = A$, $\qquad \emptyset^c = \Omega$, $\qquad \Omega^c = \emptyset$*

3. *$\emptyset \subset A$*

4. *$A \subset A$*

5. *$A \subset B$ and $B \subset C$ implies $A \subset C$*

6. $A \subset B$ if and only if $B^c \subset A^c$

7. $A \cup A = A = A \cap A$

8. $A \cup \Omega = \Omega, \qquad A \cap \Omega = A$

9. $A \cup \emptyset = A, \qquad A \cap \emptyset = \emptyset.$

Definition A.1.7. The power set of a set A is the set of all subsets of A, including the empty set $\emptyset$ and A. It is denoted by 2^A.

Proposition A.1.2. *If A is a finite set then*

$$|2^A| = 2^{|A|}.$$

Proof. We can describe each element of 2^A by a list of $|A|$ 0 or 1 digits (1 if the corresponding element is selected and 0 otherwise). The proposition follows since there are $2^{|A|}$ such lists (see Proposition 1.6.1). ∎

Example A.1.5.

$$2^{\{a,b\}} = \{\emptyset, \{a,b\}, \{a\}, \{b\}\}$$

$$|2^{\{a,b\}}| = 4$$

$$\sum_{A \in 2^{\{a,b\}}} |A| = 0 + 2 + 1 + 1 = 4.$$

The R package sets is convenient for illustrating basic concepts.

```
library(sets)
A = set("a", "b", "c")
2^A

## {{}, {"a"}, {"b"}, {"c"}, {"a", "b"},
##  {"a", "c"}, {"b", "c"}, {"a", "b",
##  "c"}}

A = set("a", "b", set("a", "b"))
2^A

## {{}, {"a"}, {"b"}, {{"a", "b"}}, {"a",
##  "b"}, {"a", {"a", "b"}}, {"b", {"a",
##  "b"}}, {"a", "b", {"a", "b"}}}

A = set(1, 2, 3, 4, 5, 6, 7, 8, 9, 10)
length(2^A)   # = 2^10

## [1] 1024
```

Proposition A.1.3 (Distributive Law of Sets).

$$\left(\bigcup_{\alpha \in Q} A_\alpha\right) \bigcap C = \bigcup_{\alpha \in Q} \left(A_\alpha \bigcap C\right)$$

$$\left(\bigcap_{\alpha \in Q} A_\alpha\right) \bigcup C = \bigcap_{\alpha \in Q} \left(A_\alpha \bigcup C\right).$$

Proof. We prove the first statement for the case of $|Q| = 2$. The proof of the second statement is similar, and the general case is a straightforward extension.

We prove the set equality $U = V$ by showing $U \subset V$ and $V \subset U$. Let x belong to the set $(A \cup B) \cap C$. This means that x is in C and also in either A or B, which implies $x \in (A \cap C) \cup (B \cap C)$. On the other hand, if x belongs to $(A \cap C) \cup (B \cap C)$, x is in $A \cap C$ or in $B \cap C$. Therefore $x \in C$ and also x is in either A or B, implying that $x \in (A \cup B) \cap C$. ∎

Proposition A.1.4.

$$S \setminus \bigcup_{\alpha \in Q} A_\alpha = \bigcap_{\alpha \in Q} (S \setminus A_\alpha)$$

$$S \setminus \bigcap_{\alpha \in Q} A_\alpha = \bigcup_{\alpha \in Q} (S \setminus A_\alpha).$$

Proof. We prove the first result. The proof of the second result is similar.

Let $x \in S \setminus \bigcup_{\alpha \in Q} A_\alpha$. This implies that x is in S but not in any of the A_α sets, which implies $x \in S \setminus A_\alpha$ for all α, and therefore $x \in \bigcap_{\alpha \in Q}(S \setminus A_\alpha)$. On the other hand, if $x \in \bigcap_{\alpha \in Q}(S \setminus A_\alpha)$, then x is in each of the $S \setminus A_\alpha$ sets. Therefore x is in S but not in any of the A_α sets, hence $x \in S \setminus \bigcup_{\alpha \in Q} A_\alpha$. ∎

Corollary A.1.1 (De-Morgan's Law).

$$\left(\bigcup_{\alpha \in Q} A_\alpha\right)^c = \bigcap_{\alpha \in Q} A_\alpha^c$$

$$\left(\bigcap_{\alpha \in Q} A_\alpha\right)^c = \bigcup_{\alpha \in Q} A_\alpha^c.$$

Proof. This is a direct corollary of Proposition A.1.4 when $S = \Omega$. ∎

Definition A.1.8. We define the sets of all natural numbers, integers, and rational numbers as follows:

$$\mathbb{N} = \{1, 2, 3, \ldots\}$$
$$\mathbb{Z} = \{\ldots, -1, 0, 1, \ldots\}$$
$$\mathbb{Q} = \{p/q : p \in \mathbb{Z}, q \in \mathbb{Z} \setminus \{0\}\}.$$

The set of real numbers[1] is denoted by $\mathbb{R}$.

Definition A.1.9. We denote the closed interval, the open interval and the two types of half-open intervals between $a, b \in \mathbb{R}$ as

$$[a, b] = \{x \in \mathbb{R} : a \le x \le b\}$$
$$(a, b) = \{x \in \mathbb{R} : a < x < b\}$$
$$[a, b) = \{x \in \mathbb{R} : a \le x < b\}$$
$$(a, b] = \{x \in \mathbb{R} : a < x \le b\}.$$

Example A.1.6.

$$\bigcup_{n \in \mathbb{N}} [0, n/(n+1)) = [0, 1),$$

$$\bigcap_{n \in \mathbb{N}} [0, n/(n+1)) = [0, 1/2).$$

Definition A.1.10. The Cartesian product of two sets A and B is

$$A \times B = \{(a, b) : a \in A, b \in B\}.$$

In a similar way we define the Cartesian product of $n \in \mathbb{N}$ sets. The repeated Cartesian product of the same set, denoted as

$$A^d = A \times \cdots \times A, \qquad d \in \mathbb{N},$$

is the set of d-tuples or d-dimensional vectors whose components are elements in A.

Example A.1.7. *$\mathbb{R}^d$ is the set of all d dimensional vectors whose components are real numbers.*

Definition A.1.11. A relation R on a set A is a subset of $A \times A$, or in other words a set of pairs of elements in A. If $(a, b) \in R$ we denote $a \sim b$ and if $(a, b) \notin R$ we denote $a \nsim b$. A relation is reflexive if $a \sim a$ for all $a \in A$. It is symmetric if $a \sim b$ implies $b \sim a$ for all $a, b \in A$. It is transitive if $a \sim b$ and $b \sim c$ implies $a \sim c$ for all $a, b, c \in A$. An equivalence relation is a relation that is reflexive, symmetric, and transitive.

Example A.1.8. *Consider relation 1 where $a \sim b$ if $a \le b$ over $\mathbb{R}$ and relation 2 where $a \sim b$ if $a = b$ over $\mathbb{Z}$. Relation 1 is reflexive and transitive but not symmetric. Relation 2 is reflexive, symmetric, and transitive and is therefore an equivalence relation.*

---

[1]One way to rigorously define the set of real numbers is as the completion of the rational numbers. The details may be found in standard real analysis textbooks, for example [37]. We do not pursue this formal definition here.

Definition A.1.12. The sets $U_\alpha, \alpha \in A$ form a partition of U if

$$\biguplus_{\alpha \in A} U_\alpha = U.$$

(see Definition A.1.6.) In other words, the union of the pairwise disjoint sets $U_\alpha, \alpha \in A$ is U. The sets U_α are called equivalence classes.

An equivalence relation $\sim$ on A induces a partition of A as follows: $a \sim b$ if and only if a and b are in the same equivalence class.

Example A.1.9. *Consider the set A of all cities and the relation $a \sim b$ if the cities a, b are in the same country. This relation is reflexive, symmetric, and transitive, and therefore is an equivalence relation. This equivalence relation induces a partition of all cities into equivalence classes consisting of all cities in the same country. The number of equivalence classes is the number of countries.*

A.2 Functions

Definition A.2.1. Let A, B be two sets. A function $f : A \to B$ assigns one element $b \in B$ for every element $a \in A$, denoted by $b = f(a)$.

Definition A.2.2. For a function $f : A \to B$, we define

$$\text{range } f = \{f(a) : a \in A\} \tag{A.1}$$

$$f^{-1}(b) = \{a \in A : f(a) = b\} \tag{A.2}$$

$$f^{-1}(H) = \{a \in A : f(a) \in H\}. \tag{A.3}$$

If range $f = B$, we say that f is onto. If for all $b \in B$, $|f^{-1}(b)| \le 1$ we say that f is 1-1 or one-to-one. A function that is both onto and 1-1 is called a bijection.

If $f : A \to B$ is one-to-one, f^{-1} is also a function $f^{-1} : B' \to A$ where $B' = \text{range } f \subset B$. If $f : A \to B$ is a bijection then $f^{-1} : B \to A$ is also a bijection.

Definition A.2.3. Let $f : A \to B$ and $f : B \to C$ be two functions. Their function composition is a function $f \circ g : A \to C$ defined as $(f \circ g)(x) = f(g(x))$.

Proposition A.2.1. *For any function f and any indexed collection of sets $U_\alpha, \alpha \in A$,*

$$f^{-1}\left(\bigcap_{\alpha \in A} U_\alpha\right) = \bigcap_{\alpha \in A} f^{-1}(U_\alpha)$$

$$f^{-1}\left(\bigcup_{\alpha \in A} U_\alpha\right) = \bigcup_{\alpha \in A} f^{-1}(U_\alpha)$$

$$f\left(\bigcup_{\alpha \in A} U_\alpha\right) = \bigcup_{\alpha \in A} f(U_\alpha).$$

Proof. We prove the result above for a union and intersection of two sets. The proof of the general case of a collection of sets is similar.

If $x \in f^{-1}(U \cap V)$ then $f(x)$ is in both U and V and therefore $x \in f^{-1}(U) \cap f^{-1}(V)$. If $x \in f^{-1}(U) \cap f^{-1}(V)$ then $f(x) \in U$ and $f(x) \in V$ and therefore $f(x) \in U \cap V$, implying that $x \in f^{-1}(U \cap V)$.

If $x \in f^{-1}(U \cup V)$ then $f(x) \in U \cup V$, which implies $f(x) \in U$ or $f(x) \in V$ and therefore $x \in f^{-1}(U) \cup f^{-1}(V)$. On the other hand, if $x \in f^{-1}(U) \cup f^{-1}(V)$ then $f(x) \in U$ or $f(x) \in V$ implying $f(x) \in U \cup V$ and $x \in f^{-1}(U \cup V)$.

If $y \in f(U \cup V)$ then $y = f(x)$ for some x in either U or V and $y = f(x) \in f(U) \cup f(V)$. On the other hand, if $y \in f(U) \cup f(V)$ then $y = f(x)$ for some x that belongs to either U or V, implying $y = f(x) \in f(U \cup V)$. ∎

Interestingly, the statement $f(U \cap V) = f(U) \cap f(V)$ is *not* true in general.

Example A.2.1. *For $f : \mathbb{R} \to \mathbb{R}$, $f(x) = x^2$, we have $f^{-1}(x) = \{\sqrt{x}, -\sqrt{x}\}$ implying that f is not one-to-one. We also have $f^{-1}([1,2]) = [-\sqrt{2}, 1] \cup [1, \sqrt{2}]$.*

Definition A.2.4. Given two functions $f, g : A \to \mathbb{R}$ we denote $f \leq g$ if $f(x) \leq g(x)$ for all $x \in A$, $f < g$ if $f(x) < g(x)$ for all $x \in A$, and $f \equiv g$ if $f(x) = g(x)$ for all $x \in A$.

Definition A.2.5. We denote a sequence of functions $f_n : A \to \mathbb{R}$, $n \in \mathbb{N}$ satisfying $f_1 \leq f_2 \leq f_3 \leq \cdots$ as $f_n \nearrow$.

Definition A.2.6. Given a set $A \subset \Omega$ we define the indicator function $I_A : \Omega \to \mathbb{R}$ as

$$I_A(\omega) = \begin{cases} 1 & \omega \in A \\ 0 & \text{otherwise} \end{cases}, \qquad \omega \in \Omega.$$

Example A.2.2. *For any set A, we have $I_A = 1 - I_{A^c}$.*

Example A.2.3. *If $A \subset B$, we have $I_A \leq I_B$.*

A.3   Cardinality

The most obvious generalization of the size of a set A to infinite sets (see Definition A.1.3) implies the obvious statement that infinite sets have infinite size. A more useful generalization can be made by noticing that two finite sets A, B have the same size if and only if there exists a bijection between them, and generalizing this notion to infinite sets.

Definition A.3.1. Two sets A and B (finite or infinite) are said to have the same cardinality, denoted by $A \sim B$ if there exists a bijection between them.

Note that the cardinality concept above defines a relation that is reflexive ($A \sim A$) symmetric ($A \sim B \Rightarrow B \sim A$), and transitive ($A \sim B$ and $B \sim C$ implies $A \sim C$) and is thus an equivalence relation. The cardinality relation thus partitions the set of all sets to equivalence classes containing sets with the same cardinality. For each natural number $k \in \mathbb{N}$, we have an equivalence class containing all finite sets of that size. But there are also other equivalence classes containing infinite sets, the most important one being the equivalence class that contains the natural numbers $\mathbb{N}$.

Definition A.3.2. Let A be an infinite set. If $A \sim \mathbb{N}$ then A is a countably infinite set. If $A \not\sim \mathbb{N}$ then A is a uncountably infinite set.

Proposition A.3.1. *Every infinite subset E of a countably infinite set A is countably infinite.*

Proof. A is countably infinite, so we can construct an infinite sequence $x_1, x_2, \ldots$ containing the elements of A. We can construct another sequence, $y_1, y_2, \ldots$, that is obtained by omitting from the first sequence the elements in $A \setminus E$. The sequence $y_1, y_2, \ldots$ corresponds to a bijection between the natural numbers and E. ∎

Proposition A.3.2. *A countable union of countably infinite sets is countably infinite.*

Proof. Let $A_n, n \in \mathbb{N}$ be a collection of countably infinite sets. We can arrange the elements of each A_n as a sequence that forms the n-row of a table with infinite rows and columns. We refer to the element at the i-row and j-column in that table as A_{ij}. Traversing the table in the following order: $A_{11}, A_{21}, A_{12}, A_{31}, A_{22}, A_{13}$, and so on (traversing south-west to north-east diagonals of the table starting at the north-west corner), we express the elements of $A = \cup_{n \in \mathbb{N}} A_n$ as an infinite sequence. That sequence forms a bijection from $\mathbb{N}$ to A. ∎

Corollary A.3.1. *If A is countably infinite then so is A^d, for $d \in \mathbb{N}$.*

Proof. If A is countably infinite, then $A \times A$ corresponds to one copy of A for each element of A, and thus we have a bijection between A^2 and a countably infinite union of countably infinite sets. The previous proposition implies that A^2 is countably infinite. The general case follows by induction. ∎

It can be shown that countably infinite sets are the "smallest" sets (in terms of the above definition of cardinality) among all infinite sets. In other words, if A is uncountably infinite, then there exists an onto function $f : A \to \mathbb{N}$ but no onto function $f : \mathbb{N} \to A$.

Proposition A.3.3. *Assuming $a < b$ and $d \in \mathbb{N}$,*

$$\mathbb{N} \sim \mathbb{Z} \sim \mathbb{Q}$$
$$\mathbb{N} \not\sim [a, b]$$
$$\mathbb{N} \not\sim (a, b)$$
$$\mathbb{N} \not\sim \mathbb{R}$$
$$\mathbb{N} \not\sim \mathbb{R}^d.$$

We first define the concept of a binary expansion of a number, which will be used in the proof of the proposition below.

Definition A.3.3. The binary expansion of a number $r \in [0, 1]$ is defined as $0.b_1 b_2, b_3 \ldots$ where $b_n \in \{0, 1\}, n \in \mathbb{N}$ and

$$r = \sum_{n \in \mathbb{N}} b_n 2^{-n}.$$

Example A.3.1. *The binary expansions of $1/4$ is 0.01 and the binary expansion of $3/4$ is $0.11 = 1/2 + 1/4$.*

Proof. Obviously there exists a bijection between the natural numbers and the natural numbers. The mapping $f : \mathbb{N} \to \mathbb{Z}$ defined as

$$f(n) = \begin{cases} (n-1)/2 & n \text{ is odd} \\ -n/2 & n \text{ is even} \end{cases}$$

maps 1 to 0, 2 to -1, 3 to 1, 4 to -2, 5 to 2, etc., and forms a bijection between $\mathbb{N}$ and $\mathbb{Z}$.

A rational number is a ratio of a numerator and a denominator integers (note however that two pairs of numerators and denominators may yield the same rational number). We thus have that $\mathbb{Q}$ is a subset of $\mathbb{Z}^2$, which is countably infinite by Proposition A.3.2. Using Proposition A.3.1, we have that $\mathbb{Q}$ is countably infinite.

We next show that there does not exist a bijection between the natural numbers and the interval $[0, 1]$. If there was such a mapping f, we could arrange the numbers in $[0, 1]$ as a sequence $f(n), n \in \mathbb{N}$ and form a table A with infinite rows and columns where column n is the binary expansion of the real number $f(n) \in [0, 1]$.

We could then create a new real number whose binary expansion is $b_n, n \in \mathbb{N}$ with $b_n \neq A_{nn}$ for all $n \in \mathbb{N}$ by traversing the diagonal of the table and choosing the alternative digits. This new real number is in $[0, 1]$ since it has a binary expansion, and yet it is different from any of the columns of A. We have thus found a real number that is different from any other number[2] in the range of f, contradicting the fact that f is onto.

Since there is no onto function from the naturals to $[0, 1]$ there can be no onto function from the natural numbers to $\mathbb{R}$ or its Cartesian products $\mathbb{R}^d$.   ∎

We extend below the definition of a Cartesian product (Definition A.1.10) to an infinite number of sets.

Definition A.3.4. Let A, T be sets. The notation A^T denotes a Cartesian product of multiple copies of A, one copy for each element of the set T. In other words, A^T is the set of all functions from T to A. The notation A^∞ denotes $A^\mathbb{N}$, a product of a countably infinite copies of A.

---

[2]The setup described above is slightly simplified. A rigorous proof needs to resolve the fact that some numbers in $[0, 1]$ have two different binary expansions, for example, $0.0111111\ldots = 0.1$. See for example [37].

Example A.3.2. *The set $\mathbb{R}^\infty$ is the set of all infinite sequences over the real line $\mathbb{R}$*

$$\mathbb{R}^\infty = \{(a_1, a_2, a_2, \ldots,) : a_n \in \mathbb{R} \text{ for all } n \in \mathbb{N}\}$$

and the set $\{0,1\}^\infty$ is the set of all infinite binary sequences. The set $R^{[0,1]}$ is a Cartesian product of multiple copies of the real numbers — one copy for each element of the interval $[0,1]$, or in other words the set of all functions from $[0,1]$ to $\mathbb{R}$.

Example A.3.3. *The set $\{0,1\}^A$ is the set of all functions from A to $\{0,1\}$, each such function implying a selection of an arbitrary subset of A (the selected elements are mapped to 1 and the remaining elements are mapped to 0). A similar interpretation may given to sets of size 2 that are different than $\{0,1\}$. Recalling Definition A.1.7, we thus have if $|B| = 2$, then B^A corresponds to the power set 2^A, justifying its notation.*

A.4 Limits of Sets

Definition A.4.1. For a sequence of sets A_n, $n \in \mathbb{N}$, we define

$$\inf_{k \geq n} A_k = \bigcap_{k=n}^{\infty} A_k$$

$$\sup_{k \geq n} A_k = \bigcup_{k=n}^{\infty} A_k$$

$$\liminf_{n \to \infty} A_n = \bigcup_{n \in \mathbb{N}} \inf_{k \geq n} A_k = \bigcup_{n \in \mathbb{N}} \bigcap_{k=n}^{\infty} A_k$$

$$\limsup_{n \to \infty} A_n = \bigcap_{n \in \mathbb{N}} \sup_{k \geq n} A_k = \bigcap_{n \in \mathbb{N}} \bigcup_{k=n}^{\infty} A_k.$$

Applying De-Morgan's law (Proposition A.1.1) we have

$$\left(\liminf_{n \to \infty} A_n\right)^c = \limsup_{n \to \infty} A_n^c.$$

Definition A.4.2. If for a sequence of sets A_n, $n \in \mathbb{N}$, we have $\liminf_{n \to \infty} A_n = \limsup_{n \to \infty} A_n$, we define the limit of A_n, $n \in \mathbb{N}$ to be

$$\lim_{n \to \infty} A_n = \liminf_{n \to \infty} A_n = \limsup_{n \to \infty} A_n,$$

The notation $A_n \to A$ is equivalent to the notation $\lim_{n \to \infty} A_n = A$.

Example A.4.1. *For the sequence of sets $A_k = [0, k/(k+1))$ from Example A.1.6* *we have*

$$\inf_{k \geq n} A_k = [0, n/(n+1))$$

$$\sup_{k \geq n} A_k = [0, 1)$$

$$\limsup_{n \to \infty} A_n = [0, 1)$$

$$\liminf_{n \to \infty} A_n = [0, 1)$$

$$\lim A_n = [0, 1).$$

We have the following interpretation for the $\liminf$ and $\limsup$ limits.

Proposition A.4.1. *Let $A_n, n \in \mathbb{N}$ be a sequence of subsets of Ω. Then*

$$\limsup_{n \to \infty} A_n = \left\{ \omega \in \Omega : \sum_{n \in \mathbb{N}} I_{A_n}(\omega) = \infty \right\}$$

$$\liminf_{n \to \infty} A_n = \left\{ \omega \in \Omega : \sum_{n \in \mathbb{N}} I_{A_n^c}(\omega) < \infty \right\}.$$

In other words, $\limsup_{n \to \infty} A_n$ is the set of $\omega \in \Omega$ that appear infinitely often (abbreviated i.o.) in the sequence A_n, and $\liminf_{n \to \infty} A_n$ is the set of $\omega \in \Omega$ that always appear in the sequence A_n except for a finite number of times.

Proof. We prove the first part. The proof of the second part is similar. If $\omega \in \limsup_{n \to \infty} A_n$ then by definition for all n there exists a k_n such that $\omega \in A_{k_n}$. For that ω we have $\sum_{n \in \mathbb{N}} I_{A_n}(\omega) = \infty$. Conversely, if $\sum_{n \in \mathbb{N}} I_{A_n}(\omega) = \infty$, there exists a sequence $k_1, k_2, \ldots$ such that $\omega \in A_{k_n}$, implying that for all $n \in \mathbb{N}$, $\omega \in \cup_{i \geq n} A_i$. ∎

Corollary A.4.1.

$$\liminf_{n \to \infty} A_n \subset \limsup_{n \to \infty} A_n.$$

Definition A.4.3. A sequence of sets $A_n, n \in \mathbb{N}$ is monotonic non-decreasing if $A_1 \subset A_2 \subset A_3 \subset \cdots$ and monotonic non-increasing if $\cdots \subset A_3 \subset A_2 \subset A_1$. We denote this as $A_n \nearrow$ and $A_n \searrow$, respectively. If $\lim A_n = A$, we denote this as $A_n \nearrow A$ and $A_n \searrow A$, respectively.

Proposition A.4.2. *If $A_n \nearrow$ then $\lim_{n \to \infty} A_n = \cup_{n \in \mathbb{N}} A_n$ and if $A_n \searrow$ then $\lim_{n \to \infty} A_n = \cap_{n \in \mathbb{N}} A_n$.*

Proof. We prove the first statement. The proof of the second statement is similar. We need to show that if A_n is monotonic non-decreasing, then $\limsup A_n = \liminf A_n = \cup_n A_n$. Since $A_i \subset A_{i+1}$, we have $\cap_{k \geq n} A_k = A_n$, and

$$\liminf_{n \to \infty} A_n = \bigcup_{n \in \mathbb{N}} \bigcap_{k \geq n} A_k = \bigcup_{n \in \mathbb{N}} A_n$$

$$\limsup_{n \to \infty} A_n = \bigcap_{n \in \mathbb{N}} \bigcup_{k \geq n} A_k \subset \bigcup_{k \in \mathbb{N}} A_k = \liminf_{n \to \infty} A_n \subset \limsup_{n \to \infty} A_n.$$

■

The following corollary of the above proposition motivates the notations lim inf and lim sup.

Corollary A.4.2. *Since $B_n = \cup_{k \geq n} A_k$ and $C_n = \cap_{k \geq n} A_k$ are monotonic sequences*

$$\liminf_{n \to \infty} A_n = \lim_{n \to \infty} \inf_{k \geq n} A_n$$

$$\limsup_{n \to \infty} A_n = \lim_{n \to \infty} \sup_{k \geq n} A_n.$$

A.5   Notes

Most of the material in this section is standard material in set theory. More information may be found in any set theory textbook. A classic textbook is [20]. Limits of sets are usually described at the beginning of measure theory or probability theory textbooks, for example [5, 1, 33].

A.6   Exercises

1. Prove the assertions in Example A.1.1.

2. Prove the assertion in Example A.1.6.

3. Prove that $f(U \cap V) = f(U) \cap f(V)$ is not true in general.

4. Let $A_0 = \{a\}$ and define $A_k = 2^{A_{k-1}}$ for $k \in \mathbb{N}$. Write down the elements of the sets A_k for all $k = 1, 2, 3$.

5. Let A, B, C be three finite sets. Describe intuitively the sets $A^{(B^C)}$ and $(A^B)^C$. What are the sizes of these two sets?

6. Let A be a finite set and B be a countably infinite set. Are the sets A^∞ and B^∞ countably infinite or uncountably infinite?

7. Find a sequence of sets A_n, $n \in \mathbb{N}$ for which $\liminf A_n \neq \limsup A_n$.

8. Describe an equivalence relation with an uncountably infinite set of equivalence classes, each of which is a set of size 2.

Appendix B

Metric Spaces

This chapter describes basic metric topology and the Euclidean spaces. It is essential for developing limit theorems, differentiation, and integration.

B.1 Basic Definitions

Definition B.1.1. Let $\mathcal{X}$ be a set. A function $d : \mathcal{X} \times \mathcal{X} \to \mathbb{R}$ is called a distance function if and only if the following properties hold for all $x, y, z \in \mathcal{X}$

- non-negativity: $d(x, y) \geq 0$

- positivity: $d(x, y) = 0$ if and only if $x = y$

- symmetry: $d(x, y) = d(y, x)$

- triangle inequality: $d(x, z) \leq d(x, y) + d(y, z)$.

The pair $(\mathcal{X}, d)$ is called a metric space (we sometimes refer to $\mathcal{X}$ as a metric space if no confusion arises).

The definitions below assume a metric space $(\mathcal{X}, d)$. The complement A^c of a set $A \subset \mathcal{X}$ is taken with respect to the metric space: $A^c = \mathcal{X} \setminus A$.

Definition B.1.2. The open ball with center $x \in \mathcal{X}$ and radius $r > 0$ is

$$B_r(x) = \{y \in \mathcal{X} : d(x, y) < r\}.$$

Definition B.1.3. A set $A \subset \mathcal{X}$ is an open set if it is a union of open balls. A set A is a closed set if it is the complement of an open set. A set A is bounded if $A \subset B_r(x)$ for some $r > 0$ and x.

Proposition B.1.1. *If $G_\alpha, \alpha \in Q$ are open sets and $F_\alpha, \alpha \in Q$ are closed sets then*

1. $\cup_{\alpha \in Q} G_\alpha$ is open

2. $\cap_{\alpha \in Q} F_\alpha$ is closed.

If $G_1, \ldots, G_n$ are open sets and $F_1, \ldots, F_n$ are closed sets then

1. $\cap_{i=1}^n G_i$ is open

2. $\cup_{i=1}^n F_i$ is closed.

Note that in the first part of the proposition above there are arbitrarily many open and closed sets, and in the second part there are only a finite number.

Proof. By definition, a union of open sets is a union of open balls. Using De-Morgan's rule (Corollary A.1.1) and is therefore open. The complement of an intersection of closed sets is a union of open sets, which is an open set. This implies that an intersection of closed sets is closed.

To prove the second part, assume that $G_1, \ldots, G_n$ are open sets and pick an arbitrary point x in their intersection. Since open sets are union of open balls, the point x is in n different open balls with different centers and radii. We can find an open ball $B_{\rho_x}(x)$ centered at x that is a subset of the intersection of the n open balls, and is therefore a subset of $\cap_{i=1}^n G_i$. It follows that

$$\bigcup_{x \in \cap_{i=1}^n G_i} B_{\rho_x}(x) = \bigcap_{i=1}^n G_i,$$

showing that $\cap_{i=1}^n G_i$ is a union of open balls and therefore an open set. The complementary result follows from De-Morgan's rule, as in the first part of the proof. ∎

In the examples below, we use the metric space $(\mathbb{R}, d)$, where $d(x, y) = |x - y|$. Verifying that $d(x, y) = |x - y|$ is indeed a distance function on $\mathbb{R}$ is straightforward. In Section B.4 we describe this metric space and its high dimensional generalization in more detail.

Example B.1.1. *In $\mathbb{R}$, the open interval (a, b) is open (open ball), as is $(a, b) \cup (c, d)$ (union of two open balls). The set $(a, +\infty)$ is a union of infinitely many open balls*

$$(a, +\infty) = \bigcup_{b : b \geq a+1} B_1(b)$$

and is therefore also an open set. Similarly, the set $(-\infty, a)$ is open. The closed interval $[a, b]$ is a closed set since its complement $(-\infty, a) \cup (b, \infty)$ is a union of two open sets and therefore is open. The half open interval $(a, b]$ is neither open nor closed.

Example B.1.2. *The set $\mathbb{R}$ is open since it is a union of open balls $\cup_{n \in \mathbb{N}} B_n(0)$, and its complement $\emptyset$ is closed. On the other hand, $\emptyset$ is an intersection of two open sets, for example $\emptyset = (1, 2) \cap (3, 4)$, which implies that $\emptyset$ is open and its complement $\mathbb{R}$ is closed. We thus have that the sets $\mathbb{R}$ and $\emptyset$ are both open and closed.*

Definition B.1.4. A point x is an interior point of a set A if $B_r(x) \subset A$ for some $r > 0$. The set of interior points of A is denoted by int A. A point x is a boundary point of a set A if for every $r > 0$, $B_r(x)$ contains at least one point in A and one point not in A. The set of boundary points of A is denoted by ∂A. The closure of a set A, denoted by $\overline{A}$, is the union of its interior and its boundary:

$$\overline{A} = (\text{int } A) \cup (\partial A).$$

Proposition B.1.2. *Let A, F, G be an arbitrary set, a closed set, and an open set, respectively, in a metric space $(\mathcal{X}, d)$. Then*

$$int\, G = G.$$
$$int\, A \subset A \subset \overline{A}$$
$$F = \overline{F}.$$

Proof. We start by proving the first property. Let x be an arbitrary element of G. Since G is a union of open balls, there exists an open ball containing x. There exists an open ball $B_r(x)$ contained in that open ball, implying that x is an interior point. The second property is obvious. To prove the third property note that $x \in F^c$ is contained in an open ball $B_{r'}(y) \subset F^c$ and choosing r small enough we have $B_r(x) \subset B_{r'}(y) \subset F^c$. It follows that x is not a boundary point, implying that $\partial F \subset F$ and therefore $\overline{F} \subset F$. The inclusion $F \subset \overline{F}$ is obvious. ∎

Example B.1.3. *In $\mathbb{R}$, we have*

$$int\,[a, b) = (a, b)$$
$$\partial [a, b) = \{a, b\}$$
$$\overline{[a, b)} = [a, b].$$

Definition B.1.5. A metric space is second countable if there exists a countably infinite collection of open sets $\mathcal{G} = \{G_n, n \in \mathbb{N}\}$ such that every open set can be expressed as a union of elements in $\mathcal{G}$.

Definition B.1.6. A set A is compact if every open covering has a finite sub-covering. In other words, if $\mathcal{U}$ is a set of open sets whose union contains A, there exists a finite number of sets in $\mathcal{U}$ that also contain A.

B.2 Limits

Definition B.2.1. A sequence $x^{(n)}$, $n \in \mathbb{N}$ in a metric space $(\mathcal{X}, d)$ converges to a limit x, denoted by $x^{(n)} \to x$ or $\lim_{n \to \infty} x^{(n)} = x$, if

$$\forall \epsilon > 0, \quad \exists N \in \mathbb{N}, \quad n > N \quad \Rightarrow \quad d(x^{(n)}, x) < \epsilon.$$

In other words, for all $\epsilon > 0$, there exists a $N \in \mathbb{N}$ such that whenever $n > N$, we have $d(x^{(n)}, x) < \epsilon$. If a sequence does not converges to a limit, we say that the sequence diverges.

It is possible that a sequence $x^{(n)}$ converges to a limit x, but the limit will never be realized: $x^{(n)} \neq x$ for all $n \in \mathbb{N}$. However, as n increases the tail subsequences

$$A_n = \{x^{(k)} : k \geq n\}, \quad n \in \mathbb{N}$$

are contained in smaller and smaller neighborhoods of x. We sometimes write $\lim x^{(n)}$ rather than $\lim_{n\to\infty} x^{(n)}$ when there is no confusion. We use superscripts $x^{(n)}$ rather than subscripts x_n since the latter will be used to denote vector components later on in this chapter.

Unless stated otherwise, real numbers in this book are associated with the metric space $(\mathbb{R}, d)$ where $d(x, y) = |x - y|$. Thus, convergence of a sequence of real numbers $x^{(n)}$ to a limit implies that

$$\forall \epsilon > 0, \quad \exists N \in \mathbb{N}, \quad n > N \quad \Rightarrow \quad |x^{(n)} - x| < \epsilon.$$

Definition B.2.2. For a sequence $s^{(n)} \in \mathbb{R}, n \in \mathbb{N}$ we write $\lim s^{(n)} = +\infty$ if

$$\forall M \in \mathbb{R}, \quad \exists N \in \mathbb{N}, \quad n > N \Rightarrow s^{(n)} \geq M.$$

Similarly we write $\lim s^{(n)} = -\infty$ if

$$\forall M \in \mathbb{R}, \quad \exists N \in \mathbb{N}, \quad n > N \Rightarrow s^{(n)} \leq M.$$

Example B.2.1. *The sequence $1, 1, 1, \ldots$ converges to 1, but the sequence $1, 2, 1, 2, 1, \ldots$ does not converge to any limit.*

Example B.2.2. *The sequence $2^{-n}, n \in \mathbb{N}$ converges to 0 since the tail sequences of $1/(2^n)$ get arbitrarily close to 0. Formally, for all ϵ there exists N such that $n > N$ implies $|2^{-n} - 0| = 2^{-n} = (1/2)^n < \epsilon$. For example, we can take N be a natural number greater than $\log_{1/2}(\epsilon)$.*

Example B.2.3. *We have $\lim_{n\to\infty} n = \lim_{n\to\infty} n^2 = +\infty$ but $1, 0, 2, 0, 3, 0, \ldots$ does not converge to a finite limit nor does it converge to infinity.*

Definition B.2.3. A sequence of real numbers $x^{(n)} \in \mathbb{R}, n \in \mathbb{N}$ is monotonic increasing, denoted by $x^{(n)} \nearrow$, if $x^{(n)} \leq x^{(n+1)}$ for all $n \in \mathbb{N}$ and is monotonic decreasing, denoted by $x^{(n)} \searrow$, if $x^{(n)} \geq x^{(n+1)}$ for all $n \in \mathbb{N}$. A sequence is monotonic if it is either monotonic increasing or monotonic decreasing. We use the notations $x^{(n)} \nearrow x$ and $x^{(n)} \searrow x$ if $\lim x^{(n)} = x$ and $x^{(n)}$ is monotonic increasing or decreasing, respectively. Replacing the $\leq$ symbol above with $<$, we get the analogous concept of strictly monotonic sequences.

Definition B.2.4. An upper bound x of a set $A \subset \mathbb{R}$ satisfies $x \geq y$ for all $y \in A$. A lower bound x of a set $A \subset \mathbb{R}$ satisfies $x \leq y$ for all $y \in A$. The smallest upper bound of a set E is the value α such that if $\beta < \alpha$ then β is not an upper bound. This value is called the supremum of E and is denoted by $\sup E$. The largest lower bound of a set E is the value α such that if $\beta > \alpha$ then β is not a lower bound. This value is called the infimum of E and is denoted by $\inf E$.

Example B.2.4. *We have* $\sup[a,b] = \sup(a,b) = b$ *and* $\inf[a,b] = \inf(a,b) = a$.

An important property of the real numbers is that the supremum of a set $A \subset \mathbb{R}$ that is bounded from above always exist. Similarly, the infimum of a set $A \subset \mathbb{R}$ that is bounded from below always exist. A proof of this property is available in [37, Chapter 1]. Interestingly, the set of rational numbers $\mathbb{Q}$ does not have this property.

Proposition B.2.1. *A monotonic increasing sequence* $x^{(n)} \in \mathbb{R}$, $n \in \mathbb{N}$ *that is bounded from above converges to a finite limit. Similarly, a monotonic decreasing sequence* $x^{(n)} \in \mathbb{R}$, $n \in \mathbb{N}$ *that is bounded from below converges to a finite limit. Monotonic sequences that are not bounded also converge to a limit, which may be a finite number or* $+\infty$ *(if it is increasing) or* $-\infty$ *(if it is decreasing).*

Proof. We assume the sequence is monotonic increasing and bounded. The sequence of distances $y^{(n)} = |x^{(n)} - \sup\{x^{(n)} : n \in \mathbb{N}\}|$, $n \in \mathbb{N}$ is monotonically decreasing and for ϵ there exists N such that $n > n$ implies $y^{(n)} < \epsilon$. This means that $x^{(n)}$ converges to $\sup\{x^{(n)} : n \in \mathbb{N}\}$. The case of monotonic decreasing is similar. ∎

Definition B.2.5. A function $f : (a,b) \to \mathbb{R}$ is monotonic increasing if for all $c < d$, we have $f(c) \le f(d)$ and monotonic decreasing if for all $c < d$, we have $f(c) \ge f(d)$. If the inequalities hold without equality the function is strictly monotonic increasing or strictly monotonic decreasing.

The following definitions are analogous the the $\limsup$ and $\liminf$ definitions in Section A.4.

Definition B.2.6. For a sequence $s^{(n)} \in \mathbb{R}, n \in \mathbb{N}$ we define

$$\liminf_{n \to \infty} s^{(n)} = \lim_{n \to \infty} \inf\{s^{(k)} : k \ge n\}$$
$$\limsup_{n \to \infty} s^{(n)} = \lim_{n \to \infty} \sup\{s^{(k)} : k \ge n\}.$$

Proposition B.2.2. *The values* $\liminf_{n \to \infty} s^{(n)}$ *and* $\liminf_{n \to \infty} s^{(n)}$ *always exist, either as a finite number or as* $+\infty$ *or* $-\infty$.

Proof. The sequence $v^{(n)} = \inf\{s^{(k)} : k \ge n\}$ is monotonic increasing, which converges by Proposition B.2.1 either to a finite limit or to an infinite limit. The proof in the case of $\limsup s^{(n)}$ is analogous. ∎

Definition B.2.7. A subsequence of a sequence $x^{(n)}, n \in \mathbb{N}$ is a sequence obtained from $x^{(n)}, n \in \mathbb{N}$ by omitting some indices.

Definition B.2.8. If a sub-sequence of $x^{(n)}$ converges to α then α is called a sub-sequential limit of the sequence $x^{(n)}$.

Note that sub-sequential limits may exist even when a limit may not. For example, the sequence $1, 2, 1, 2, 1, \ldots$ in Example B.2.1 has no limit, but it has two sub-sequential limits: 1 and 2.

Proposition B.2.3. *For a sequence $s^{(n)} \in \mathbb{R}, n \in \mathbb{N}$ whose set of sub-sequential limits is E, we have*

$$\limsup_{n \to \infty} s^{(n)} = \sup E$$

$$\liminf_{n \to \infty} s^{(n)} = \inf E.$$

Proof. If $\liminf_{n \to \infty} s^{(n)} < \inf E$, then for all $\epsilon > 0$ there exists $N > 0$ such that $\inf\{s^{(k)} : k \geq N\} \leq \inf E - \epsilon$. This contradicts the definition of E as the set of sub-sequential limits. The case $\liminf_{n \to \infty} s^{(n)} > \inf E$ leads to a similar contradiction. The proof of the first statement is similar. ∎

Corollary B.2.1.

$$-\infty \leq \liminf s^{(n)} \leq \limsup s^{(n)} \leq +\infty.$$

Proposition B.2.4. *For $s^{(n)} \in \mathbb{R}, n \in \mathbb{N}$, we have $\lim s^{(n)} = s$ if and only if*

$$\liminf s^{(n)} = \limsup s^{(n)} = s.$$

Proof. If $\liminf s^{(n)} = \limsup s^{(n)} = \alpha$ then $v^{(n)} = \inf_{k \geq n} s^{(k)}$ and $w^{(n)} = \sup_{k \geq n} s^{(k)}$ converge to α, implying that for all $\epsilon > 0$ there exists $N > 0$ such $n > N$ implies $d(s^{(n)}, \alpha) < \epsilon$. This implies that $s^{(n)} \to \alpha$.

If $s^{(n)} \to \alpha$ then for all $\epsilon > 0$ there exists $N > 0$ such that $n > N$ implies $d(s^{(n)}, \alpha) < \epsilon$. It follows that $v^{(n)} = \inf_{k \geq n} s^{(k)}$ and $w^{(n)} = \sup_{k \geq n} s^{(k)}$ converge to α. ∎

Example B.2.5. *We have*

$$\limsup_{n \to \infty} 2^{-n} = \liminf_{n \to \infty} 2^{-n} = \lim_{n \to \infty} 2^{-n} = 0$$

$$\limsup_{n \to \infty} n = \liminf_{n \to \infty} n = \lim_{n \to \infty} n = +\infty,$$

and for $x^{(n)} = 1, 0, 2, 0, 3, 0, 4, 0, \ldots$ we have that $\lim_{n \to \infty} x^{(n)}$ does not exist and

$$0 = \liminf x^{(n)} < \limsup x^{(n)} = +\infty.$$

Definition B.2.9. Let $f : A \to B$ be a function from one metric space to another. The limit $\lim_{x \to y} f(x)$ is the value v for which

$$\forall \epsilon > 0, \quad \exists \delta > 0, \quad d(x, y) < \delta \quad \Rightarrow \quad d(f(x), v) < \epsilon.$$

When $B = \mathbb{R}$ we write $\lim_{x \to y} f(x) = +\infty$ if

$$\forall M \in \mathbb{R}, \quad \exists \delta > 0 \quad d(x, y) < \delta \quad \Rightarrow \quad f(x) > M$$

and $\lim_{x \to y} f(x) = -\infty$ if

$$\forall M \in \mathbb{R}, \quad \exists \delta > 0 \quad d(x, y) < \delta \quad \Rightarrow \quad f(x) < M.$$

Definition B.2.10. For a function $f : \mathbb{R} \to B$, where B is a metric space, we write $\lim_{x \to +\infty} f(x) = v$ if

$$\forall \epsilon > 0 \quad \exists N \in \mathbb{N}, \quad x > N \quad \Rightarrow \quad d(f(x), v) < \epsilon$$

and $\lim_{x \to -\infty} f(x) = v$ if

$$\forall \epsilon > 0 \quad \exists N \in \mathbb{N}, \quad x < N \quad \Rightarrow \quad d(f(x), v) < \epsilon.$$

There are two limit concepts defined above, one for sequences and one for functions. The following proposition connects these two concepts.

Proposition B.2.5. $\lim_{x \to y} f(x) = v$ *if and only if the following statement holds*

$$\lim_{n \to \infty} x^{(n)} = y \text{ and } \forall n \in \mathbb{N}, x^{(n)} \neq y \quad implies \quad \lim_{n \to \infty} f(x^{(n)}) = v. \quad (*)$$

Proof. We assume $\lim_{x \to y} f(x) = v$ and $\lim x^{(n)} = y$ and for all n, $x^{(n)} \neq y$, and show that $\lim_{n \to \infty} f(x^{(n)}) = v$. For all $\delta > 0$ there exists N such that for $n > N$, $d(x^{(n)}, y) < \delta$. Also, for every $\epsilon > 0$ there exists $\delta > 0$ such that if $d(x, y) < \delta$ then $d(f(x), v) < \epsilon$. Combining these two results shows that $\lim_{n \to \infty} f(x^{(n)}) = v$.

Conversely, we assume that $\lim_{x \to y} f(x) \neq v$ and show that Equation (*) does not hold. In this case $\exists \epsilon > 0$ such that for all $\delta > 0$ there exists a point x for which $d(x, y) < \delta$ and $d(f(x), v) > \epsilon$. Repeating this argument for $\delta_n = 1/n$, we find a sequence $x^{(n)}$ converging to y but for which $\lim f(x^{(n)}) \neq v$. ∎

One sided limits of $f : \mathbb{R} \to \mathbb{R}$ are similar to limits, except that the approach $x \to a$ is restricted to be from a single side. In some cases, the limit $\lim_{x \to a} f(x)$ does not exist but the one sided limits do.

Definition B.2.11. The right-sided limit of $f : \mathbb{R} \to \mathbb{R}$ is $L = \lim_{x \to a^+} f(x)$ if

$$\forall \epsilon > 0 \quad \exists \delta > 0, \quad a < x < a + \delta \quad \Rightarrow \quad |f(x) - L| < \epsilon.$$

The left-sided limit is $L = \lim_{x \to a^-} f(x)$ if

$$\forall \epsilon > 0 \quad \exists \delta > 0, \quad a - \delta < x < a \quad \Rightarrow \quad |f(x) - L| < \epsilon.$$

Definition B.2.12. Let $f_n : A \to B, n \in \mathbb{N}$ be a sequence of functions for which $f_n(x) \to f(x)$ for all $x \in A$. We then say that f_n converge pointwise to f, and denote it by $f_n \to f$. We denote $f_n \nearrow f$ if in addition to $f_n \to f$ we also have $B = \mathbb{R}$ and $f_1 \leq f_2 \leq f_3 \leq \cdots$ (recall Definition A.2.5).

Definition B.2.13. A function $f : A \to \mathbb{R}$ is bounded if $\sup_{x \in A} |f(x)| < \infty$, bounded from above if $\sup_{x \in A} f(x) < \infty$, and bounded from below if $\inf_{x \in A} f(x) > -\infty$.

B.3   Continuity

Definition B.3.1. A function from one metric space to another, $f : A \to B$, is continuous at p if for all $\epsilon > 0$ there exists $\delta > 0$ such that

$$d(x,p) < \delta \qquad \text{implies} \qquad d(f(x), f(p)) < \epsilon.$$

If f is continuous at all $p \in A$ then we say that f is continuous on A or simply continuous.

Note that there are two distance functions in the definition above: one on the metric space A, and the other on the metric space B. For simplicity, we denote them both using d, but nevertheless they may correspond to different functions.

Proposition B.3.1. *Let $f : A \to B$ be a function and p be a limit point of A (there exists a sequence of points in A converging to p). Then*

$$f \text{ is continuous at } p \qquad \text{if and only if} \qquad \lim_{x \to p} f(x) = f(p).$$

Proof. The proof follows from comparing the definitions of a continuous function and the limit of a function. ∎

Proposition B.3.2 (Continuity Preserves Limits). *Let $f : A \to B$ be a continuous function and $x^{(n)}, n \in \mathbb{N}$ be a convergent sequence. Then*

$$\lim_{n \to \infty} f(x^{(n)}) = f\left(\lim_{n \to \infty} x^{(n)} \right).$$

Proof. The proof follows from the previous proposition. ∎

Proposition B.3.3. *If $f : A \to B$ and $g : B \to C$ are continuous functions then their composition $g \circ f : A \to C$ is also a continuous function.*

Proof. Since g is continuous, for $\epsilon > 0$ there exists α such that

$$d(u,w) < \alpha' \quad \Rightarrow \quad d(g(u), g(w)) < \epsilon.$$

Since f is continuous, for $\alpha > 0$ there exists $\delta > 0$ such that

$$d(x,p) < \delta \quad \Rightarrow \quad d(f(x), f(p)) < \alpha.$$

Combining the two results above with $\alpha = \alpha'$, we see that for all $\epsilon > 0$ there exists $\delta > 0$ such that

$$d(x,p) < \delta \quad \Rightarrow \quad d(g(f(x)), g(f(p))) < \epsilon.$$

∎

Proposition B.3.4. *A function $f : A \to B$ is continuous if and only if*

$$G \text{ is open} \qquad \text{implies} \qquad f^{-1}(G) \text{ is open.}$$

Proof. We assume that f is continuous and consider an open set $G \subset B$ and a point $x \in A$ such that $f(x) \in G$. Since G is open, every point in G is an interior point (see previous subsection). Therefore, there $\epsilon > 0$ such that $B_\epsilon(f(x)) \subset G$. Since f is continuous at x, there exists $\delta > 0$ such that $d(y, x) < \delta$ implies $d(f(y), f(x)) < \epsilon$. It follows that x is lies in an open ball that is contained in $f^{-1}(G)$. Repeating this argument for all x such that $f(x) \in G$, we get that $f^{-1}(G)$ is a union of open balls, and therefore $f^{-1}(G)$ is open.

Conversely, we assume that $f^{-1}(G)$ is open whenever G is open and consider an arbitrary $x \in A$. The set $B_\epsilon(f(x))$ is an open ball, and therefore $f^{-1}(B_\epsilon(f(x)))$ is open as well. This implies that $f^{-1}(B_\epsilon(f(x)))$ contains an open ball $B_\delta(x)$ for which $f(B_\delta(x)) \subset B_\epsilon(f(x))$. The continuity of the function f follows. ∎

The above proposition is sometimes used to define continuous functions. This leads to a more general definition of continuous functions $f : A \to B$ between sets that are not necessarily metric spaces.

Definition B.3.2. A function $f : A \to B$ between two metric spaces is uniformly continuous on A if for all $\epsilon > 0$ there exists δ such that

$$d(x, y) < \delta \quad \text{implies} \quad d(f(x), f(y)) < \epsilon.$$

The difference between a function that is continuous on A and uniformly continuous on A is that, in the former case, δ may vary depending on x in addition to ϵ, and in the latter case, there is a single δ for each ϵ.

The following two propositions connect compactness and continuity. Proofs are available in most real analysis textbook, for example [37].

Proposition B.3.5. *Let $f : A \to B$ be a continuous function between two metric spaces. If A is compact, then f is uniformly continuous on A.*

Proposition B.3.6. *A continuous function $f : A \to \mathbb{R}$ on a compact set A attains $\inf_{x \in A} f(x)$, $\sup_{x \in A} f(x)$ and all values in between. In this case we refer to the supremum and infimum as the maximum and minimum of f over A, and denote them $\max_{x \in A} f(x)$ and $\min_{x \in A} f(x)$, respectively.*

B.4   The Euclidean Space

As described in Chapter A, the Euclidean space $\mathbb{R}^d = R \times \cdots \times \mathbb{R}$ is the set of all ordered d-tuples or vectors over the real numbers $\mathbb{R}$ (when $d = 1$ we refer to the vectors as scalars). We denote a vector in $\mathbb{R}^d$ when $d > 1$ in bold and refer to its scalar components via subscripts. For example the vector $\boldsymbol{x} = (x_1, \ldots, x_d)$ has d scalar components $x_i \in \mathbb{R}$, $i = 1, \ldots, d$. Together with the Euclidean distance

$$d(\boldsymbol{x}, \boldsymbol{y}) = \sqrt{\sum_{i=1}^{d} (x_i - y_i)^2},$$

the Euclidean space is a metric space $(\mathbb{R}, d)$ (Proposition B.4.4 later in this chapter proves that the Euclidean distance above is a valid distance function).

Definition B.4.1. Multiplication of a vector by a scalar $c \in \mathbb{R}$ and the addition of two vectors of the same dimensionality are defined as

$$c\boldsymbol{x} \stackrel{\text{def}}{=} (cx_1, \dots, cx_d) \in \mathbb{R}^d$$
$$\boldsymbol{x} + \boldsymbol{y} \stackrel{\text{def}}{=} (x_1 + y_1, \dots, x_d + y_d) \in \mathbb{R}^d.$$

Example B.4.1. For all vectors $\boldsymbol{x}$, the vector $\boldsymbol{0} = (0, \dots, 0)$ satisfies $\boldsymbol{x} + \boldsymbol{0} = \boldsymbol{x}$ and the scalar 1 satisfies $1\boldsymbol{x} = \boldsymbol{x}$.

Definition B.4.2. A function $T : \mathbb{R}^m \to \mathbb{R}^n$ is a linear transformation if

$$T(\alpha \boldsymbol{u} + \beta \boldsymbol{v}) = \alpha T(\boldsymbol{u}) + \beta T(\boldsymbol{v})$$

for all $\boldsymbol{u}, \boldsymbol{v} \in \mathbb{R}^m$ and all $\alpha, \beta \in \mathbb{R}$.

Example B.4.2. The function $T : \mathbb{R}^d \to \mathbb{R}$, defined by $T(\boldsymbol{x}) = \sum_{i=1}^d c_i x_i$, is a linear transformation.

Definition B.4.3. An inner product is a function $g(\cdot, \cdot) : \mathbb{R}^d \times \mathbb{R}^d \to \mathbb{R}$ that satisfies the following for all $\boldsymbol{u}, \boldsymbol{v}, \boldsymbol{w} \in \mathbb{R}^d$ and for all $\alpha, \beta \in \mathbb{R}$.

- symmetry: $g(\boldsymbol{v}, \boldsymbol{u}) = g(\boldsymbol{u}, \boldsymbol{v})$

- bilinearity: $g(\alpha \boldsymbol{u} + \beta \boldsymbol{v}, \boldsymbol{w}) = \alpha g(\boldsymbol{u}, \boldsymbol{w}) + \beta g(\boldsymbol{v}, \boldsymbol{w})$

- positivity: $g(\boldsymbol{v}, \boldsymbol{v}) \geq 0$.

Note that as a consequence of the first two properties above

$$g(\alpha \boldsymbol{u} + \beta \boldsymbol{v}, \gamma \boldsymbol{w} + \delta \boldsymbol{z}) = \alpha \gamma g(\boldsymbol{u}, \boldsymbol{w}) + \alpha \delta g(\boldsymbol{u}, \boldsymbol{z}) + \beta \gamma g(\boldsymbol{v}, \boldsymbol{w}) + \beta \delta g(\boldsymbol{v}, \boldsymbol{z}).$$

Proposition B.4.1 (Cauchy Schwartz Inequality). *For any inner product g and for all $\boldsymbol{x}, \boldsymbol{y} \in \mathbb{R}^d$*

$$g^2(\boldsymbol{x}, \boldsymbol{y}) \leq g(\boldsymbol{x}, \boldsymbol{x}) \, g(\boldsymbol{y}, \boldsymbol{y}).$$

Proof. Using the positivity and bi-linearity properties of the inner product,

$$0 \leq g(g(\boldsymbol{y}, \boldsymbol{y})\boldsymbol{x} - g(\boldsymbol{x}, \boldsymbol{y})\boldsymbol{y}, g(\boldsymbol{y}, \boldsymbol{y})\boldsymbol{x} - g(\boldsymbol{x}, \boldsymbol{y})\boldsymbol{y})$$
$$= g^2(\boldsymbol{y}, \boldsymbol{y})g(\boldsymbol{x}, \boldsymbol{x}) - 2g(\boldsymbol{y}, \boldsymbol{y})g^2(\boldsymbol{x}, \boldsymbol{y}) + g^2(\boldsymbol{x}, \boldsymbol{y})g(\boldsymbol{y}, \boldsymbol{y})$$
$$= g(\boldsymbol{y}, \boldsymbol{y})(g(\boldsymbol{x}, \boldsymbol{x})g(\boldsymbol{y}, \boldsymbol{y}) - g^2(\boldsymbol{x}, \boldsymbol{y})).$$

Since $g(\boldsymbol{y}, \boldsymbol{y}) \geq 0$, the proposition follows.   ∎

Proposition B.4.2. *The Euclidean product*

$$\langle \boldsymbol{x}, \boldsymbol{y} \rangle \overset{\text{def}}{=} \sum_{i=1}^{d} x_i y_i$$

is an inner product

Proof. It is easy to verify the three properties above by direct substitution. ∎

Definition B.4.4. A norm is a function $h : \mathbb{R}^d \to \mathbb{R}$ satisfying the following for all $\boldsymbol{x}, \boldsymbol{y} \in \mathbb{R}^d$, and for all $c \in \mathbb{R}$.

- non-negativity: $h(\boldsymbol{x}) \geq 0$

- positivity: $h(\boldsymbol{x}) = 0$ if and only if $\boldsymbol{x} = \boldsymbol{0}$

- homogeneity: $h(c\,\boldsymbol{x}) = |c|\,h(\boldsymbol{x})$

- triangle inequality: $h(\boldsymbol{x} + \boldsymbol{y}) \leq h(\boldsymbol{x}) + h(\boldsymbol{y})$.

Proposition B.4.3. *The function*

$$\|\boldsymbol{x}\| \overset{\text{def}}{=} \sqrt{\langle \boldsymbol{x}, \boldsymbol{x} \rangle} = \left(\sum_{i=1}^{d} x_i^2 \right)^{1/2}$$

is a norm, commonly known as the Euclidean norm.

Proof. The first three properties are obvious. The fourth property follows from the Cauchy Schwartz inequality applied to the Euclidean inner product

$$\begin{aligned}
\|\boldsymbol{x} + \boldsymbol{y}\|^2 &= \langle \boldsymbol{x} + \boldsymbol{y}, \boldsymbol{x} + \boldsymbol{y} \rangle \\
&= \|\boldsymbol{x}\|^2 + 2\langle \boldsymbol{x}, \boldsymbol{y} \rangle + \|\boldsymbol{y}\|^2 \\
&\leq \|\boldsymbol{x}\|^2 + 2\|\boldsymbol{x}\|\|\boldsymbol{y}\| + \|\boldsymbol{y}\|^2 \\
&= (\|\boldsymbol{x}\| + \|\boldsymbol{y}\|)^2.
\end{aligned}$$

∎

The Euclidean norm is the most popular norm. The more general L_p norm

$$\|\boldsymbol{x}\|_p = \left(\sum_{i=1}^{d} |x_i|^p \right)^{1/p}, \qquad p \geq 1,$$

has the following special cases

$$\|\boldsymbol{x}\|_2 = \|\boldsymbol{x}\| \qquad \text{(the Euclidean norm)}$$

$$\|\boldsymbol{x}\|_1 = \sum_{i=1}^{d} |x_i|$$

$$\|\boldsymbol{x}\|_\infty = \max\{|x_1|, \ldots, |x_d|\} \qquad \text{(achieved by letting } p \to \infty\text{)}.$$

The L_p norm can be further generalized as follows.

Definition B.4.5. Assuming W is a non-singular matrix[1], the weighted $L_{p,W}$ norm is

$$\|x\|_{p,W} = \|Wx\|_p.$$

Weighted norms are convenient for emphasizing some dimensions over others. For example, if $\text{diag}(w)$ is an all zero matrix, except for its diagonal

$$\text{diag}(w) = \begin{pmatrix} w_1 & 0 & \cdots & 0 \\ 0 & w_2 & \cdots & 0 \\ \vdots & \vdots & \ddots & \vdots \\ 0 & \cdots & 0 & w_d \end{pmatrix},$$

the corresponding weighted L_2 norm is

$$\|x\|_{2,\text{diag}(w)} = \sqrt{\sum_{i=1}^{d} w_i^2 x_i^2}.$$

Definition B.4.6. The angle between $x, y \in \mathbb{R}^d$ is the value θ for which

$$\cos\theta = \frac{\langle x, y \rangle}{\|x\| \cdot \|y\|}.$$

Definition B.4.7. Vectors $x^{(i)}, i = 1, \ldots, k$ are said to be orthogonal if

$$i \neq j \qquad \text{implies} \qquad \langle x^{(i)}, x^{(j)} \rangle = 0.$$

If, in addition, $\|x^{(i)}\| = 1$, $i = 1, \ldots, k$, the vectors are said to be orthonormal.

Note that the orthogonality condition above corresponds to a a 90 degree angle (perpendicularity) between every two distinct vectors.

Proposition B.4.4. *For any norm $h(\cdot)$, the function $d(x,y) = h(x - y)$ is a valid distance function. Accordingly, we may interpret the norm $h(x)$ as the corresponding distance of x from the origin $d(x, 0)$.*

Proof. It is straightforward to verify that $d(x,y) = h(x - y)$ satisfies the first three properties of a distance function. The fourth follows from applying the triangle inequality property of norms to $x - y$ and $y - z$:

$$h(x - z) \leq h(x - y) + h(y - z).$$

∎

---

[1] Matrices, and the matrix product notation Wx, are defined at the beginning of Chapter C.

The propositions above confirm that the Euclidean space $(\mathbb{R}^d, d)$, where

$$d(\boldsymbol{x}, \boldsymbol{y}) = \sqrt{\sum_{i=1}^{d}(x_i - y_i)^2},$$

is a metric space.

Figure B.1 shows contour lines of four L_p norms in the left column and four $L_{p,W}$ norms in the right column, where $W = \text{diag}(2, 1)$. A non-diagonal matrix W would result in rotated versions of the figures in the right column.

The R code below generates the left and right columns of the figure.

```
s = seq(-1, 1, length.out = 50)
R = expand.grid(x1 = s, x2 = s)
# generate left column (Lp norms)
D = rbind(R, R, R, R)
D$Norm[1:2500] = abs(D$x1[1:2500]) + abs(D$x2[1:2500])
D$Norm[2501:5000] = ((abs(D$x1[2501:5000]))^(1.5) +
    (abs(D$x2[2501:5000]))^(1.5))^(2/3)
D$Norm[5001:7500] = ((abs(D$x1[5001:7500]))^2 +
    (abs(D$x2[5001:7500]))^2)^(1/2)
D$Norm[7501:10000] = pmax(abs(D$x1[7501:10000]),
    abs(D$x2[7501:10000]))
D$p = c(rep("Lp norm, p=1", 2500), rep("Lp norm p=1.5",
    2500), rep("Lp norm p=2", 2500), rep("Lp norm p=inf",
    2500))
ggplot(D, aes(x1, x2, z = Norm)) + facet_grid(p ~
    .) + stat_contour(bins = 4)


# generate right column (weighted Lp norms)
D = rbind(R, R, R, R)
D$Norm[1:2500] = abs(2 * D$x1[1:2500]) + abs(D$x2[1:2500])
D$Norm[2501:5000] = ((2 * abs(D$x1[2501:5000]))^(1.5) +
    (abs(D$x2[2501:5000]))^(1.5))^(2/3)
D$Norm[5001:7500] = ((2 * abs(D$x1[5001:7500]))^2 +
    (abs(D$x2[5001:7500]))^2)^(1/2)
D$Norm[7501:10000] = pmax(abs(2 * D$x1[7501:10000]),
    abs(D$x2[7501:10000]))
D$p = c(rep("weighted Lp norm p=1", 2500), rep("weighted Lp norm p=1.5",
    2500), rep("weighted Lp norm p=2", 2500),
    rep("weighted Lp norm p=inf", 2500))
ggplot(D, aes(x1, x2, z = Norm)) + facet_grid(p ~
    .) + stat_contour(bins = 4)
```

Example B.4.3. *Consider the set $\mathbb{R}^\infty$ defined in Example A.3.2. The proofs of the propositions above can be extended to show that $\langle \boldsymbol{x}, \boldsymbol{y} \rangle = \sum_{n \in \mathbb{N}} x_n y_n$ is an inner product on $\mathbb{R}^\infty$, $\|\boldsymbol{x}\|_2 = \sqrt{\sum_{n \in \mathbb{N}} x_n^2}$ is a norm on $\mathbb{R}^\infty$ and $d(\boldsymbol{x}, \boldsymbol{y}) =$*

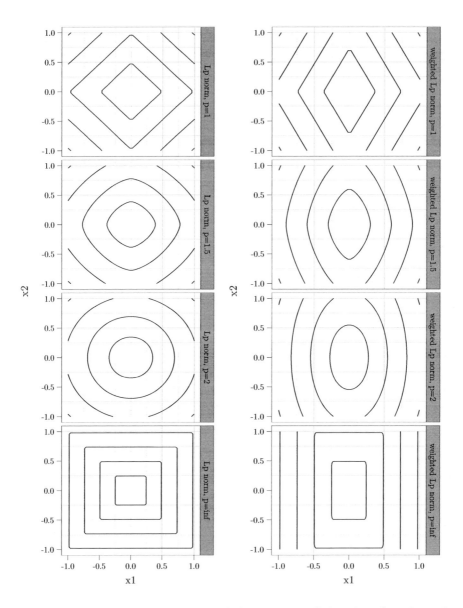

Figure B.1: Equal height contours of the L_p norm (left column) and weighted $L_{p,W}$ norm with $W = \text{diag}(2,1)$ (right column), in the two dimensional case $d = 2$. Each row corresponds to a different value of p. As p increases from 1 to ∞ the contours change their shape from diamond-shape to square-shape.

$\sqrt{\sum_{n \in \mathbb{N}}(x_n - y_n)^2}$ is a distance on $\mathbb{R}^\infty$, all provided that the infinite sums converge. In general, however, the infinite sums may not converge, making the generalizations above inappropriate.

Example B.4.4. Consider the set $\mathbb{R}^\infty$ defined in Example A.3.2 and the distance function

$$\bar{d}(\boldsymbol{x}, \boldsymbol{y}) = \sup\left\{\frac{\min(|x_n - y_n|, 1)}{n} : n \in \mathbb{N}\right\}. \tag{B.1}$$

Note that $\bar{d}$ is well-defined and finite, is symmetric, is non-negative, and is zero if and only if $\boldsymbol{x} = \boldsymbol{y}$. It also satisfies the triangle inequality

$$\bar{d}(\boldsymbol{x}, \boldsymbol{z}) = \sup\left\{\frac{\min(|x_n - y_n + y_n - z_n|, 1)}{n} : n \in \mathbb{N}\right\}$$

$$\leq \sup\left\{\frac{\min(|x_n - y_n| + |y_n - z_n|, 1)}{n} : n \in \mathbb{N}\right\}$$

$$\leq \sup\left\{\frac{\min(|x_n - y_n|, 1)}{n} + \frac{\min(|y_n - z_n|, 1)}{n} : n \in \mathbb{N}\right\}$$

$$\leq \bar{d}(\boldsymbol{x}, \boldsymbol{y}) + \bar{d}(\boldsymbol{y}, \boldsymbol{z})$$

and is therefore a distance function.

If $\boldsymbol{x} \in B_\epsilon(\boldsymbol{0})$ for a given ϵ then for all $n \in \mathbb{N}$ we have $\min(|x_n|, 1) < n\epsilon$. There exists some N such that $n > N$ corresponds to $n\epsilon > 1$, implying that the components $x_{N+1}, x_{N+2}, \ldots$ of vectors $\boldsymbol{x} \in B_\epsilon(\boldsymbol{0})$ are unrestricted. Similarly, for $\boldsymbol{x} \in B_\epsilon(\boldsymbol{0})$ the components x_n, where $n < N$, satisfy $|x_n| < n\epsilon$. It follows that points in $B_\epsilon(\boldsymbol{0})$ are an intersection of a finite number of sets of the form

$$\mathbb{R} \times \cdots \times \mathbb{R} \times (a, b) \times \mathbb{R} \times \cdots . \tag{*}$$

(we refer to sets of the form (*) as simple cylinders.) Since $\bar{d}(\boldsymbol{x} + \boldsymbol{c}, \boldsymbol{z} + \boldsymbol{c}) = \bar{d}(\boldsymbol{x}, \boldsymbol{z})$, we also have that $B_\epsilon(\boldsymbol{y})$ is an intersection of a finite number of simple cylinders or sets of the form expressed in (*).

On the other hand, let A be an intersection of a finite number of simple cylinders. Then for each $\boldsymbol{x} \in A$ we can construct $B_\epsilon(\boldsymbol{y})$ such that $\boldsymbol{x} \in B_\epsilon(\boldsymbol{y}) \subset A$ (taking ϵ to be sufficiently small). This implies that A is a union of open balls and is therefore an open set. A union of intersections of a finite number of simple cylinders is a union of open sets and therefore is open also.

In summary, we have thus demonstrated that in the space $(\mathbb{R}^\infty, \bar{d})$, the set of open sets is equivalent to the set of unions of intersections of a finite number of simple cylinders.

The convergence problems mentioned in Example B.4.3 leads to the common practice of defining the metric structure on $\mathbb{R}^\infty$ using the distance function $\bar{d}$ in Example B.4.4 rather than the Euclidean distance. In fact, whenever we refer to the metric structure of $\mathbb{R}^\infty$ we will assume the metric structure of $\bar{d}$ derived above. This metric structure is commonly referred to in the literature as the product topology of $\mathbb{R}^{\infty}$.

The metric structure of the Euclidean space simplifies some of the properties described in the previous chapter.

Proposition B.4.5. *The Euclidean space $\mathbb{R}^d$ is second countable, and in particular one choice for $\mathcal{G}$ in Definition B.1.5 is the set of all open balls with rational centers and radii. Similarly, the space $(\mathbb{R}^{\infty}, \bar{d})$ (see Example B.4.4) is second countable.*

Proof. We define $\mathcal{G}$ to be the set of all open balls with rational centers and rational radii. Since $\mathbb{Q}$ and $\mathbb{Q}^d$ are countably infinite sets, the set $\mathcal{G}$ is countably infinite.

Let G be an open set and $\boldsymbol{x} \in G$. By Proposition B.1.2 there exists $r > 0$ for which $B_r(\boldsymbol{x}) \subset G$. Since for every real number there is a rational number that is arbitrarily close, we can select $B_{r'}(\boldsymbol{x}')$ where $r', \boldsymbol{x}'$ are rationals such that $\boldsymbol{x} \in B_{r'}(\boldsymbol{x}') \subset B_r(\boldsymbol{x}) \subset G$. Repeating this for every $\boldsymbol{x} \in G$ and taking the union of the resulting rational balls completes the proof.

In the case of R^{∞}, second countability is demonstrated by taking all simple cylinders (see Example B.4.4) whose base (a, b) has rational endpoints and noting that a countable union of sets that are countably infinite is countably infinite. ∎

Proposition B.4.6. *In the Euclidean space $\boldsymbol{x}^{(n)} \to \boldsymbol{x}$ if and only if $x_j^{(n)} \to x_j$ for all $j = 1, \ldots, d$.*

Proof. We recall Proposition B.3.2, which states that continuous functions such as $f(x) = x^2$, $f(x) = \sqrt{x}$, and $f(x, y) = x + y$ preserve limits. It follows that $\|\boldsymbol{v}^{(n)}\| \to 0$ if and only if $\|\boldsymbol{v}^{(n)}\|^2 = \sum_{j=1}^d (v_j^{(n)})^2 \to 0$, which occurs if and only if $(v_j^{(n)})^2 \to 0$ for all $j = 1, \ldots, d$. This occurs if and only if $v_j^{(n)} \to 0$ for all $j = 1, \ldots, d$ ∎

Example B.4.5. *The sequence of vectors $(2^{-n}, 3^{-n}), n \in \mathbb{N}$ converges as $n \to \infty$ to $\boldsymbol{0} = (0, 0)$.*

Proposition B.4.7. *The function $\boldsymbol{f} : \mathbb{R}^d \to \mathbb{R}^k$ defined by $\boldsymbol{f}(\boldsymbol{x}) = (f_1(\boldsymbol{x}), \ldots, f_k(\boldsymbol{x}))$ is continuous if and only if the functions $f_1, \ldots, f_k$ are continuous.*

Note that above we are denoting the function using bold-face $\boldsymbol{f}$ to indicate that $\boldsymbol{f}(\boldsymbol{x})$ is a vector.

Proof. We assume that f_i are continuous for all $i = 1, \ldots, k$ and consider an arbitrary $\epsilon > 0$. Then for $\epsilon' = \epsilon/\sqrt{k} > 0$ there exists $\delta > 0$ such that if $\|\boldsymbol{x} - \boldsymbol{y}\|^2 < \delta^2$ then $|f_i(\boldsymbol{x}) - f_i(\boldsymbol{y})|^2 < \epsilon^2/k$, $i = 1, \ldots, k$. Note that we choose the same δ for all $i = 1, \ldots, k$; for example, by setting $\delta = \min(\delta_1, \ldots, \delta_k)$. It follows that if $\|\boldsymbol{x} - \boldsymbol{y}\|^2 < \delta$ then

$$\|\boldsymbol{f}(\boldsymbol{x}) - \boldsymbol{f}(\boldsymbol{y})\|^2 = \sum_{j=1}^k |f_j(\boldsymbol{x}) - f_j(\boldsymbol{y})|^2 \leq k\epsilon^2/k = \epsilon^2,$$

showing that $\boldsymbol{f}$ is continuous.

Conversely, if $\boldsymbol{f}$ is continuous, for all $\epsilon > 0$ there exists $\delta > 0$ such that whenever $\|\boldsymbol{x} - \boldsymbol{y}\|^2 \le \delta^2$ we have

$$\|\boldsymbol{f}(\boldsymbol{x}) - \boldsymbol{f}(\boldsymbol{y})\|^2 = \sum_{j=1}^{k} |f_j(\boldsymbol{x}) - f_j(\boldsymbol{y})|^2 < \epsilon^2.$$

This implies $|f_j(\boldsymbol{x}) - f_j(\boldsymbol{y})|^2 < \epsilon^2$ and $|f_j(\boldsymbol{x}) - f_j(\boldsymbol{y})| < \epsilon$, implying the continuity of $f_1, \ldots, f_k$. ∎

Proposition B.4.8. *The following functions* $f : \mathbb{R}^d \to \mathbb{R}^k$ *are continuous:*

$$f(\boldsymbol{x}) = cx_i$$
$$f(\boldsymbol{x}, \boldsymbol{y}) = \boldsymbol{x} + \boldsymbol{y}$$
$$f(\boldsymbol{x}, \boldsymbol{y}) = \langle \boldsymbol{x}, \boldsymbol{y} \rangle$$
$$f(x) = 1/x.$$

Note that in the first and third cases above, we have $k = 1$, and in the last case above, we have $k = d = 1$.

Proof. To prove the first assertion, let $\epsilon > 0$ be given, and define $\delta = \epsilon/|c|$. Whenever

$$\|\boldsymbol{x} - \boldsymbol{y}\|^2 = \sum_{j=1}^{d} |x_j - y_j|^2 < \epsilon^2/|c|^2$$

we also have

$$|cx_i - cy_i| \le |c| \cdot |x_i - y_i| < |c|\epsilon/|c| = \epsilon.$$

To prove the second assertion note that whenever $\|(\boldsymbol{u}, \boldsymbol{w}) - (\boldsymbol{x}, \boldsymbol{y})\| < \epsilon/\sqrt{2d}$, then for all $j = 1, \ldots, d$, $|x_j - u_j| < \epsilon/\sqrt{2d}$ and $|y_j - w_j| < \epsilon/\sqrt{2d}$. Using the triangle inequality property of the Euclidean norm, we have

$$\|(\boldsymbol{x} + \boldsymbol{y}) - (\boldsymbol{u} + \boldsymbol{w})\| = \|(\boldsymbol{x} - \boldsymbol{u}) + (\boldsymbol{y} - \boldsymbol{w})\| \le \|\boldsymbol{x} - \boldsymbol{u}\| + \|\boldsymbol{y} - \boldsymbol{w}\| \le \sqrt{2d\epsilon^2/(2d)} = \epsilon.$$

The proofs of the other propositions are similar. ∎

Corollary B.4.1. *All polynomials are continuous functions.*

The concept of compactness (Definition B.1.6) has some important consequences (Propositions B.3.6 and B.3.5). The general definition of a compact space (Definition B.1.6) is hard to verify, but the following simple condition is very useful for verifying compactness in $\mathbb{R}^d$. A proof is available in [37].

Proposition B.4.9. *If a set* $A \subset \mathbb{R}^d$ *is closed and bounded, it is compact.*

B.5   Growth of Functions

The big-Oh and other growth notations are useful in comparing the asymptotic behavior of two functions $f, g : \mathbb{R} \to \mathbb{R}$. We list below their definitions.

Definition B.5.1. For two functions $f, g : \mathbb{R} \to \mathbb{R}$ we define the following notations:

$$f = O(g) \text{ as } x \to L \qquad \text{if} \qquad \limsup_{x \to L} \frac{|f(x)|}{|g(x)|} < \infty$$

$$f = o(g) \text{ as } x \to L \qquad \text{if} \qquad \lim_{x \to L} \frac{f(x)}{g(x)} = 0$$

$$f \sim g \text{ as } x \to L \qquad \text{if} \qquad \lim_{x \to L} \frac{f(x)}{g(x)} = 1$$

$$f \asymp g \text{ as } x \to L \qquad \text{if} \qquad f = O(g) \text{ and } g = O(f).$$

Above, L may correspond to a finite limit or an infinite limit: $+\infty, -\infty$. If L is omitted, it is typically assumed to be $+\infty$.

Intuitively, $f = O(g)$ implies that f grows as fast as g or slower, and $f = o(g)$ implies that f grows slower than g does. The definitions above assume that $g(x)$ is non-zero, or at least non-zero as $x \to L$.

Example B.5.1. *We have* $f = O(g)$ *if and only if* $f(x)/g(x) = O(1)$ *and* $f(x) = o(g(x))$ *if and only if* $f(x)/g(x) = o(1)$. *In this case, the notation 1 corresponds to the constant function* $1(x) = 1$ *for all* x.

Example B.5.2. *Assuming* $n, k > 0$, *we have* $x^n = O(x^k)$ *if and only if* $k \geq n$ *and* $x^{-n} = o(x^{-k})$ *if and only if* $k < n$ *(both as* $x \to \infty$*). We also have* $x^n = O(\exp(x))$ *and* $\exp(-x) = o(x^{-n})$ *as* $x \to \infty$ *for all* $n > 0$.

B.6   Notes

More details on metric spaces and the Euclidean space are available in standard real analysis textbooks. A classic example is [37]. A more general presentation of open sets and continuity that does not assume the presence of a metric space is available in topology textbooks, for example [30]. Much of the material presented in Section B.4 holds also for infinite dimensional vector spaces. More details on analysis in infinite dimensional spaces are available in standard functional analysis textbooks. A classic example is [27].

B.7   Exercises

1. Verify the claims made in Example B.4.3 regarding the Euclidean norm and Euclidean distance in $\mathbb{R}^\infty$.

2. Construct a subset of $\mathbb{R}^\infty$ on which the Euclidean distance generalization is a distance function.

3. Consider the space $\mathbb{R}^{[0,1]}$ of functions from $[0,1]$ to $\mathbb{R}$ (see Definition A.3.4). Is $\langle f, g \rangle = \int_0^1 f(x)g(x)\, dt$ a valid inner product? Is the natural generalization of the Euclidean distance a valid distance function on $\mathbb{R}^{[0,1]}$?

4. Show that the L_1 and L_2 distance functions result in the same collection of open sets in the Euclidean space $\mathbb{R}$.

5. What is the condition for a sequence $\boldsymbol{x}^{(n)}, n \in \mathbb{N}$ to converge to $\boldsymbol{x}$ in $(\mathbb{R}^\infty, \bar{d})$. Is it identical to the condition that $x_k^{(n)} \to x_k$ for all $k \in \mathbb{N}$?

6. Generalize the distance function $\bar{d}$ in Example B.4.4 to the set $\mathbb{R}^{[0,1]}$. Is the resulting space $(\mathbb{R}^{[0,1]}, \bar{d})$ a metric space?

Appendix C

Linear Algebra

This chapter describes some basic concepts in linear algebra, including determinants, eigenvalues and eigenvectors, and the singular value decomposition.

C.1 Basic Definitions

Definition C.1.1. A matrix A is a two dimensional array of real numbers. The elements of A are indexed by their row and column: A_{ij} is the element at the i-row and j-column. The set of all n by m matrices is denoted $\mathbb{R}^{n \times m}$. The transpose of a matrix $A \in \mathbb{R}^{n \times m}$ is $A^\top \in \mathbb{R}^{m \times n}$, defined as $[A^\top]_{ij} \stackrel{\text{def}}{=} A_{ji}$. Vectors are a special case of matrices with either one row or one column.

```
# define a 2 by 2 matrix and a 2x1 vector
A = matrix(c(1, 2, 3, 4), nrow = 2, ncol = 2)
A

##      [,1] [,2]
## [1,]    1    3
## [2,]    2    4

t(A)   # matrix transpose of A

##      [,1] [,2]
## [1,]    1    2
## [2,]    3    4

v = c(1, -1)   # row vector
v

## [1]  1 -1
```

257

When indexing vectors we sometimes omit the redundant index. For example, if v is a column vector we write $[v]_{j1}$ as simply v_j. We similarly consider 1×1 matrices as scalars and refer to them without redundant indices.

We follow convention and write vectors in bold lowercase, for example x, y and scalars in non-bold lowercase, for example x, y. We denote matrices in non-bold uppercase, such as A, B. Important exceptions are random variables, denoted by non-bold uppercase, for example X, Y and random vectors, denoted by bold uppercase, for example $\boldsymbol{X}, \boldsymbol{Y}$.

Definition C.1.2. Multiplying a matrix A by a scalar $c \in \mathbb{R}$ is defined by

$$[cA]_{ij} \overset{\text{def}}{=} cA_{ij}.$$

The addition of two matrices A, B of the same size is defined as

$$[A + B]_{ij} \overset{\text{def}}{=} A_{ij} + B_{ij}.$$

```
B = matrix(2, nrow = 2, ncol = 2)
A
```

```
##      [,1] [,2]
## [1,]    1    3
## [2,]    2    4
```

```
B
```

```
##      [,1] [,2]
## [1,]    2    2
## [2,]    2    2
```

```
2 * A
```

```
##      [,1] [,2]
## [1,]    2    6
## [2,]    4    8
```

```
A + B
```

```
##      [,1] [,2]
## [1,]    3    5
## [2,]    4    6
```

Definition C.1.3. A matrix having the same number of rows and columns is called a square matrix. The diagonal elements of a square matrix A are $A_{11}, \ldots, A_{nn}$. A diagonal matrix is a square matrix whose off diagonal elements are zero. A diagonal matrix whose diagonal elements are 1 is called the identity matrix and is denoted by I. Given a vector of diagonal values v we denote the corresponding diagonal matrix as diag(v).

```
A
```

```
##      [,1] [,2]
## [1,]    1    3
## [2,]    2    4
```

```
diag(A)   # diagonal of the matrix A
```

```
## [1] 1 4
```

```
diag(c(1, -1))   # construct a diagonal matrix from a vector
```

```
##      [,1] [,2]
## [1,]    1    0
## [2,]    0   -1
```

```
diag(5)   # 5x5 identity matrix
```

```
##      [,1] [,2] [,3] [,4] [,5]
## [1,]    1    0    0    0    0
## [2,]    0    1    0    0    0
## [3,]    0    0    1    0    0
## [4,]    0    0    0    1    0
## [5,]    0    0    0    0    1
```

Definition C.1.4. An upper triangular matrix is a square matrix for which all elements below the diagonal are zero. A lower triangular matrix is a square matrix for which all elements above the diagonal are zero. A triangular matrix is a lower triangular matrix or an upper triangular matrix.

```
# 5x5 upper triangular matrix of ones
round(upper.tri(matrix(1, 5, 5), diag = TRUE))
```

```
##      [,1] [,2] [,3] [,4] [,5]
## [1,]    1    1    1    1    1
## [2,]    0    1    1    1    1
## [3,]    0    0    1    1    1
## [4,]    0    0    0    1    1
## [5,]    0    0    0    0    1
```

Definition C.1.5. The product of a matrix $A \in \mathbb{R}^{n \times m}$ and a matrix $B \in \mathbb{R}^{m \times k}$ is the matrix $AB \in \mathbb{R}^{n \times k}$ defined as

$$[AB]_{ij} = \sum_{r=1}^{m} A_{ir} B_{rj}.$$

Note that while matrix multiplication by a scalar and matrix addition are commutative, matrix product is not: $AB \neq BA$ in general. All three operations are associative: $A + (B + C) = (A + B) + C$, $c(AB) = (cA)B$, and $A(BC) = (AB)C$. This implies we can omit parenthesis and write $A + B + C$, cAB, and ABC without any ambiguity.

A

```
##         [,1] [,2]
## [1,]    1    3
## [2,]    2    4
```

B

```
##         [,1] [,2]
## [1,]    2    2
## [2,]    2    2
```

```
# matrix product is not commutatitve
A %*% B
```

```
##         [,1] [,2]
## [1,]     8    8
## [2,]    12   12
```

```
B %*% A
```

```
##         [,1] [,2]
## [1,]     6   14
## [2,]     6   14
```

```
# matrix product is associative
(A %*% B) %*% (A + 1)  == A %*% (B %*% (A + 1))
```

```
##         [,1] [,2]
## [1,] TRUE TRUE
## [2,] TRUE TRUE
```

When we use the matrix product notation AB we assume implicitly that it is defined, implying that A has the same number of columns as B has rows. Repeated multiplications is denoted using an exponent, for example $AA = A^2$ and $AA^k = A^{k+1}$.

Example C.1.1. *Using the above definition of matrix multiplication, we see that for all matrices A, we have $AI = IA = A$. Similarly, we have $AI^k = I^k A = A$ and $AIA = A^2$.*

Example C.1.2. *For $A \in \mathbb{R}^{n \times m}$, $\boldsymbol{x} \in \mathbb{R}^{m \times 1}$, and $\boldsymbol{y} \in \mathbb{R}^{1 \times n}$, we have*

$$A\boldsymbol{x} \in \mathbb{R}^{n \times 1}, \qquad [A\boldsymbol{x}]_i = \sum_{r=1}^{m} A_{ir} x_r$$

$$\boldsymbol{y}A \in \mathbb{R}^{1 \times m}, \qquad [\boldsymbol{y}A]_i = \sum_{r=1}^{n} y_r A_{ri}$$

$$\boldsymbol{y}A\boldsymbol{x} \in \mathbb{R}^{1 \times 1}, \qquad \boldsymbol{y}A\boldsymbol{x} = \sum_{i=1}^{n} \sum_{j=1}^{m} y_i A_{ij} x_j.$$

If $n = m$, we also have

$$\boldsymbol{x}^\top A \boldsymbol{x} \in \mathbb{R}^{1 \times 1} \qquad \boldsymbol{x}^\top A \boldsymbol{x} = \sum_{i=1}^{n} \sum_{j=1}^{n} x_i A_{ij} x_j.$$

Definition C.1.6. For two vectors $\boldsymbol{x} \in \mathbb{R}^{n \times 1}, \boldsymbol{y} \in \mathbb{R}^{m \times 1}$ we define the outer product (see also Section B.4) as

$$\boldsymbol{x} \otimes \boldsymbol{y} \stackrel{\text{def}}{=} \boldsymbol{x}\boldsymbol{y}^\top \in \mathbb{R}^{n \times m},$$

and if $n = m$ we define the inner product (see also Section B.4) as

$$\langle \boldsymbol{x}, \boldsymbol{y} \rangle \stackrel{\text{def}}{=} \boldsymbol{x}^\top \boldsymbol{y} = \boldsymbol{y}^\top \boldsymbol{x} \in \mathbb{R}.$$

```
x = c(1, 2, 3)
y = c(-1, 0, 1)
x %*% y   # inner product

##      [,1]
## [1,]    2

x %o% y   # outer product

##      [,1] [,2] [,3]
## [1,]   -1    0    1
## [2,]   -2    0    2
## [3,]   -3    0    3
```

Definition C.1.7. We refer to the vector $\sum_{i=1}^{n} \alpha_i v^{(n)}$, where $\alpha_1, \ldots, \alpha_n \in \mathbb{R}$, as a linear combination of the vectors $v^{(1)}, \ldots, v^{(n)}$. The α_i, $i = 1, \ldots, n$ values are called the combination coefficients.

Definition C.1.8. The vectors $v^{(1)}, \ldots, v^{(n)}$ are said to be linearly dependent if the expression $\sum_i \alpha_i v^{(i)} = 0$ implies that $\alpha_i = 0$ for all i. Equivalently this implies that $v^{(i)}$ may be written as a linear combination of the remaining vectors. If the vectors are not linearly dependent they are said to be linearly independent.

Example C.1.3. *The vectors $(0,1)$ and $(0,2)$ are linearly dependent, but the vectors $(0,1)$ and $(1,0)$ are linearly independent. In fact, the set of all linear combinations of $(0,1)$ and $(1,0)$ coincides with $\mathbb{R}^2$ (the set of all possible two dimensional vectors).*

Example C.1.4. *Generalizing the previous example, the vectors $e^{(i)} \in \mathbb{R}^{n \times 1}$ defined by $e_j^{(i)} = 1$ if $i = j$ and 0 otherwise, $i = 1, \ldots, n$, are linearly independent. Since for every vector v,*

$$v = \sum v_i e^{(i)},$$

the set of all their linear combinations of $e^{(i)}, i = 1, \ldots, n$ is $\mathbb{R}^n$ (the set of all possible n dimensional vectors). The vectors $e^{(i)}, i = 1, \ldots, n$ are called unit vectors and are together called the standard basis for $\mathbb{R}^n$.

Examining Definition C.1.5 we see that multiplying a matrix by a column vector Av yields a column vector that is a linear combination of the columns of A. Similarly the matrix multiplication AB is a matrix whose columns are each a linear combination of the column vectors of A. We will use these important observations in several proofs later on.

The proposition below helps to explain why matrices and linear algebra are so important.

Proposition C.1.1. *Any linear transformation $T : \mathbb{R}^n \to \mathbb{R}^m$ (Definition B.4.2) is realized by matrix multiplication $T(v) = Av$ for some $A \in \mathbb{R}^{m \times n}$. Conversely, the matrix multiplication mapping $v \mapsto Av$ is a linear transformation.*

Proof. Let T be a linear transformation, and denote by $e^{(i)}$, $i = 1, \ldots, n$ the standard basis from Example C.1.4. Given a vector v we express it as a linear combination of the standard basis $v = \sum_i v_i e^{(i)}$, and apply linearity

$$T(v) = T\left(\sum v_i e^{(i)}\right) = \sum v_i u^{(i)},$$

where $u^{(i)} = T(e^{(i)})$. Setting the columns of A to $u^{(i)}$, $i = 1, \ldots, m$ shows that T is a matrix multiplication mapping realized by the matrix A: $T(v) = Av$. The second statement follows from the fact that for matrix multiplication $A(c_1 u + c_2 v) = c_1 Au + c_2 Av$. ∎

Definition C.1.9. The inverse of a square matrix A is the matrix A^{-1} such that $AA^{-1} = A^{-1}A = I$. A matrix A for which an inverse exists is called non-singular. Otherwise it is called singular.

Proposition C.1.2. *The inverse of a non-singular matrix is unique.*

Proof. If A has two inverses B and C, then $AB = I$ and $C = CI = CAB = B$. ∎

One of the applications of matrix inversion is in solving linear systems of equations. Given the system of equations $A\boldsymbol{x} = \boldsymbol{y}$ where $A, \boldsymbol{y}$ are known and $\boldsymbol{x}$ is not known, we can solve for $\boldsymbol{x}$ by inverting the matrix

$$\boldsymbol{x} = A^{-1}A\boldsymbol{x} = A^{-1}\boldsymbol{y}.$$

Proposition C.1.3. *For any two matrices A, B we have $(AB)^\top = B^\top A^\top$. Similarly, for any non-singular matrices A, B we have $(AB)^{-1} = B^{-1}A^{-1}$.*

Proof. The first statement follows from

$$[(AB)^\top]_{ij} = \sum_k A_{jk}B_{ki} = \sum_k B_{ki}A_{jk} = [B^\top A^\top]_{ij}$$

and the second statement follows from $AB(B^{-1}A^{-1}) = I = B^{-1}A^{-1}AB$. ∎

Proposition C.1.4. *For any non-singular matrix A we have*

$$(A^\top)^{-1} = (A^{-1})^\top.$$

We denote the matrix above as $A^{-\top}$.

Proof. By the previous proposition

$$A^\top (A^{-1})^\top = (A^{-1}A)^\top = I^\top = I$$
$$(A^{-1})^\top A^\top = (AA^{-1})^\top = I^\top = I.$$

∎

Definition C.1.10. An orthogonal matrix A is a square matrix whose columns are orthonormal vectors, or equivalently A satisfies $AA^\top = I$.

Proposition C.1.5. *Multiplication by an orthogonal matrix preserves inner products and norms*

$$\langle \boldsymbol{v}, \boldsymbol{w} \rangle = \langle A\boldsymbol{v}, A\boldsymbol{w} \rangle$$
$$\|A\boldsymbol{x}\| = \|\boldsymbol{x}\|.$$

Proof.
$$\langle A\boldsymbol{v}, A\boldsymbol{w} \rangle = \boldsymbol{v}^\top A^\top A\boldsymbol{w} = \boldsymbol{v}^\top \boldsymbol{w}.$$

∎

As a consequence of the above proposition, mapping vectors by multiplying with an orthogonal matrix preserves the angle between the two vectors (see Section B.4). Moreover norm preservation implies that multiplying a vector by an orthogonal matrix preserves the distance between the vector and the origin. We thus interpret the mapping $\boldsymbol{x} \mapsto A\boldsymbol{x}$ for an orthogonal A as rotation or reflection in n-dimension.

```
# a 2x2 orthogonal matrix
A = matrix(c(1, 0, 0, -1), nrow = 2)
A
```

```
##      [,1] [,2]
## [1,]    1    0
## [2,]    0   -1
```

```
# verify that it is orthogonal
A %*% t(A)
```

```
##      [,1] [,2]
## [1,]    1    0
## [2,]    0    1
```

```
# the corresponding linear transformation v
# - > Av corresponds to reflection across x
# axis
A %*% c(2, 3)
```

```
##      [,1]
## [1,]    2
## [2,]   -3
```

Definition C.1.11. The Kronecker product of two matrices $A \in \mathbb{R}^{m \times m}$ and $B \in \mathbb{R}^{n \times n}$ is

$$A \otimes B = \begin{pmatrix} A_{11}B & A_{12}B & \cdots & A_{1m}B \\ A_{21}B & A_{22}B & \cdots & A_{2m}B \\ \vdots & \vdots & \vdots & \vdots \\ A_{m1}B & A_{22}B & \cdots & A_{mm}B \end{pmatrix}$$

where $A_{ij}B$ is the matrix corresponding to the product of a scalar and a matrix.

Note that the Kronecker product is consistent with the earlier definition of an outer product of two vectors $v \otimes w$:

$$u \otimes v = uv^{\top}.$$

Example C.1.5. *The matrix $I \otimes B$ is the block diagonal matrix*

$$I \otimes B = \begin{pmatrix} B & 0 & \cdots & 0 \\ 0 & B & \cdots & 0 \\ \vdots & \vdots & \vdots & \vdots \\ 0 & 0 & \cdots & B \end{pmatrix}.$$

C.2 Rank

Definition C.2.1. The linear space spanned by the vectors $S = \{\boldsymbol{v}^{(1)}, \ldots, \boldsymbol{v}^{(n)}\}$ is the set of all linear combinations of vectors in S. A basis of a linear space S is any set of linearly independent vectors that span it. The dimension $\dim S$ of a linear space S is the size of its basis.

Example C.2.1. *The space $\mathbb{R}^n$ is spanned by the standard basis $\boldsymbol{e}^{(i)}, i = 1, \ldots, n$ from Example C.1.4. Since the standard basis vectors are linearly independent, they are a basis for $\mathbb{R}^n$ in the sense of the previous definition. Another possible basis for $\mathbb{R}^n$ is $\{\boldsymbol{u}^{(i)} : i = 1, \ldots, n\}$, where $\boldsymbol{u}^{(i)}$ is defined by $u_j^{(i)} = 1$ if $j \leq i$ and 0 otherwise.*

Definition C.2.2. The column space of a matrix $A \in \mathbb{R}^{n \times m}$ is the space $\mathsf{col}(A)$ spanned by the columns of A or equivalently

$$\mathsf{col}(A) = \{A\boldsymbol{v} : \boldsymbol{v} \in \mathbb{R}^{m \times 1}\} \subset \mathbb{R}^n.$$

We refer to $\dim \mathsf{col}(A)$ as the rank of the column space of A. The row space of $A \in \mathbb{R}^{n \times m}$ is the space $\mathsf{row}(A)$ spanned by the rows of A. The null space of $A \in \mathbb{R}^{n \times m}$ is

$$\mathsf{null}(A) = \{\boldsymbol{v} : A\boldsymbol{v} = \boldsymbol{0}\} \subset \mathbb{R}^m.$$

Proposition C.2.1. *For any matrix A,*

$$\dim \mathsf{col}(A) = \dim \mathsf{row}(A).$$

We denote that number as $\mathsf{rank}(A)$.

Proof. Consider a matrix A with $\dim \mathsf{col}(A) = r$ and arrange the r vectors spanning $\mathsf{col}(A)$ as columns of a matrix C. Since the columns of A are linear combination of the columns of C we have $A = CR$ for some matrix R with r rows. Every row of $A = CR$ is a linear combination of the rows of R and thus the row space of A is a subset of the row space of R whose dimension is bounded by r. The reverse inequality is obtained by applying the same argument to $A^\top$. ∎

Definition C.2.3. A matrix $A \in \mathbb{R}^{n \times m}$ has full rank is a matrix for which $\mathsf{rank}(A) = \min(n, m)$. Otherwise the matrix has low rank.

Proposition C.2.2.

$$\mathsf{rank}(AB) \leq \min(\mathsf{rank}\, A, \mathsf{rank}\, B).$$

Proof. Since the columns of AB are linear combinations of the columns of A, $\mathsf{rank}(AB) \leq \mathsf{rank}(A)$. Similarly, since the rows of AB are linear combinations of the rows of B, $\mathsf{rank}(AB) \leq \mathsf{rank}(B)$. ∎

Proposition C.2.3. *For non-singular matrices P, Q we have*

$$\text{rank}(PAQ) = \text{rank}(A).$$

Proof. Applying the above proposition, we have

$$\text{rank}(PAQ) \leq \text{rank}(AQ) \leq \text{rank}(A),$$
$$\text{rank}(A) = \text{rank}(IAI) = \text{rank}(P^{-1}PAQQ^{-1}) \leq \text{rank}(PAQ).$$

∎

Proposition C.2.4. *For a matrix $A \in \mathbb{R}^{m \times n}$,*

$$\text{rank}(A) + \text{dim}(\text{null}(A)) = n.$$

Proof. For $s = \text{dim}(\text{null}(A))$, let $\boldsymbol{\alpha}^{(i)}, i = 1, \ldots, s$ be a basis for $\text{null}(A)$, and extend it with the vectors $\boldsymbol{\beta}^{(i)}$ $i = 1, \ldots, n - s$ that together span $\mathbb{R}^n$. Given $\boldsymbol{v} \in \text{col}(A)$, we have

$$\boldsymbol{v} = A\boldsymbol{x} = A \left(\sum_{i=1}^{s} a_i \boldsymbol{\alpha}^{(i)} + \sum_{i=1}^{n-s} b_i \boldsymbol{\beta}^{(i)} \right) = \sum_{i=1}^{n-s} b_i A\boldsymbol{\beta}^{(i)}.$$

(The second equality above follows from representing $\boldsymbol{x}$ as a linear combination of the basis composed by $\boldsymbol{\alpha}^{(i)}, i = 1, \ldots, s$ and $\boldsymbol{\beta}^{(i)}$ $i = 1, \ldots, n - s$.) We thus have that every vector in $\text{col}(A)$ can be written as a linear combination of $n - s$ vectors $\boldsymbol{\gamma}^{(i)} = A\boldsymbol{\beta}^{(i)}$. This shows that $\text{rank}(A) \leq n - \text{dim}(\text{null}(A))$. Assuming that $\sum_{i=1}^{n-s} c_i \boldsymbol{\gamma}^{(i)} = 0$ we have $A \sum_{i=1}^{n-s} c_i \boldsymbol{\beta}^{(i)} = 0$ implying $\sum_{i=1}^{n-s} c_i \boldsymbol{\beta}^{(i)} \in \text{null}(A)$. But by construction of $\boldsymbol{\beta}^{(i)}$ this is possible only if $c_1 = \cdots = c_{n-s} = 0$ which implies that $\boldsymbol{\gamma}^{(i)}, i = 1, \ldots, n - s$ are linearly independent. Since the column space of A is spanned by $n - s$ linearly independent vectors, $\text{rank}(A) = n - s$. ∎

Proposition C.2.5.

$$\text{rank}(A) = \text{rank}(A^{\top}) = \text{rank}(AA^{\top}) = \text{rank}(A^{\top}A).$$

Proof. Since $A\boldsymbol{x} = 0$ implies $A^{\top}A\boldsymbol{x} = 0$, and since $A^{\top}A\boldsymbol{x}$ implies $\boldsymbol{x}A^{\top}A\boldsymbol{x} = 0$, which implies $A\boldsymbol{x} = 0$, we have $\text{null}(A) = \text{null}(A^{\top}A)$. Since both matrices have the same number of columns, the previous proposition implies that $\text{rank}(A) = \text{rank}(A^{\top}A)$. Applying the same argument to $A^{\top}$ shows that $\text{rank}(A^{\top}) = \text{rank}(AA^{\top})$. We conclude the proof by noting that $\text{rank}(A) = \text{rank}(A^{\top})$ as implied by Proposition C.2.1. ∎

C.3 Eigenvalues, Determinant, and Trace

Definition C.3.1. An eigenvector-eigenvalue pair of a square matrix A is a pair of a vector and scalar $(\boldsymbol{v}, \lambda)$ for which $A\boldsymbol{v} = \lambda \boldsymbol{v}$. The spectrum of a matrix A is the set of all its eigenvalues.

We make the following observations.

- The above definition implies that the eigenvectors and eigenvalues of A are solutions of the vector equation $(A - \lambda I)v = 0$.

- The vector v can only be a solution of $(A - \lambda I)v = 0$ if $\dim \text{null}(A - \lambda I) \geq 1$, implying that $(A - \lambda I)$ is a singular matrix.

- If v is an eigenvector of A, then so is cv (with the same eigenvalue).

Definition C.3.2. A permutation of order n is a one to one and onto function from $\{1, \ldots, n\}$ to $\{1, \ldots, n\}$. The product of two permutations $\pi\sigma$ is defined to be their function composition $\pi \circ \sigma$. The set of all permutations of order n is denoted $\mathfrak{S}_n$.

Definition C.3.3. A pair $a, b \in \{1, \ldots, n\}$ is an inversion of a permutation $\sigma \in \mathfrak{S}_n$ if $a < b$ but $\sigma(b) < \sigma(a)$. In other words, the function σ reverses the natural order of a, b. The number of inversions of a permutation is called its parity. Permutations with an even parity are called even and permutations with an odd parity are called odd. The sign of a permutation σ, denoted $s(\sigma)$, equals 1 if σ is even and -1 if σ is odd.

Definition C.3.4. The determinant of a square matrix $A \in \mathbb{R}^{n \times n}$ is defined as

$$\det A = \sum_{\sigma \in \mathfrak{S}_n} s(\sigma)\, A_{1,\sigma(1)} A_{2,\sigma(2)} \cdots A_{n,\sigma(n)}.$$

The definition above states that the determinant is a sum of many terms, each a product of matrix elements from each row and with differing columns. The sum alternates between adding and subtracting these products, depending on the parity of the permutation.

Example C.3.1. *The permutation of a 2×2 matrix A is*

$$\det A = A_{11} A_{22} - A_{12} A_{21}.$$

There are precisely two permutations in $\mathfrak{S}_2$: the identity σ_1 ($\sigma_1(i) = i$) and the non-identity σ_2 ($\sigma_2(1) = 2$ and $\sigma_2(2) = 1$). The identity permutation has zero inversions and is therefore even. The permutation σ_2 has one inversion (the pair (1,2) and is therefore odd).

Example C.3.2. *The permutation of a 3×3 matrix A is*

$$\det A = A_{11} A_{22} A_{33} - A_{11} A_{23} A_{32} - A_{12} A_{21} A_{33}$$
$$+ A_{1,2} A_{2,3} A_{3,1} + A_{1,3} A_{2,1} A_{3,2} - A_{1,3} A_{2,2} A_{3,1}.$$

The first term corresponds to the identity permutation, which is even. The next two terms correspond to permutations with a single inversion ((2,3) in the first case and (1,2) in the second case), making them odd. The next two terms have two inversions making them even ((2,3) and (1,3) in the first case and (1,2) and (1,3) in the second case). The last term has three inversions making it odd ((1,2), (1,3), and (2,3)).

Proposition C.3.1. *The determinant of a triangular matrix is the product of its diagonal elements.*

Proof. All products in the definition of the determinant zero out except for the single product containing all diagonal elements.   ∎

Note that the above proposition applies in particular to diagonal matrices.

Proposition C.3.2. *The following properties hold*

1. $\det I = 1$.

2. $\det A$ *is a linear function of the j-column vector $v = (A_{1j}, \ldots, A_{nj})$ assuming other columns are held fixed.*

3. *If A' is obtained from A by interchanging two columns then $\det A = -\det A'$.*

4. *If A has two equal columns then $\det A = 0$.*

5. $\det(A) = \det(A^\top)$.

Proof. The first property follows from Proposition C.3.1. To prove the second property we note that multiplying the j column by a c multiplies every product in the sum by c and therefore it multiplies the determinant itself by c. Adding a vector w to the j-column v of A results in a new matrix A' whose determinant has product terms containing a sum $(v_i + w_i)$ that may be expanded into a sum of the two determinants – one with v as the j-column and one with w as the j-column. To prove the third property note that each product term in the sum defining the determinant of A' corresponds to a product term in the sum defining the determinant of A but with the sign inverted (since the parity function changes sign). The fourth property if a corollary of the third property. Examining the definition of the determinant as a sum of products, we see that $\det(A) = \det(A^\top)$.
∎

Proposition C.3.3. *For any two square matrices A, B we have*

$$\det(BA) = \det B \det A$$
$$\det(A^{-1}) = (\det A)^{-1}.$$

Proof. The proof of $\det(BA) = \det B \det A$ follows from the previous proposition applied to $f(A) = \det(BA)$. For a detailed proof of this result see most linear algebra textbook or [37, Chapter 9]. The statement $\det(A^{-1}) = (\det A)^{-1}$ follows from $\det(BA) = \det B \det A$ with $B = A^{-1}$ and the fact that $\det I = 1$.   ∎

Proposition C.3.4. *A square matrix is singular if and only if its determinant is zero.*

Proof. If A is non-singular $1 = \det I = \det AA^{-1} = \det A \det A^{-1}$ so that $\det A$ is non zero. If A is singular then its columns are linearly dependent and there is one column, say $\boldsymbol{v}^{(j)}$ that is a linear combination of the other columns $\boldsymbol{v}^{(j)} = \sum_{k \neq j} a_k \boldsymbol{v}^{(k)}$. Since the determinant of a matrix with two identical columns is zero, we can replace $\boldsymbol{v}^{(j)}$ by $\boldsymbol{v}^{(k)}$, $k \neq j$ and get a matrix whose determinant is zero. By the second and fourth properties of Proposition C.3.2, replacing $\boldsymbol{v}^{(j)}$ by $\boldsymbol{v}^{(j)} - \sum_{k \neq j} a_k \boldsymbol{v}^{(k)}$ results in a matrix whose determinant is the same as the original matrix. Since doing so results in a determinant of a matrix with a zero column, $\det A = 0$. ∎

Definition C.3.5. The characteristic polynomial of a square matrix A is the function

$$f(\lambda) = \det(\lambda I - A), \qquad \lambda \in \mathbb{R}.$$

Substituting the definition of the determinant in the equation above, we see that $f(\lambda)$ is indeed a polynomial function in λ.

Recall from the previous section that for $(\lambda, \boldsymbol{v})$ to be eigenvalue-eigenvector of A the matrix $(A - \lambda I)$ must be singular. Combining this with the proposition above, we get that the eigenvalues are the roots of the characteristic polynomial:

$$f(\lambda) = \det(\lambda I - A) = 0.$$

This observation leads to a simple procedure for finding the eigenvalues of a given square matrix A by finding the roots of $f(\lambda)$ (either analytically or numerically). Once the eigenvalues $\lambda_1, \ldots, \lambda_k$ are known, we can obtain the eigenvectors by solving the linear equations

$$(A - \lambda_i I)\boldsymbol{v}^{(i)} = \boldsymbol{0}, \qquad i = 1, \ldots, k.$$

Since the eigenvalues $\lambda_1, \ldots, \lambda_n$ are the roots of the characteristic polynomial $f(\lambda)$, we can write it as the following product

$$f(\lambda) = \det(\lambda I - A) = \prod_i (\lambda - \lambda_i). \tag{C.1}$$

This factorization applies for any polynomial $f(x) = \prod(x - x_i)$ where x_i are the roots. For example $f(x) = x^2 - 3x + 2 = (x - 1)(x - 2)$.

Example C.3.3. *The characteristic polynomial corresponding to the matrix* $\begin{pmatrix} 1 & 1 \\ 0 & 1 \end{pmatrix}$ *is*

$$0 = \det \begin{pmatrix} \lambda - 1 & -1 \\ 0 & \lambda - 1 \end{pmatrix} = (\lambda - 1)^2,$$

whose solution over $\lambda \in \mathbb{R}$ is $\lambda = 1$. There is thus a single eigenvalue $\lambda = 1$. To find the eigenvector or eigenvectors we solve the linear system of equations

$$\mathbf{0} = \begin{pmatrix} 1 - \lambda & 1 \\ 0 & 1 - \lambda \end{pmatrix} \mathbf{v} = \begin{pmatrix} 0 & 1 \\ 0 & 0 \end{pmatrix} \mathbf{v}$$

whose solution is $\mathbf{v} = (a, 0)$ for all a, for example $\mathbf{v} = (1, 0)$.

Definition C.3.6. The trace of a square matrix $\text{tr}(A)$ is the sum of its diagonal elements.

Proposition C.3.5. *The following properties hold:*

$$\text{tr}(A + B) = \text{tr}(A) + \text{tr}(B)$$
$$\text{tr}(AB) = \text{tr}(BA).$$

Proof. The first property is trivial. The second follows from

$$\text{tr}(AB) = \sum_k \sum_j A_{kj} B_{jk} = \sum_k \sum_j B_{kj} A_{jk} = \text{tr}(BA).$$

■

Definition C.3.7. The Frobenius norm of a matrix A is the Euclidean norm of the vector obtained by concatenating the matrix columns into one long vector

$$\|A\|_F \overset{\text{def}}{=} \sqrt{\sum_i \sum_j A_{ij}^2}.$$

Since the diagonal elements of $AA^\top$ are the sum of squares of the rows of A, and the diagonal elements of $A^\top A$ are the sum of squares of the columns of A, we have

$$\|A\|_F = \sqrt{\text{tr}(A^\top A)} = \sqrt{\text{tr}(AA^\top)}.$$

Proposition C.3.6. *For any matrix A and any orthogonal matrix U,*

$$\|UA\|_F = \|A\|_F.$$

Proof.

$$\|UA\|_F = \sqrt{\text{tr}((UA)^\top UA)} = \sqrt{\text{tr}(A^\top A)} = \|A\|_F.$$

■

Proposition C.3.7. *If A is a square matrix with eigenvalues λ_i, $i = 1, \ldots, n$*

$$\text{tr}\, A = \sum_{i=1}^n \lambda_i$$

$$\det A = \prod_{i=1}^n \lambda_i.$$

Proof. Expressing the characteristic polynomial $f(\lambda)$ as a product $\prod_i (\lambda - \lambda_i)$, we get

$$0 = f(\lambda) = \prod_{i=1}^n (\lambda - \lambda_i) = \lambda^n - \lambda^{n-1} \sum_{i=1}^n \lambda_i + \cdots + (-1)^n \prod_{i=1}^n \lambda_i. \qquad \text{(C.2)}$$

Recall from the second statement of Proposition C.3.2 that multiplying a column by -1 inverts the sign of the determinant. It follows that $\det A = (-1)^n \det(-A)$, and since $f(0) = \det(-A) = (-1)^n \prod \lambda_i$, we have $\det A = \prod_{i=1}^n \lambda_i$.

Substituting the definition of the determinant in $\det(\lambda I - A)$, we see that the only terms of power λ^{n-1} result from a multiplication of the diagonal terms $\prod_{i=1}^n (\lambda - A_{ii})$. More specifically, there are n terms containing a power λ^{n-1} in the determinant expansion: $-A_{11}\lambda^{n-1}, \ldots, -A_{11}\lambda^{n-1}$. Collecting these terms, we get that the coefficient associated with λ^{n-1} in the characteristic polynomial is $-\operatorname{tr}(A) = \sum_{i=1}^n A_{ii}$. Comparing this to the coefficient of λ^{n-1} in the equation above we get that $\operatorname{tr}(A) = \sum \lambda_i$. ∎

The proposition below is one of the central results in linear algebra. A proof is available in most linear algebra textbooks.

Proposition C.3.8 (Spectral Decomposition). *For a symmetric matrix A we have*

$$U^\top A U = \operatorname{diag}(\boldsymbol{\lambda}) \quad or \quad A = U \operatorname{diag}(\boldsymbol{\lambda}) U^\top,$$

where $\boldsymbol{\lambda}$ is the vector of eigenvalues and U is an orthogonal matrix whose columns are the corresponding eigenvectors.

Proposition C.3.9. *A square orthogonal matrix is non-singular and has determinant +1 or -1.*

Proof. Examining the definition of the determinant, we see that $\det(A) = \det(A^\top)$. It then follows that

$$1 = \det(I) = \det(AA^\top) = \det A \det A^\top = (\det A)^2.$$

∎

Proposition C.3.10. *A square matrix is orthogonal if and only if $A^{-1} = A^\top$. Moreover, A is orthogonal if and only if $A^\top$ is orthogonal.*

Proof. Since an orthogonal matrix is non-singular it has an inverse and since the inverse is unique, it must coincide with $A^\top$ and we have $I = A^{-1}A = A^\top A$. ∎

Recall that every linear transformation T is realized by a matrix multiplication operation: $T(\boldsymbol{x}) = A\boldsymbol{x}$. If A is orthogonal, the mapping $\boldsymbol{x} \mapsto A\boldsymbol{x}$ may be interpreted as a geometric rotation or reflection around the axis and the mapping $\boldsymbol{x} \mapsto A^\top \boldsymbol{x} = A^{-1}\boldsymbol{x}$ is the inverse rotation or reflection. If A is diagonal, the mapping $\boldsymbol{x} \mapsto A\boldsymbol{x}$ may be interpreted as stretching some dimensions and compressing other dimensions. Applying the spectral decomposition to a symmetric A, we get a decomposition of A as a product of three matrices $U \operatorname{diag}(\boldsymbol{\lambda}) U^\top$. This implies that the linear transformation $T(\boldsymbol{x}) = A\boldsymbol{x}$ can be viewed as a sequence of three linear transformations: the first begin a rotation or reflection, the second being scaling of the dimensions, and the third being another rotation or reflection.

Proposition C.3.11. *For a symmetric A, $\operatorname{rank}(A)$ equals the number of non-zero eigenvalues.*

Proof. By the spectral decomposition we have

$$\operatorname{rank}(A) = \operatorname{rank}(U^\top \operatorname{diag}(\boldsymbol{\lambda}) U) = \operatorname{rank}(\operatorname{diag}(\boldsymbol{\lambda})).$$

∎

Proposition C.3.12. *A symmetric $A \in \mathbb{R}^{n \times n}$ has n orthonormal eigenvectors and $\operatorname{col}(A)$ is spanned by the eigenvectors corresponding to non-zero eigenvalues.*

Proof. The spectral decomposition $AU^\top = UA$ implies that the columns of U are orthonormal eigenvectors. If $\boldsymbol{v} \in \operatorname{col}(A)$ then $\boldsymbol{v} = A\boldsymbol{x}$ for some $\boldsymbol{x}$. Expressing $\boldsymbol{x}$ as a linear combination of the columns $\boldsymbol{\alpha}^{(i)}$, $i = 1, \ldots, n$ of U we have that $\boldsymbol{v} = A\boldsymbol{x} = A\sum a_i \boldsymbol{\alpha}_i = \sum a_i \lambda_i \boldsymbol{\alpha}^{(i)}$ where the second sum is over the indices corresponding to non-zero eigenvalues. ∎

C.4   Positive Semi-Definite Matrices

Definition C.4.1. A square symmetric matrix $H \in \mathbb{R}^{n \times n}$ is positive semi-definite (psd) if

$$\boldsymbol{v}^\top H \boldsymbol{v} \geq 0, \qquad \forall \boldsymbol{v} \in \mathbb{R}^n$$

and positive definite (pd) if the inequality holds with equality only for vectors $\boldsymbol{v} = \boldsymbol{0}$. A square symmetric matrix $H \in \mathbb{R}^{n \times n}$ is negative semi-definite (nsd) if

$$\boldsymbol{v}^\top H \boldsymbol{v} \leq 0, \qquad \forall \boldsymbol{v} \in \mathbb{R}^n$$

and negative definite (nd) if the inequality holds with equality only for vectors $\boldsymbol{v} = \boldsymbol{0}$.

We make the following observations.

- The matrix A is psd if any only if $-A$ is nsd, and similarly a matrix A is pd if and only if $-A$ is nd.

- The psd and pd concepts are denoted by $0 \preceq A$ and $0 \prec A$, respectively. The nsd and nd concepts are denoted by $A \preceq 0$ and $A \prec 0$, respectively.

- The notations above can be extended to denote a partial order on matrices: $A \preceq B$ if and only if $A - B \preceq 0$ and $A \prec B$ if any only if $A - B \prec 0$. Note that $A \prec B$ does not imply that all entries of A are smaller than all entries of B.

Proposition C.4.1. *A symmetric matrix is psd if and only if all eigenvalues are non-negative. It is nsd if and only if all eigenvalues are non-positive. It is pd if and only if all eigenvalues are positive. It is nd if and only if all eigenvalues are negative.*

Proof. Let v be an arbitrary vector. Using the spectral decomposition, we have

$$v^\top A v = (v^\top U) \, \text{diag}(\lambda)(U^\top v) = \sum_{i=1}^n \lambda_i ([v^\top T]_i)^2,$$

where U is a matrix containing the n orthogonal eigenvectors of A. The above expression is non-negative for all v if and only if $\lambda_i \geq 0$ for all $i = 1, \ldots, n$. The rest of the proof is similar. ∎

Proposition C.4.2. *Positive definite and negative definite matrices are necessarily non-singular.*

Proof. Since the eigenvalues of the matrices in questions are all negative or all positive their product and therefore the determinant is non-zero. ∎

Proposition C.4.3. *A symmetric matrix of rank r is psd if and only if there exists a square matrix R of rank r such that $A = R^\top R$. If A is pd then R is singular.*

Proof. Let A be a psd matrix A of rank r. Then it has r non-zero eigenvalues and we can write its spectral decomposition as

$$A = V \, \text{diag}(\lambda_1, \ldots, \lambda_r) V^\top = V \, \text{diag}(\sqrt{\lambda_1}, \ldots, \sqrt{\lambda_r}) \, \text{diag}(\sqrt{\lambda_1}, \ldots, \sqrt{\lambda_r}) V^\top,$$

where V is a matrix whose columns contain the r eigenvectors corresponding to non-zero eigenvalues. Conversely, if $A = RR^\top$ then $\text{rank}(A) = \text{rank}(R) = r$ (see Proposition C.2.5) and

$$(v^\top R)(R^\top v) = w^\top w \geq 0.$$

Finally, if $A = R^\top R$ is pd then it is non-singular and therefore of full rank. It follows that R is also non-singular and of full rank (see Proposition C.2.5). ∎

Proposition C.4.4. *If A is pd then so is A^{-1}*

Proof. sLet A be a pd matrix. Using the previous proposition, we have

$$A^{-1} = (RR^\top)^{-1} = R^{-1}(R^\top)^{-1} = R^{-1}(R^{-1})^\top$$

where R is a non-singular square matrix. Using the previous proposition again (the converse part), we get that A^{-1} is pd. ∎

Proposition C.4.5. *The diagonal elements of a pd matrix are all positive.*

Proof. Using the standard basis $e^{(i)}$ (see Example C.1.4), we have

$$e^{(i)^\top} A e^{(i)} = A_{ii}, \qquad i = 1, \ldots, n.$$

It follows that if A is pd, $A_{ii} > 0, i = 1, \ldots, n$. ∎

Proposition C.4.6. *If A is positive definite there exists a square root matrix $A^{1/2}$ for which $A^{1/2} A^{1/2} = A$.*

Proof. Let A be a pd matrix with positive eigenvalues. Using the spectral decomposition, we have

$$
\begin{aligned}
A &= U^\top \operatorname{diag}(\lambda_1, \ldots, \lambda_n) U \\
&= (U^\top \operatorname{diag}(\sqrt{\lambda_1}, \ldots, \sqrt{\lambda_n}) U)(U^\top \operatorname{diag}(\sqrt{\lambda_1}, \ldots, \sqrt{\lambda_n}) U) \\
&= A^{1/2} A^{1/2}.
\end{aligned}
$$

∎

C.5   Singular Value Decomposition

The singular value decomposition (SVD) generalizes the spectral decomposition for non-symmetric matrices.

Proposition C.5.1 (Singular Value Decomposition). *For any matrix $A \in \mathbb{R}^{n \times m}$, we have*

$$A = U \Sigma V^\top, \qquad \text{where}$$

- $U \in \mathbb{R}^{n \times n}$ *is an orthogonal matrix whose columns are the eigenvectors of $AA^\top$*

- $V \in \mathbb{R}^{m \times m}$ *is an orthogonal matrix whose columns are the eigenvectors of $A^\top A$*

- $\Sigma \in \mathbb{R}^{n \times m}$ *is an all zero matrix except for the first r diagonal elements $\sigma_i = \Sigma_{ii}, i = 1, \ldots, r$ (called singular values) that are the square roots of the eigenvalues of $A^\top A$ and of $AA^\top$ (these two matrices have the same eigenvalues).*

We assume above that the singular values are sorted in descending order and the eigenvectors are sorted according to descending order of their eigenvalues.

Proof. We assume in the proof below (without loss of generality) that $n \geq m$. The case $m > n$ can then be established by transposing the SVD of $A^\top$:

$$A = (A^\top)^\top = (U' \Sigma' V'^\top)^\top = V' (U' \Sigma')^\top = V' \Sigma'^\top U'^\top.$$

Note that since $n \geq m$,

$$\Sigma = \begin{pmatrix} \operatorname{diag}(\sigma_1, \ldots, \sigma_m) \\ \mathbf{0} \end{pmatrix}, \qquad \sigma_i = 0 \text{ for } r < i \leq m.$$

Assuming $\mathsf{rank}(A) = r$ we have that also $\mathsf{rank}(A^\top A) = r$ and the spectral decomposition of $A^\top A$ implies

$$A^\top AV = V \, \mathsf{diag}(\sigma_1^2, \ldots, \sigma_r^2, 0, \ldots, 0)$$

where σ_i^2 are the eigenvalues of $A^\top A$ and the columns of V, denoted $\boldsymbol{v}^{(i)}$, are the corresponding orthonormal eigenvectors. Defining $\boldsymbol{u}^{(i)} = A\boldsymbol{v}^{(i)}/\sigma_i$ we note that and

$$A^\top \boldsymbol{u}^{(i)} = A^\top A \boldsymbol{v}^{(i)}/\sigma_i = \sigma_i \boldsymbol{v}^{(i)}$$
$$AA^\top \boldsymbol{u}^{(i)} = \sigma_i A\boldsymbol{v}^{(i)} = \sigma_i^2 \boldsymbol{u}^{(i)},$$

implying that $\boldsymbol{u}^{(i)}$ are eigenvectors of $AA^\top$ corresponding to eigenvalues σ_i^2. Since the eigenvectors $\boldsymbol{v}^{(i)}$ are orthonormal, then so are the eigenvectors $\boldsymbol{u}^{(i)}$

$$(\boldsymbol{u}^{(i)})^\top \boldsymbol{u}^{(j)} = \frac{(\boldsymbol{v}^{(i)})^\top A^\top A\boldsymbol{v}^{(j)}}{\sigma_i^2} = (\boldsymbol{v}^{(i)})^\top \boldsymbol{v}^{(j)} = \begin{cases} 1 & i = j \\ 0 & i \neq j \end{cases}.$$

We have thus far a matrix V whose columns are eigenvectors of $A^\top A$ with eigenvalues σ_i^2, and a matrix U whose columns are r eigenvectors of $AA^\top$ corresponding to eigenvalues σ_i^2. We augment the eigenvectors $\boldsymbol{u}^{(i)}$, $i = 1, \ldots, r$ with orthonormal vectors $\boldsymbol{u}^{(i)}$, $i = r+1, \ldots, n$ that span $\mathsf{null}(AA^\top)$ and together $\boldsymbol{u}^{(i)}$, $i = 1, \ldots, n$ are a full orthonormal set of eigenvectors of $AA^\top$ with eigenvalues σ_i^2 (with $\sigma_i = 0$ for $i > r$).

Since

$$[U^\top AV]_{ij} = (\boldsymbol{u}^{(i)})^\top A\boldsymbol{v}^{(j)} = \begin{cases} \sigma_j \, (\boldsymbol{u}^{(i)})^\top \boldsymbol{u}^{(j)} & i \leq r \\ 0 & i > r \end{cases}$$

we get $U^\top AV = \Sigma$ and consequentially the desired decompositions $A = U\Sigma V^\top$. ∎

When A has full rank $(r = \min(m,n))$ the singular values $\boldsymbol{\sigma} = (\sigma_1, \ldots, \sigma_r)$ are all strictly positive and $\Sigma = \begin{pmatrix} \mathsf{diag}(\boldsymbol{\sigma}) \\ \mathbf{0} \end{pmatrix}$ or $\Sigma = \begin{pmatrix} \mathsf{diag}(\boldsymbol{\sigma}) & \mathbf{0} \end{pmatrix}$.

Corollary C.5.1. *The SVD decomposition of a rank r matrix A may be written as*

$$A = \sum_{i=1}^{r} \sigma_i \boldsymbol{u}^{(i)} \otimes \boldsymbol{v}^{(j)}$$

where $\boldsymbol{u}^{(i)}, \boldsymbol{v}^{(j)}$ are the columns of U and V respectively. We thus have a decomposition of A as a sum of r rank-one matrices.

Proof. The statement may be easily proved by writing the SVD with Σ as a sum of $\Sigma_i = \mathsf{diag}(0, \ldots, 0, \sigma_i, , 0 \ldots, 1)$ matrices. ∎

The following corollary shows that the SVD is almost identical to the spectral decomposition for symmetric matrices.

Corollary C.5.2. *If A is a symmetric matrix the singular values are the absolute values of the eigenvalues of A: $\sigma_i = |\lambda_i|$ and the columns of $U = V$ are the eigenvectors of A. If in addition A is a symmetric positive definite matrix then U, V, Σ are square non-singular matrices.*

Proof. If A is symmetric then $AA^\top = A^\top A = A^2$ and U, V, Σ are square matrices. The eigenvectors of A are also eigenvectors of A^2 with squared corresponding eigenvalues and the singular values are the absolute values of the eigenvalues of A. If A is pd then it has full rank and a complete set of non-zero eigenvalues. ∎

Corollary C.5.3. *Denoting the vector of singular values by $\boldsymbol{\sigma}$ we have*

$$\|A\|_F = \|\boldsymbol{\sigma}\|.$$

Proof. This is a direct consequence of the SVD and the fact that multiplying by an orthogonal matrix preserves the Frobenius norm. ∎

Recall the intuitive description of the spectral decomposition earlier in this chapter. It states that any linear transformation $T : \mathbb{R}^d \to \mathbb{R}^d$ corresponding to a multiplication by a symmetric matrix is equivalent to rotation or reflection, followed by axis scaling followed by an inverse rotation or reflection. The SVD generalizes this interpretation to non-symmetric and even non-square matrices. Any linear transformation $T : R^m \to \mathbb{R}^n$ may be decomposed as a sequence of a rotation or reflection, followed by axis scaling, followed by another rotation or scaling. Geometrically, this implies that the mapping $\boldsymbol{x} \mapsto A\boldsymbol{x}$ transforms a sphere into an ellipse. The orientations of the ellipse axes are provided by the columns of U, V and the elongation of the different ellipse axes are determined by the singular values.

C.6   Notes

The material concerning basic definitions, rank, eigenvalues and eigenvectors, trace, and determinant are standard in most introductory linear algebra textbooks. An example of a popular textbook is [43]. More details on the more advanced material such as positive definite matrices and matrix decompositions may be found in matrix analysis textbooks, for example [23, 24]. Another useful resource on linear algebra and its role in probability theory is [39].

C.7   Exercises

1. What is the column space of the matrix $\begin{pmatrix} 1 & 1 & 1 \\ 0 & 1 & 1 \\ 0 & 0 & 0 \end{pmatrix}$? (Identify a set of orthogonal vectors that span it).

2. What is the column space, null space, and rank of the matrix whose entries are all ones.

3. Find the eigenvalues and eigenvectors of the matrix $\begin{pmatrix} 2 & 1 \\ 1 & 2 \end{pmatrix}$.

4. Consider the vectors $(1, 1, 0, 0)$ and $(0, 0, 1, 1)$. Show that they are linearly independent, and find two additional vectors that together with the two vectors above form a basis for $\mathbb{R}^4$.

5. Find a matrix of size 4×4 whose null space is spanned by the vector $(1, 0, 0, 0)$.

6. Compute analytically the spectral decomposition and the SVD of the matrix $\begin{pmatrix} 1 & 1 \\ 0 & 1 \end{pmatrix}$.

Appendix D

Differentiation

This chapter covers univariate differentiation (differentiation of a function $f : \mathbb{R} \to \mathbb{R}$) and one dimensional Taylor series. It also includes some standard power series results that are used elsewhere in the book. Multivariate differentiation and Taylor series are covered in Chapter F.

D.1   Univariate Differentiation

Definition D.1.1. Given a function $f : \mathbb{R} \to \mathbb{R}$, the limit (if it exists)

$$f'(x) = \lim_{t \to x} \frac{f(t) - f(x)}{t - x}$$

is called the derivative of f at x. If the derivative exists over a set $A \subset \mathbb{R}$, f is said to be differentiable over that set.

The derivative $f'(x)$ is sometimes denoted $\frac{df(x)}{dx}$ or $df(x)/dx$.

Intuitively, the derivative of a function $f'(x)$ is the slope of the line segment connecting the point $(t, f(t))$ in $\mathbb{R}^2$ to the point $(x, f(x))$ as $t \to x$ (see Figure D.1). If $f'(x)$ is positive, the slope is increasing. If $f'(x)$ is negative, the slope is decreasing. The second derivative $f^{(2)}(x)$ is the derivative of the function $f'(x)$: $f^{(2)}(x) = (f')'(x)$. The third and higher derivatives are similarly defined.

Proposition D.1.1. *If f is differentiable at a point x then it is also continuous at x.*

Proof.

$$\lim_{t \to x} f(t) - f(x) = \lim_{t \to x} \frac{f(t) - f(x)}{t - x}(t - x) = \lim_{t \to x} \frac{f(t) - f(x)}{t - x} \lim_{t \to x}(t - x) = f'(x) \cdot 0 = 0$$

where the last equality follows from the fact that a product is continuous and continuous functions preserves limits (see Chapter B). It follows that $\lim_{t \to x} f(t) = f(x)$, implying that f is continuous. ∎

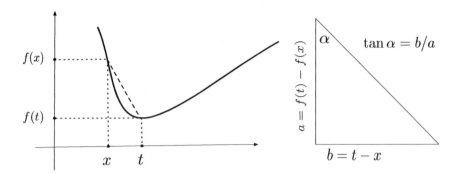

Figure D.1: The ratio of $f(t) - f(x)$ to $t - x$ is the slope of the line connecting $(t, f(t))$ and $(x, f(x))$ ($\tan \alpha = b/a$, right panel). The derivative is the limit of that slope $\tan \alpha$ as $t \to x$.

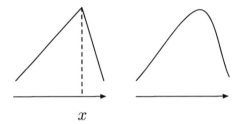

Figure D.2: A function is differentiable if it is smooth rather than angular. The function on the right is differentiable everywhere. The function on the left is differentiable everywhere except at the angular position x.

Intuitively, a function is differentiable at x when it is smooth rather than having an angle. See Figure D.2 for an illustration as well as an example for a function that is continuous but not differentiable.

Proposition D.1.2. *For two differentiable functions f, g we have*

$$(f + g)'(x) = f'(x) + g'(x)$$
$$(fg)'(x) = f'(x)g(x) + f(x)g'(x)$$
$$(f/g)'(x) = \frac{g(x)f'(x) - g'(x)f(x)}{g^2(x)}.$$

Proof. The first statement follows from the facts that the ratio in $(f + g)'(x)$ separates into a sum of two ratios and that $\lim f + g = \lim f + \lim g$. The second statement follows form

$$\lim_{t \to x} \frac{f(t)g(t) - f(x)g(x)}{t - x} = \lim_{t \to x} \frac{f(t)(g(t) - g(x)) + g(x)(f(t) - f(x))}{t - x}$$

$$= f(x) \lim_{t \to x} \frac{g(t) - g(x)}{t - x} + g(x) \lim_{t \to x} \frac{f(t) - f(x)}{t - x},$$

where we used the fact that $\lim uv = \lim u \cdot \lim v$ and Proposition D.1.1. Similarly, the third statement follows form

$$(f/g)'(x) = \lim_{t \to x} \frac{1}{g(t)g(x)} \left(g(x) \frac{f(t) - f(x)}{t - x} - f(x) \frac{g(t) - g(x)}{t - x} \right)$$

$$= \frac{1}{g^2(x)} \lim_{t \to x} \left(g(x) \frac{f(t) - f(x)}{t - x} - f(x) \frac{g(t) - g(x)}{t - x} \right).$$

∎

The derivative of the constant function $f(x) = c$ is zero since the ratio is zero regardless of t. The derivative of $f(x) = x$ is 1 since the ratio is always 1 regardless of t (the numerator and denominator cancel out). Repeated applications of the previous proposition shows that $(x^2)' = 2x$, $(ax^2)' = 2ax$ and by induction $(ax^n)' = anx^{n-1}$. It follows that every polynomial is differentiable, and its derivative may be obtained by applying the rule $(ax^n)' = anx^{n-1}$ to each of the polynomial terms separately and adding up the derivatives.

Lemma D.1.1 (e.g., [46]). *If f is differentiable at y_0 then there exists $r > 0$ such that*

$$f(y) = f(y_0) + (f'(y_0) + E(y))(y - y_0), \qquad |y - y_0| < r$$

for some function E that is continuous at y_0 and satisfies $E(y_0) = 0$.

Proof. The function

$$E(y) = \begin{cases} 0 & y = y_0 \\ \frac{f(y) - f(y_0)}{y - y_0} - f'(y_0) & y \neq y_0 \end{cases}$$

is continuous at $y = y_0$ and satisfies the statement in the proposition. ∎

Proposition D.1.3 (Chain Rule). *For two differentiable function $f, g : \mathbb{R} \to \mathbb{R}$*

$$(f \circ g)'(x_0) = f'(g(x_0))g'(x_0).$$

Proof.

$$\lim_{x \to x_0} \frac{f(g(x)) - f(g(x_0))}{x - x_0} = \lim_{x \to x_0} \frac{(f'(g(x_0)) + E(g(x)))(g(x) - g(x_0))}{x - x_0}$$

$$= f'(g(x_0))g'(x_0)$$

where we used Lemma D.1.1 in the first equality (applied f and $y_0 = g(x_0)$), the continuity of g and of E at $g(x_0)$, and $E(g(x_0)) = 0$ in the second equality. ∎

Definition D.1.2. A function $f : A \subset \mathbb{R} \to \mathbb{R}$ has a local maximum at $x \in A$ if $f(x)$ is higher or equal than all values in $\{f(y) : y \in B_r(x)\}$ for some $r > 0$. A function $f : A \subset \mathbb{R} \to \mathbb{R}$ has a local minimum at $x \in A$ if $f(x)$ is lower or equal than all values in $\{f(y) : y \in B_r(x)\}$ for some $r > 0$.

Proposition D.1.4. *If a differentiable function $f : [a, b] \to \mathbb{R}$ has a local maximum or a local minimum at $x \in (a, b)$, then $f'(x) = 0$.*

Proof. Assume f has a local maximum at x and $f(x)$ is greater or equal than all values $f(y)$ where $|y - x| < r$ for some $r > 0$. Setting $\delta = r$, if $x - \delta < y < x$ then $(f(y) - f(x))/(y - x) \geq 0$, implying that $f'(x) \geq 0$. Similarly if $x < y < x + \delta$, then $(f(y) - f(x))/(y - x) \leq 0$, implying that $f'(x) \leq 0$. We conclude that $f'(x) = 0$. A similar proof applies to the local minimum case. ∎

Proposition D.1.5 (Mean Value Theorem). *For any two differentiable functions $f, g : [a, b] \to \mathbb{R}$ there exists a point $x \in (a, b)$ such that*

$$(f(b) - f(a))g'(x) = (g(b) - g(a))f'(x).$$

In particular, setting $g(x) = x$ we get that there exists a point $x \in (a, b)$ such that

$$f'(x) = \frac{f(b) - f(a)}{b - a}.$$

Proof. The function

$$h(x) = (f(b) - f(a))g(x) - (g(b) - g(a))f(x)$$

is continuous, differentiable, and satisfies

$$h(a) = f(b)g(a) - f(a)g(b) = h(b).$$

Since $[a, b]$ is compact, by Proposition B.3.6, h attains its maximum and minimum on $[a, b]$. Since $h(a) = h(b) = 0$, if $\max_{x \in [a,b]} h(x) \neq 0$, the maximum occurs in (a, b) and by Proposition D.1.4 $h'(x) = 0$. Similarly, if $\max_{x \in [a,b]} h(x) \neq 0$, then $h'(x) = 0$ as well. If both the minimum and maximum are 0 then $h(y)$ equals 0 for all y, and in particular $h'(x) = 0$. ∎

Proposition D.1.6. *For a differentiable function $f : (a, b) \to \mathbb{R}$, we have*

- *If $f'(x) \geq 0$ for all $x \in (a, b)$ then f is monotonic increasing,*

- *if $f'(x) \leq 0$ for all $x \in (a, b)$ then f is monotonic decreasing,*

- *if $f'(x) > 0$ for all $x \in (a, b)$ then f is strictly monotonic increasing,*

- *if $f'(x) < 0$ for all $x \in (a, b)$ then f is strictly monotonic decreasing,*

- *if $f'(x) = 0$ for all $x \in (a, b)$ then f is constant,*

Proof. The mean-value theorem above implies that for all $(u, v) \subset (a, b)$ there exists $x \in (u, v)$ such that

$$f(v) - f(u) = (v - u)f'(x) \geq 0.$$

The rest of the proof is similar. ∎

In many cases it is easy to find the limit of a ratio $\lim(f(x)/g(x))$ using Proposition B.4.8 that implies $\lim(f(x)/g(x)) = (\lim f(x))/(\lim g(x))$, assuming that the (a) denominator is not zero, and (b) the numerator and denominator are finite. For example,

$$\lim_{x \to 0} \frac{x^2 + 4x}{\cos x} = \frac{\lim_{x \to 0} x^2 + 4x}{\lim_{x \to 0} \cos x} = \frac{0}{1} = 0.$$

L'Hospital's rule below may be used to compute $\lim f(x)/g(x)$ when either (a) or (b) fails.

Proposition D.1.7 (L'Hospital's Rule). *For two differentiable functions $f, g : [\alpha, \beta] \to \mathbb{R}$, if*

$$(a): \qquad \lim_{x \to a} f(x) = 0 = \lim_{x \to a} g(x)$$

or

$$(b): \qquad \lim_{x \to a} f(x) = \pm\infty = \lim_{x \to a} g(x)$$

(where the value a may be finite, ∞, or $-\infty$) then

$$\lim_{x \to a} \frac{f(x)}{g(x)} = \lim_{x \to a} \frac{f'(x)}{g'(x)},$$

assuming that the limit in the right hand side above exists.

Proof. We assume that f', g' are continuous and $\lim_{x \to a} f(x) = \lim_{x \to a} g(x) = 0$ (condition (a)). In this case $f(a) = g(a) = 0$ and

$$\lim_{x \to a} \frac{f(x)}{g(x)} = \lim_{x \to a} \frac{\frac{f(x) - f(a)}{x - a}}{\frac{g(x) - g(a)}{x - a}} = \frac{\lim_{x \to a} \frac{f(x) - f(a)}{x - a}}{\lim_{x \to a} \frac{g(x) - g(a)}{x - a}} = \frac{f'(a)}{g'(a)} = \lim_{x \to a} \frac{f'(x)}{g'(x)}.$$

A similar proof applies if condition (b) holds. Proofs without requiring continuity of f', g' appear in [37] or [46]. ∎

Example D.1.1. *We have*

$$\lim_{x \to 0} \frac{x^2 + 4x}{\sin x} = \frac{\lim_{x \to 0} (x^2 + 4x)'}{\lim_{x \to 0} (\sin x)'} = \frac{\lim_{x \to 0} 2x + 4}{\lim_{x \to 0} \cos x} = \frac{4}{1}.$$

Note that an application of the ratio rule $\lim(u/v) = \lim u / \lim v$ does not work here since the limit of the numerator and the limit of the denominator both equal 0.

D.2   Taylor Expansion and Power Series

Taylor's important theorem approximates a smooth function with a polynomial.

Proposition D.2.1 (Taylor Series Theorem). *Let $f : [a, b] \to \mathbb{R}$ be a function whose higher order derivatives exist and define the k-order Taylor approximation at $x \in [a, b]$ around $\alpha \in [a, b]$ to be*

$$P_k(x) = \sum_{l=0}^{k} \frac{f^{(l)}(\alpha)}{l!} (x - \alpha)^l.$$

Then there exists a point y between x and α such that

$$f(x) - P_k(x) = \frac{f^{(k+1)}(y)}{(k+1)!} (x - \alpha)^{k+1}. \tag{D.1}$$

If in addition $\sup_{k \in \mathbb{N}} |f^{(k)}(y)|$ is finite for all $y \in (a, b)$, then $f(x) = \lim_{k \to \infty} P_k(x)$.

Proof. For an arbitrary $x, \alpha \in [a, b]$, we define

$$M \overset{\text{def}}{=} (f(x) - P_k(x))/(x - \alpha)^{k+1},$$
$$g(y) \overset{\text{def}}{=} f(y) - P_k(y) - M(y - \alpha)^{k+1}, \qquad y \in [a, b],$$

and observe that

$$P^{(l)}(\alpha) = f^{(l)}(\alpha) \qquad\qquad l = 0, \ldots, k$$
$$g^{(l)}(\alpha) = 0 \qquad\qquad l = 0, \ldots, k.$$

Since

$$g^{(k+1)}(y) = f^{(k+1)}(y) - (k+1)! M,$$

we will prove (D.1) if we can show that

$$g^{(k+1)}(y) = 0, \qquad \text{for some } y \in B_x(\alpha).$$

Since $g(\alpha) = 0$ (since $g^{(0)}(\alpha) = 0$ – see above) and $g(x) = 0$ (by definition of M), the mean value theorem implies that $g'(z_1) = 0$ for some z_1 that lies between α and x. Similarly, since $g'(\alpha) = 0 = g'(z_1)$ we have $g^{(2)}(z_2) = 0$ for some z_2 lying between α and z_1, and so on until $g^{(k+1)}(z_{k+1}) = 0$ for some z_{k+1} lying between α and x. This concludes the proof of (D.1).

If the higher order derivatives at x are bounded as indicated in the second part of the proposition, then

$$\lim_{k \to \infty} \left| f^{(k+1)}(y) \frac{(x - \alpha)^{k+1}}{(k+1)!} \right| \leq C \lim_{k \to \infty} \left| \frac{(x - \alpha)^{k+1}}{(k+1)!} \right| = 0,$$

where the last equality follows from Equation (1.1) (the factorial function grows faster to ∞ than the exponential function).   ■

If the conditions in the proposition above hold, we can express the Taylor approximation using the big-oh and little-oh notations (see Section B.5) at the limit $x \to \alpha$:

$$f(x) = \sum_{l=0}^{n} \frac{f^{(l)}(\alpha)}{l!}(x - \alpha)^l + O(|x - \alpha|^{n+1})$$

and

$$f(x) = \sum_{l=0}^{n} \frac{f^{(l)}(\alpha)}{l!}(x - \alpha)^l + o(|x - \alpha|^l).$$

The case $\alpha = 0$ is sometimes of particular interest: we have

$$f(x) = \sum_{l=0}^{n} \frac{f^{(l)}(0)}{l!}x^l + O(|x|^{l+1})$$

and

$$f(x) = \sum_{l=0}^{n} \frac{f^{(l)}(0)}{l!}x^l + o(|x|^n).$$

The Taylor approximation P_k approximates f at the point x using higher order derivatives of f, evaluated at α. Thus, functions $f : \mathbb{R} \to \mathbb{R}$ whose higher order derivatives exist and are bounded are completely specified for all values by their derivatives at a single point (any point).

The following proposition is sometimes useful. In many cases, the third statement below is taken to be the definition of the exponential function. We prove it in order to demonstrates the Taylor series proposition above.

Proposition D.2.2.

$$\lim_{n \to \infty} \left(1 + \frac{x}{n}\right)^n = \exp(x) \tag{D.2}$$

$$\lim_{n \to \infty} (1 + a_n)^n = \exp\left(\lim_{n \to \infty} na_n\right) \tag{D.3}$$

$$\lim_{n \to \infty} 1 + \frac{x^1}{1!} + \cdots + \frac{x^n}{n!} = \exp(x). \tag{D.4}$$

(where we assume in the second statement that $y = \lim na_n$ exists and is finite.)

Proof.

$$\log \lim_{n \to \infty} (1 + x/n)^n = \lim_{n \to \infty} \log(1 + x/n)^n = \lim_{n \to \infty} n \log(1 + x/n)$$

$$= \lim_{h \to 0} x \frac{\log(1 + h)}{h} = x \frac{d \log x}{dx}\bigg|_{x=1} = x \cdot 1,$$

where we used the fact that log is a continuous function and the substitution $h = x/n$. The proof of the second statement is similar:

$$\log \lim_{n \to \infty} (1 + a_n)^n = \lim_{n \to \infty} \log(1 + a_n)^n = \lim_{n \to \infty} n \log(1 + a_n)$$

$$= \lim_{h \to 0} n a_n \frac{\log(1 + h)}{h} = \lim_{n \to \infty} n a_n \cdot \lim_{h \to 0} \frac{\log(1 + h)}{h}$$

$$= \lim_{n \to \infty} n a_n \frac{d \log x}{dx}\Big|_{x=1} = y \cdot 1,$$

where we used the substitution $h = a_n$ and used the fact that $a_n \to 0$. The third statement follows from the Taylor series proposition above, using the fact that $\exp(x)^{(k)} = \exp(x)$, and that convergence holds since $\sup_{k \in \mathbb{N}} (\exp(x))^{(k)}$ is finite. ∎

Proposition D.2.3 (Finite Geometric Series).

$$\sum_{k=0}^{n-1} ar^k = a \frac{1 - r^n}{1 - r}, \qquad r \neq 1.$$

Proof. Denoting $R = \sum_{k=0}^{n-1} ar^k$, we have that

$$Rr = ar + ar^2 + \cdots + ar^n$$

$$R(1 - r) = R - Rr = a - ar^n$$

$$R = \frac{a - ar^n}{1 - r}.$$

∎

Corollary D.2.1 (Infinite Geometric Series).

$$\sum_{k=0}^{\infty} ar^k = \begin{cases} a(1 - r)^{-1} & |r| < 1 \\ \text{series does not converge} & \text{otherwise} \end{cases}$$

Proof. Take the limit $n \to \infty$ in the previous proposition and note that we need $|r| < 1$ to ensure convergence. ∎

Example D.2.1. *Using the results above for $x \in B_1(0)$ we have*

$$\frac{1}{1 - x} = 1 + x + x^2 + x^3 + \cdots \tag{D.5}$$

which is also the Taylor series expansion. Differentiating both sides of the equation above we also get for $x \in B_1(0)$,

$$\frac{1}{(1 - x)^2} = \sum_{n=0}^{\infty} (n + 1)x^n = \sum_{n=0}^{\infty} nx^n + \frac{1}{1 - x},$$

implying the following expression

$$\sum_{n=0}^{\infty} nx^n = \frac{1}{(1-x)^2} - \frac{1}{1-x}. \tag{D.6}$$

Differentiating the equation above again, we get the expression

$$\sum_{n=0}^{\infty} n^2 x^n = \left(\frac{1}{(1-x)^2} - \frac{1}{1-x} \right)' x = \frac{2x}{(1-x)^3} - \frac{x}{(1-x)^2}. \tag{D.7}$$

Proposition D.2.4. *Let a_n be a non-increasing sequence of positive numbers. Then the sequence $A = \sum_{n \in \mathbb{N}} a_n$ converges if and only if the sequence $B = \sum_{n=0}^{\infty} 2^n a_{2^n}$ converges.*

Proof. Writing the sequence A as

$$A = (a_1) + (a_2 + a_3) + (a_4 + a_5 + a_6 + a_7) + \cdots$$

we note that each parenthetical term above is less than or equal the first summand in that term times the number of summands: $(a_1) \le a_1$, $(a_2 + a_3) \le a_2 + a_2$, $(a_4 + a_5 + a_6 + a_7) \le a_4 + a_4 + a_4 + a_4$, and so on. This establishes the inequality $A \le B$ and implies that A converges if B converges.

It remains to show that B converges if A converges. To do that we write

$$B = (a_1 + a_2) + (a_2 + a_4 + a_4 + a_4) + (a_4 + a_8 + \cdots) + \cdots$$

and note that $(a_1 + a_2) \le (a_1 + a_1)$, $(a_2 + a_4 + a_4 + a_4) \le (a_2 + a_2 + a_3 + a_3)$, and so on. This implies the inequality

$$B \le a_1 + a_1 + a_2 + a_2 + a_3 + a_3 + a_4 + a_4 + \cdots = 2A,$$

which implies that B converges if A converges. ∎

Proposition D.2.5. *The sequence $\sum_{n \in \mathbb{N}} n^{-\alpha}$ converges if and only if $\alpha > 1$.*

Proof. By Proposition D.2.4, the series $\sum_{n \in \mathbb{N}} n^{-\alpha}$ converges if and only if the series

$$\sum_{n \in \mathbb{N}} 2^n (2^{-n\alpha}) = \sum_{n \in \mathbb{N}} 2^{n - n\alpha} = B$$

converges. This occurs by Corollary D.2.1 if and only if $\alpha > 1$. ∎

Section F.6 describes the multivariate generalization of differentiation and Taylor expansions.

D.3   Notes

The material in this chapter is standard material in any calculus or real analysis textbook. Our exposition follows [37] and [46]. A more elementary description is available in [45].

D.4   Exercises

1. Finish the proof of Proposition D.1.6.

2. Find the Taylor expansion of $\sin(x)$ and $\cos(x)$ around $\alpha = 0$. Use your Taylor expansions to prove that $(\sin(x))' = \cos(x)$ and $(\cos(x))' = -\sin(x)$.

3. Find a concrete example where the Taylor polynomial P_k of f does not converge to f as $k \to \infty$.

4. Find a function whose derivative is the polynomial $Q(x) = a_n x^n + a_{n-1} x^{n-1} + \cdots + a_1 x + a_0$. Are there other functions whose derivatives equal Q?

5. The chapter describes differentiable functions as smooth, and non-differentiable functions as non-smooth. The function $f(x) = |x|$ is smooth everywhere except at $x = 0$ where it has a sharp corner. Prove that $|x|$ is differentiable everywhere except at 0. Argue informally why the characterization of differentiable functions as smooth is appropriate.

6. Express $(f \circ g \circ h)'(x)$ in terms of $f(x), g(x), h(x)$ and their derivatives. Can you generalize your result to a composition of an arbitrary number of differentiable functions?

Appendix E

Measure Theory*

We attempt in this book to circumvent the use of measure theory as much as possible. However, in several places where measure theory is essential we make an exception (for example the limit theorems in Chapter 8 and Kolmogorov's extension theorem in Chapter 6). Sections containing such exceptions are marked by an asterisk.

This chapter contains a introduction to the parts of measure theory that are the most essential to probability theory. The next chapter covers Lebesgue integration and the Lebesgue measure.

E.1 σ-Algebras*

Definition E.1.1. An algebra $\mathcal{C}$ is a non-empty set of sets that satisfies: (i) $\Omega \in \mathcal{C}$, (ii) if $A \in \mathcal{C}$ then $A^c \in \mathcal{C}$, and (iii) if $A, B \in \mathcal{C}$ then $A \cup B \in \mathcal{C}$.

We make the following comments.

- By induction, if $\mathcal{C}$ is an algebra then $A_1, \ldots, A_k \in \mathcal{C}$ implies $A_1 \cup \cdots \cup A_n \in \mathcal{C}$.

- Using De-Morgan's rule and the second property above, $A_1, \ldots, A_k \in \mathcal{C}$ implies $A_1 \cap \cdots \cap A_n \in \mathcal{C}$,

- A more compact definition of an algebra is a set containing Ω that is closed under finite unions, finite intersections, and complements.

Definition E.1.2. A σ-algebra is a non-empty set of sets that is closed under countable unions, countable intersections, and complements.

In other words, if $A_n, n \in \mathbb{N}$ reside in a σ-algebra $\mathcal{A}$ then we also have $\cup_{n \in \mathbb{N}} A_n \in \mathcal{A}$, $\cap_{n \in \mathbb{N}} A_n \in \mathcal{A}$ and $A_n^c \in \mathcal{A}$.

Definition E.1.3. A measurable space $(\Omega, \mathcal{F})$ consists of a set Ω and a σ-algebra of subsets of Ω.

Example E.1.1. *The power set 2^{Ω} is a σ-algebra. It contains all subsets and is therefore closed under complements and countable unions and intersections.*

Note that every σ-algebra necessarily includes $\emptyset$ and Ω since $A_n \cap A_n^c = \emptyset$ and $A_n \cup A_n^c = \Omega$. As a consequence, a σ-algebra is also closed under finite unions and intersections (define A_k above for $k \geq c$ to be either $\emptyset$ or Ω), implying that a σ algebra is also an algebra.

Proposition E.1.1. *A set of sets $\mathcal{A}$ is a σ-algebra if and only if (i) $\Omega \in \mathcal{A}$, (ii) $A \in \mathcal{A}$ implies $A^c \in \mathcal{A}$, and (iii) if $A_n \in \mathcal{A}$ for $n \in \mathbb{N}$ then $\cup_n A_n \in \mathcal{A}$.*

Proof. If $\mathcal{A}$ is a σ-algebra then it obviously satisfies the three properties. If $\mathcal{A}$ satisfies the three properties, it is obviously closed under union and under complements. It remains to show that it is closed under intersections. Let $B_n, n \in \mathbb{N}$ be a sequence of sets in $\mathcal{A}$. Since $\mathcal{A}$ is closed under complements, we have $A_n \stackrel{\text{def}}{=} B_n^c \in \mathcal{A}$ and also $\cup_n A_n \in \mathcal{A}$. Taking the complement again and applying De-Morgan's theorem we get $(\cup_n A_n)^c = \cap_n A_n^c = \cap_n B_n \in \mathcal{A}$. ∎

Proposition E.1.2. *An intersection of multiple σ-algebras is also a σ-algebra.*

Proof. Since each σ algebra contains Ω their intersection is non-empty and it contains Ω as well. If A is a member of the intersection then it is a member of all the σ-algebras and therefore A^c is also a member of all the σ-algebras. It follows that A^c is also in the intersection. The intersections of the σ algebras is closed under countable unions and intersections for the same reason. ∎

Proposition E.1.3. *Given a set of sets $\mathcal{A}$, there exists a unique minimal σ-algebra containing $\mathcal{A}$. We refer to this σ-algebra as the σ-algebra generated by $\mathcal{A}$ and denote it as $\sigma(\mathcal{A})$.*

In other words, there is a σ-algebra containing $\mathcal{A}$ that is a subset of all other σ-algebras containing $\mathcal{A}$.

Proof. Since the power set 2^{Ω} is a σ-algebra containing $\mathcal{A}$ there exists at least one σ-algebra containing $\mathcal{A}$. We define the smallest σ-algebra to be the intersection of all σ-algebras containing $\mathcal{A}$. It is a σ-algebra by Proposition E.1.2 and by construction it is minimal in the sense that is a subset of all other σ-algebras. ∎

Corollary E.1.1. *If $\mathcal{C}$ is a σ-algebra then $\sigma(\mathcal{C}) = \mathcal{C}$.*

Proof. This is a direct corollary of the definition of $\sigma(\mathcal{C})$ as the smallest σ-algebra containing $\mathcal{C}$ and the fact that it is uniquely defined. ∎

It is desirable to have an efficient mechanism for producing the smallest σ-algebra containing a set of sets $\mathcal{C}$. A conceptual mechanism is as follows: (i) create a set $\mathcal{C}'$ containing the sets in $\mathcal{C}$, their complements, and their countable unions and intersections, (ii) repeat step (i) with $\mathcal{C}'$ substituting $\mathcal{C}$. Unfortunately, the mechanism above is not computationally efficient as it may never terminate. The proposition above asserts the existence of a smallest σ-algebra but does not offer any assistance into constructing it.

Definition E.1.4. Let $\mathcal{C}$ be a set of sets and U be a set. We denote

$$\mathcal{C} \cap U = \{C \cap U : C \in \mathcal{C}\}.$$

Proposition E.1.4. *Let $\mathcal{C}$ be a σ-algebra over Ω and $U \subset \Omega$. Then $\mathcal{C} \cap U$ is a σ-algebra over U.*

Proof. Since $U = \Omega \cap U$, we have $U \in \mathcal{C} \cap U$. If $B \in \mathcal{C} \cap U$ then $B = C \cap U$ for some $C \in \mathcal{C}$ and therefore

$$U \setminus B = (\Omega \setminus C) \cap U \in \mathcal{C} \cap U$$

(since $\Omega \setminus C = C^c \in \mathcal{C}$). If $B_n = C_n \cap U, n \in \mathbb{N}$ is a sequence of sets in $\mathcal{C} \cap U$ then $\cup C_n \in \mathcal{C}$ and

$$\bigcup_{n \in \mathbb{N}} B_n = \bigcup_{n \in \mathbb{N}} (U \cap C_n) = \left(\bigcup_{n \in \mathbb{N}} C_n \right) \cap U \in \mathcal{C} \cap U.$$

∎

Proposition E.1.5. *Let $\mathcal{C}$ be a set of sets and U be a set. Then*

$$\sigma(\mathcal{C} \cap U) = \sigma(\mathcal{C}) \cap U.$$

Proof. By the previous proposition $\sigma(\mathcal{C}) \cap U$ is a σ-algebra and

$$\sigma(\mathcal{C} \cap U) \subset \sigma(\sigma(\mathcal{C}) \cap U) = \sigma(\mathcal{C}) \cap U.$$

It remains to show that $\sigma(\mathcal{C}) \cap U \subset \sigma(\mathcal{C} \cap U)$.
    We show that the set

$$\mathcal{G} = \{A : A \cap U \in \sigma(\mathcal{C} \cap U)\}$$

is a σ-algebra. The first requirement holds since $\Omega \in \mathcal{G}$. If $A \in \mathcal{G}$ then $A \cap U \in \sigma(\mathcal{C} \cap U)$, implying that

$$A^c \cap U = U \setminus (A \cap U) \in \sigma(\mathcal{C} \cap U)$$

as well (the set difference of two sets in a σ algebra is in the σ algebra as well). It follows that $A^c \in \mathcal{G}$, showing that the second requirement holds as well. To show the third requirement, note that if $A_n \in \mathcal{G}$ for $n \in \mathbb{N}$ then

$$(\cup_n A_n) \cap U = \cup_n (A_n \cap U) \in \sigma(\mathcal{C} \cap U)$$

(since $A_n \cap U \in \sigma(\mathcal{C} \cap U)$) and thus $\cup_n A_n \in \mathcal{G}$.

Since $\mathcal{G}$ is a σ-algebra and $\mathcal{C} \subset \mathcal{G}$ (if $A \in \mathcal{C}$ then $A \cap U \in \sigma(\mathcal{C} \cap U)$ implying that $A \in \mathcal{G}$) we have

$$\sigma(\mathcal{C}) \subset \sigma(\mathcal{G}) = \mathcal{G}.$$

(The last equality holds since $\mathcal{G}$ is a σ-algebra.) If $A \in \sigma(\mathcal{C})$ then $A \in \mathcal{G}$ and $A \cap U \in \sigma(\mathcal{C} \cap U)$ implying that $A \in \sigma(\mathcal{C} \cap U)$. It follows that $\sigma(\mathcal{C}) \cap U \subset \sigma(\mathcal{C} \cap U)$.
∎

E.2   The Measure Function*

Definition E.2.1. Let $\mathcal{F}$ be an algebra. A measure is a function $\mu : \mathcal{F} \to [0, \infty) \cup \{+\infty\}$ that satisfies:

(Empty set) $\mu(\emptyset) = 0$

(Countable additivity) If $E_n, n \in \mathbb{N}$ is a sequence of disjoint events ($E_i \cap E_j = \emptyset$ whenever $i \neq j$)

$$\mu\left(\bigcup_{i=1}^{\infty} E_i \right) = \sum_{i=1}^{\infty} \mu(E_i).$$

If $\mathcal{F}$ is a σ-algebra, the triplet $(\Omega, \mathcal{F}, \mu)$ is called a measure space.

Contrasting this with Definition 1.2.1, we see that a probability is a measure function that satisfies $\mu(\Omega) = 1$.

Proposition E.2.1 (The Continuity of Measure). *Any measure with $\mu(\Omega) < \infty$ satisfies the following properties*

1 Finite Additivity. *For a finite sequence of disjoint sets $E_1, \ldots, E_k$,*

$$\mu\left(\sum_{i=1}^{k} E_k \right) = \sum_{i=1}^{k} \mu(E_i).$$

2 Continuity from Below. *If $E_n \nearrow E$, $E \in \mathcal{F}$, and $E_n \in \mathcal{F}$ for all $n \in \mathbb{N}$, then $\mu(E_n) \nearrow \mu(E)$.*

3 Continuity from above. *If $E_n \searrow E$, $E \in \mathcal{F}$, $E_n \in \mathcal{F}$ for all $n \in \mathbb{N}$, and $\mu(E_1) < \infty$, then $\mu(E_n) \searrow \mu(E)$.*

4 Countable sub-additivity. *If $E_n \in \mathcal{F}$ and $\cup_n E_n \in \mathcal{F}$ then*

$$\mu\left(\bigcup_{i=1}^{\infty} E_i \right) \leq \sum_{i=1}^{\infty} \mu(E_i).$$

Note that the notations $\nearrow$, $\searrow$ mean convergence of monotonic sequences (Definition B.2.3) and convergence of monotonic sets (Definition A.4.3), depending on whether they are associated with sequences of numbers or sets.

Proof. The first property follows from the countable additivity of the measure function and setting $E_l = \emptyset$ for $l > k$.

We prove next the second property. We define the following sequence of disjoint sets $B_1 = E_1$ and $B_k = E_k \backslash E_{k-1}$ where $E = \cup_{n \in \mathbb{N}} B_n$ and $E_n = \cup_{k=1}^{n} B_k$. By the countable and finite additivity of μ

$$\mu(E) = \sum_{n \in \mathbb{N}} \mu(B_n) = \lim_{n \to \infty} \sum_{k=1}^{n} \mu(B_k) = \lim_{n \to \infty} \mu(E_n)$$

proving (2). To prove property (3) note that $E_1 \backslash E_n \nearrow E_1 \backslash E$, which together with property (2) implies

$$\mu(E_1) - \mu(E_n) = \mu(E_1 \backslash E_n) \nearrow \mu(E_1 \backslash E) = \mu(E_1) - \mu(E).$$

To prove (4), we denote $B = E_1$ and $B_n = E_n \cap E_1^c \cap \cdots \cap E_{n-1}^c$ for $n \in \mathbb{N}$. Note that B_k is a disjoint sequence of sets and $\cup_{n=1}^{k} B_n = \cup_{n=1}^{k} E_n$ for all $k \in \mathbb{N}$. Since $B_k \subset E_k$ we have

$$\mu\left(\bigcup_{n=1}^{k} E_k\right) = \mu\left(\bigcup_{n=1}^{k} B_k\right) = \sum_{n=1}^{k} \mu(B_k) \le \sum_{n=1}^{k} \mu(E_k).$$

It remains to let $k \to \infty$ and use property (2) on the left hand side, applied to the sets $E_k' = \cup_{n=1}^{k} E_n \nearrow \cup_{n \in \mathbb{N}} E_k$.   ∎

E.3   Caratheodory's Extension Theorem*

In many cases it is hard to define a probability function on all sets A in a σ-algebra. Caratheodory's extension theorem shows that it is sufficient to define the probability measure on an algebra $\mathcal{C}$. The probability measure is then uniquely defined on $\sigma(\mathcal{C})$, in a way consistent with its definition on $\mathcal{C}$. The proof of this important result is somewhat lengthy. The existence of the extension is based on Dynkin's theorem, an important result from set theory. The uniqueness of the extension is based on the concept of an outer measure.

E.3.1   Dynkin's Theorem*

Definition E.3.1. A set of sets $\mathcal{C}$ is a π-system if the following holds:

$$A, B \in \mathcal{C} \quad \text{implies} \quad A \cap B \in \mathcal{C}.$$

It follows by induction that for a π-system $\mathcal{C}$, if $A_1, \ldots, A_k \in \mathcal{C}$, we have $A_1 \cap \cdots \cap A_k \in \mathcal{C}$.

Definition E.3.2. A set of sets $\mathcal{C}$ is a λ-system if it satisfies: (i) $\Omega \in \mathcal{C}$, (ii) $A \in \mathcal{C}$ implies $A^c \in \mathcal{C}$, and (iii) $A_n \in \mathcal{C}, n \in \mathbb{N}$ and $A_n \cap A_m = \emptyset$ for $n \ne m$ implies $\cup_n A_n \in \mathcal{C}$.

Note that for a λ-system $\mathcal{C}$, we have $\emptyset = \Omega^c \in \mathcal{C}$, which together with property (iii) implies also that if $A_n \in \mathcal{C}, n = 1, \ldots, k$ and $A_n \cap A_m = \emptyset$ for $n \neq m$ then $\cup_{n=1}^{k} A_n \in \mathcal{C}$ (take $A_n = \emptyset$ for all $n > k$).

Proposition E.3.1. *Assume that $\mathcal{C}$ satisfies properties (i) and (iii) above. Then property (ii) holds if and only if the following property holds*

$$(ii'): \quad A, B \in \mathcal{C}, \quad A \subset B \quad implies \quad B \setminus A \in \mathcal{C}. \quad \text{(E.1)}$$

Proof. We assume that $\mathcal{C}$ satisfies properties (i), (ii), and (iii). If $A, B \in \mathcal{C}$ and $A \subset B$, then $B^c \in \mathcal{C}$, $A \cup B^c \in \mathcal{C}$, and consequentially $(A \cup B^c)^c = A^c \cap B = B \setminus A \in \mathcal{C}$.

Conversely, we assume that $\mathcal{C}$ satisfies properties (i), and (iii), and that property (ii') holds. Then $A \in \mathcal{C}$ implies $A^c = \Omega \setminus A \in \mathcal{C}$ establishing property (ii). ∎

Proposition E.3.2. *A set of sets $\mathcal{C}$ that is both a π-system and a λ-system is a σ-algebra.*

Proof. The property $\Omega \in \mathcal{C}$ holds since $\mathcal{C}$ is a λ-system. The set $\mathcal{C}$ is closed under finite union (as a π-system) and intersections (using the fact that it is closed under finite union and under complements and De-Morgan's theorem). If $A_n \in \mathcal{C}, n \in \mathbb{N}$ then the sets $B_n = A_n \cap A_1^c \cap A_2^c \cdots A_{n-1}^c$, $n \in \mathbb{N}$ are disjoint, and by property (iii) of a λ-system their union is also in $\mathcal{C}$. Noting that $\cup_{n \in \mathbb{N}} A_n = \cup_{n \in \mathbb{N}} B_n$ establishes that $\mathcal{C}$ is closed under countable unions. ∎

Proposition E.3.3. *The intersection of several λ-systems is a lambda system. Thus, given several λ systems we have a unique minimal λ system containing them.*

Proof. The proof is very similar to the proof of Proposition E.1.2. ∎

Proposition E.3.4 (Dynkin's Theorem). *If $\mathcal{C}$ is a π-system and $\mathcal{D}$ is a λ-system. Then $\mathcal{C} \subset \mathcal{D}$ implies $\sigma(\mathcal{C}) \subset \mathcal{D}$.*

Proof. We denote by $\mathcal{D}'$ the smallest λ-system containing $\mathcal{C}$ (defined as the intersection of all λ-systems containing $\mathcal{C}$, much like the definition of the σ-algebra generated by a set of sets). We thus have $\mathcal{C} \subset \mathcal{D}' \subset \mathcal{D}$. If we show that $\mathcal{D}'$ is also a π-system, Proposition E.3.2 implies that it is a σ-algebra, which in turn implies $\mathcal{C} \subset \sigma(C) \subset \mathcal{D}' \subset \mathcal{D}$ and concludes the proof.

We denote by $\mathcal{D}_A$ to be the set of sets B for which $A \cap B \in \mathcal{D}'$ and show that if $A \in \mathcal{D}'$ then $\mathcal{D}_A$ is a λ-system. Property (i) holds since $A \cap \Omega = \Omega \in \mathcal{D}'$ (since $\mathcal{D}'$ is a λ-system). If $B, B' \in \mathcal{D}_A$ with $B \subset B'$ then $A \cap B, A \cap B' \in \mathcal{D}'$ and using property (ii') $A \cap B' \setminus A \cap B = A \cap (B' \setminus B) \in \mathcal{D}'$. This shows that $B' \setminus B \in \mathcal{D}_A$ and that property (ii') (E.1) holds for $\mathcal{D}_A$. If the disjoint sets $B_n, n \in \mathbb{N}$ are in $\mathcal{D}_A$ then $A \cap B_n \in \mathcal{D}'$. Using property (iii) for the λ system $\mathcal{D}'$, we have $A \cap (\cup_n B_n) = \cup_n (A \cap B_n) \in \mathcal{D}'$ implying that $\cup_n B_n \in \mathcal{D}_A$, showing that $\mathcal{D}_A$ is a λ-system.

If $A, B \in \mathcal{C}$ then $A \cap B \in \mathcal{C} \subset \mathcal{D}'$ (since $\mathcal{C}$ is a π-system) and so $B \in \mathcal{D}_A$. This implies that if $A \in \mathcal{C}$ then $\mathcal{C} \subset \mathcal{D}_A$. We thus have the following sequence of implications

$$
\begin{aligned}
A \in \mathcal{C} &\Rightarrow & \mathcal{C} \subset \mathcal{D}' \subset \mathcal{D}_A \quad (\mathcal{D}' \text{ is the minimal } \lambda \text{ system}) \\
A \in \mathcal{C}, B \in \mathcal{D}' &\Rightarrow & B \in \mathcal{D}_A &\Rightarrow & A \in \mathcal{D}_B \\
B \in \mathcal{D}' &\Rightarrow & \mathcal{C} \subset \mathcal{D}_B \\
B \in \mathcal{D}' &\Rightarrow & \mathcal{D}' \subset \mathcal{D}_B \quad (\mathcal{D}' \text{ is the minimal } \lambda \text{-system containing } \mathcal{C}).
\end{aligned}
$$

Using the implications above, we conclude the proof by showing that $\mathcal{D}'$ is a π-system:

$$
A, B \in \mathcal{D}' \quad \Rightarrow \quad A \in \mathcal{D}_B \quad \Rightarrow \quad A \cap B \in \mathcal{D}'.
$$

∎

The following corollary establishes the uniqueness part of Caratheodory's extension theorem.

Corollary E.3.1. *Let* $\mathsf{P}_1, \mathsf{P}_2$ *be two probability measures on* $\sigma(\mathcal{C})$, *where* $\mathcal{C}$ *is a* π-*system. If* $\mathsf{P}_1, \mathsf{P}_2$ *agree on* $\mathcal{C}$ *then they also agree on* $\sigma(\mathcal{C})$.

Proof. We denote the set $\mathcal{A}$ on which P_1 and P_2 agree and note that $\mathcal{C} \subset \mathcal{A} \subset \sigma(\mathcal{C})$. Since $\mathsf{P}_1, \mathsf{P}_2$ are probability measures their agreement on A implies their agreement on A^c, thus $A \in \mathcal{A}$ implies $A^c \in \mathcal{A}$. If $A_n, n \in \mathbb{N}$ is a sequence of disjoint sets in $\mathcal{A}$ then $\cup_n A_n \in \mathcal{A}$ (by countable additivity of P_1 and P_2). It follows that $\mathcal{A}$ is a λ-system. Dynkin's theorem (Proposition E.3.4) then states that $\sigma(\mathcal{C}) \subset \mathcal{A}$. ∎

E.3.2 Outer Measure*

Definition E.3.3. Let P be a probability measure on an algebra $\mathcal{C}$. For each set $A \subset \Omega$ we define its outer measure to be

$$
\mathsf{P}^*(A) = \inf \sum_n \mathsf{P}(A_n)
$$

where the infimum ranges over all finite and infinite sequences of sets A_n in $\mathcal{C}$ that cover A ($A \subset \cup_n A_n$).

Definition E.3.4. Given a probability measure P on an algebra $\mathcal{C}$ we define a set A to be P^*-measurable if it satisfies the following equation

$$
\mathsf{P}^*(A \cup E) + \mathsf{P}^*(A^c \cup E) = \mathsf{P}^*(E), \qquad \forall E \subset \Omega. \tag{E.2}
$$

We denote the set of all P^*-measurable sets by $\mathcal{M}$.

Proposition E.3.5. *The outer probability measure function* P^* *satisfies the following properties.*

(empty set) $P^*(\emptyset) = 0$.

(non-negativity) $P^*(A) \geq 0$ *for all $A \subset \Omega$.*

(monotonicity) $A \subset B$ *implies* $P^*(A) \leq P^*(B)$.

(countable subadditivity) $P^*(\cup_{n \in \mathbb{N}} A_n) \leq \sum_{n \in \mathbb{N}} P^*(A_n)$.

Proof. The first property follows from the fact that $\emptyset$ is covered by itself, and has probability measure $P(\emptyset) = 1 - P(\Omega) = 0$. The second property holds since the function P is non-negative. The third property holds since a covering of A is also a covering of B.

We prove the fourth property below. For any ϵ and any n we construct a sequence of sets $B_{nk}, k \in \mathbb{N}$ such that $A_n \subset \cup_k B_{nk}$ and $\sum_k P(B_{nk}) \leq P^*(A_n) + \epsilon 2^{-n})$. Such a construction is possible since P^* is defined as the infimum over all possible coverings. The fourth property holds since $\cup_n A_n \subset \cup_n \cup_k B_{nk}$ and

$$P^* \left(\bigcup_n A_n \right) = P^* \left(\bigcup_n \bigcup_k B_{nk} \right) \leq \sum_n \sum_k P(B_{nk}) < \sum_n P^*(A_n) + \epsilon 2^{-n}$$

$$= \sum_n P^*(A_n) + \epsilon.$$

∎

Corollary E.3.2. *A set A is in $\mathcal{M}$ if and only if for all $E \subset \Omega$*

$$P^*(A \cap E) + P^*(A^c \cap E) \leq P^*(E). \tag{E.3}$$

Proof. By the countable sub-additivity property of the proposition above, $P^*(A \cap E) + P^*(A^c \cap E) \geq P^*(E)$ always hold. Thus equality for a specific A holds for all E if and only if (E.3) holds for the same A and all E. ∎

Proposition E.3.6. *The set $\mathcal{M}$ is an algebra.*

Proof. We have $\Omega \in \mathcal{M}$ since $P^*(\Omega \cap E) + P^*(\Omega^c \cap E) = P^*(E) + 0$. If $A \in \mathcal{M}$, we also have $A^c \in \mathcal{M}$. If $A, B \in \mathcal{M}$, then for all $E \subset \Omega$

$$P^*(E) = P^*(B \cap E) + P^*(B^c \cap E)$$
$$= P^*(A \cap B \cap E) + P^*(A^c \cap B \cap E) + P^*(A \cap B^c \cap E) + P^*(A^c \cap B^c \cap E)$$
$$\geq P^*(A \cap B \cap E) + P^*((A^c \cap B \cap E) \cup (A \cap B^c \cap E) \cup (A^c \cap B^c \cap E))$$
$$= P^*((A \cap B) \cap E) + P^*((A \cap B)^c \cap E),$$

implying that $A \cap B \in \mathcal{M}$. The inequality above follows from the fourth property of Proposition E.3.5. ∎

Proposition E.3.7. *For a finite or infinite sequence A_n of disjoint sets in $\mathcal{M}$,*

$$P^* \left(E \cap \left(\bigcup_n A_n \right) \right) = \sum_n P^*(E \cap A_n) \quad \forall E \subset \Omega. \tag{E.4}$$

Proof. We start with the case of a finite sequence of sets A_n in $\mathcal{M}$. We prove by induction on the number of sets n. The case $n = 1$ hold trivially. We consider the case $n = 2$. If $A_1 \cup A_2 = \Omega$, (E.4) is a restatement of the condition (E.2). Otherwise, using the fact that $A_1 \in \mathcal{M}$ and (E.2) we have

$$\mathsf{P}^*(A_1 \cap E \cap (A_1 \cup A_2)) + \mathsf{P}^*(A_1^c \cap E \cap (A_1 \cup A_2)) = \mathsf{P}^*(E \cap (A_1 \cup A_2)).$$

Since A_1, A_2 are disjoint, the left hand side above simplifies to the right side of (E.4). For a general n, we prove the induction argument using the $n = 2$ and $n - 1$ induction hypotheses

$$\mathsf{P}^*\left(E \cap \left(\bigcup_{k=1}^n A_k\right)\right) = \mathsf{P}^*\left(E \cap \left(\bigcup_{k=1}^{n-1} A_k\right)\right) + \mathsf{P}^*(E \cap A_k) = \sum_{k=1}^n \mathsf{P}^*(E \cap E_k).$$

For the case of an infinite sequence of disjoint sets $A_n, n \in \mathbb{N}$, we observe that

$$\mathsf{P}^*\left(E \cap \left(\bigcup_{k \in \mathbb{N}} A_k\right)\right) \geq \mathsf{P}^*\left(E \cap \left(\bigcup_{k=1}^n A_k\right)\right) = \sum_{k=1}^n \mathsf{P}^*(E \cap E_k).$$

Letting $n \to \infty$ in the equation above we conclude that

$$\mathsf{P}^*\left(E \cap \left(\bigcup_{k \in \mathbb{N}} A_k\right)\right) \geq \sum_{k \in \mathbb{N}} \mathsf{P}^*(E \cap E_k).$$

The reverse inequality holds by property 4 of Proposition E.3.5. ∎

Proposition E.3.8. *The set $\mathcal{M}$ is a σ-algebra, and P^* restricted to $\mathcal{M}$ is countable additive.*

Proof. Let $A_n, n \in \mathbb{N}$ be a sequence of disjoint sets in $\mathcal{M}$ with union A. Since $\mathcal{M}$ is an algebra, $F_n = \cup_{k=1}^n A_k \in \mathcal{M}$ and

$$\mathsf{P}^*(E) = \mathsf{P}^*(E \cap F_n) + \mathsf{P}^*(E \cap F_n^c)$$

$$= \sum_{k=1}^n \mathsf{P}^*(E \cap A_k) + \mathsf{P}^*(E \cap F_n^c)$$

$$\geq \sum_{k=1}^n \mathsf{P}^*(E \cap A_k) + \mathsf{P}^*(E \cap A^c).$$

Above, we used (E.4) to derive the second equality and the fact that $A^c \subset F_n^c$ to derive the inequality. Letting $n \to \infty$ in the equation above, and using (E.4) again, we have

$$\mathsf{P}^*(E) \geq \sum_{k \in \mathbb{N}} \mathsf{P}^*(E \cap A_k) + \mathsf{P}^*(E \cap A_k^c)$$

$$= \mathsf{P}^*(E \cap A) + \mathsf{P}^*(E \cap A^c).$$

Together with (E.3), the equation above proves that $\mathcal{M}$ is closed under a countable union of disjoint sets.

To show that $\mathcal{M}$ is a σ-algebra, it remains to show that it is closed under countable unions of arbitrary sets $B_n \in \mathcal{M}, n \in \mathbb{N}$ (not necessarily disjoint). Denoting $A_1 = B_1$, $A_k = B_k \cap A_1^c \cap \cdots \cap A_{k-1}^c$, we have that A_n is a sequence of disjoint sets in $\mathcal{M}$, and since $A = \cup_n A_n = \cup_n B_n = B$, we have

$$\mathsf{P}^*(E) \geq \mathsf{P}^*(E \cap A) + \mathsf{P}^*(E \cap A^c) = \mathsf{P}^*(E \cap B) + \mathsf{P}^*(E \cap B^c).$$

Together with (E.3), the equation above proves that $\mathcal{M}$ is closed under a countable union of arbitrary sets.

The countable additivity of P^* on $\mathcal{M}$ follows from (E.4) with Ω substituting E.   ■

Proposition E.3.9. *Let* P *be a probability measure on an algebra* $\mathcal{C}$, P^* *the associated outer measure, and* $\mathcal{M}$ *the set of all* P^*-*measurable sets. Then (i) for all* $A \in \mathcal{C}$, $\mathsf{P}^*(A) = \mathsf{P}(A)$, *and (ii)* $\mathcal{C} \subset \mathcal{M}$.

Proof. By definition $\mathsf{P}^*(A) = \inf \sum_n \mathsf{P}(A_n)$ where the infimum ranges over all covering of A within $\mathcal{C}$. If $A \in \mathcal{C}$, the covering of A achieving the infimum is precisely $A_1 = A$ implying that $\mathsf{P}^* = \mathsf{P}$ on $\mathcal{C}$ (for any covering A_n of A, $\mathsf{P}(A) \leq \mathsf{P}(\cup_n A_n) \leq \sum_n \mathsf{P}(A_n)$).

We now prove that $\mathcal{C} \subset \mathcal{M}$. Let $A \in \mathcal{C}$. For each set E and $\epsilon > 0$ select a sequence of sets $A_n, n \in \mathbb{N}$ in $\mathcal{C}$ such that

$$\sum_n \mathsf{P}(A_n) \leq \mathsf{P}^*(E) + \epsilon. \tag{E.5}$$

This is possible since $\mathsf{P}^*(E)$ is defined as the infimum of such sums over all coverings of E. Since $\mathcal{C}$ is an algebra, $B_n = A_n \cap A \in \mathcal{C}$ and $C_n = A_n \cap A^c \in \mathcal{C}$. The following inequality

$$\mathsf{P}^*(E \cap A) + \mathsf{P}^*(E \cap A^c) \leq \mathsf{P}^*\left(\bigcup_n B_n\right) + \mathsf{P}^*\left(\bigcup_n C_n\right) \leq \sum_n \mathsf{P}^*(B_n) + \mathsf{P}^*(C_n) \tag{E.6}$$

$$= \sum_n \mathsf{P}(B_n) + \sum_n \mathsf{P}(C_n) = \sum_n \mathsf{P}(A_n) \leq \mathsf{P}^*(E) + \epsilon.$$

holds for all $\epsilon \geq 0$ indicating that $A \in \mathcal{M}$. The first inequality in (E.6) follows from $E \cap A \subset \cup_n B_n$ and $E \cap A^c \subset \cup_n C_n$ and monotonicity of P^*. The second inequality follows from the countable subadditivity of P^*. The first equality follows from $B_n, C_n \in \mathcal{C}$ and the first part of this proposition. The second equality follows from (E.5).   ■

E.3.3   The Extension Theorem*

Proposition E.3.10 (Caratheodory's extension theorem). *A probability measure* P *defined on an algebra* $\mathcal{C}$ *has a unique extension to a probability measure on* $\sigma(\mathcal{C})$.

Proof. We first prove the existence of the extension. Since $\mathcal{M}$ is a σ-algebra (Proposition E.3.8) and $\sigma(\mathcal{C})$ is the smallest σ-algebra, we have

$$\mathcal{C} \subset \sigma(\mathcal{C}) \subset \mathcal{M}.$$

Furthermore, Proposition E.3.8 states that P^* is countable additive on $\mathcal{M}$ implying that its restriction to $\mathcal{M}$ is a probability measure. It follows that P^* is also a probability measure on $\sigma(\mathcal{C})$ that agrees with P on $\mathcal{C}$ (Proposition E.3.9).

The uniqueness of the extension follows from Proposition E.3.1 and the fact that the algebra $\mathcal{C}$ is also a π-system. ∎

Caratheodory's extension theorem holds also for measures that are not probability measures, including measures for which $\mu(\Omega) = +\infty$. A proof is available in [5].

E.4 Independent σ-Algebras*

Definition E.4.1. Let $(\Omega, \mathcal{F}, \mu)$ be a measure space. Two events $A, B \in \mathcal{F}$ are independent if $\mu(A \cap B) = \mu(A)\mu(B)$. A finite number of events $A_1, \ldots, A_n \in \mathcal{F}$ are independent if

$$\mu(A_1 \cap \cdots A_n) = \mu(A_1) \cdots \mu(A_n).$$

An infinite collection of events $A_\theta, \theta \in \Theta$ is independent if for every $k \in \mathbb{N}$ and for every finite subset of distinct events $A_{\theta_1}, \ldots, A_{\theta_k}, \theta_1, \ldots, \theta_k \in \Theta$ we have

$$\mu(A_{\theta_1} \cap \ldots \cap A_{\theta_k}) = \mu(A_{\theta_1}) \cdots \mu(A_{\theta_k}).$$

The definitions above of set independence extends to independence of sets of sets.

Definition E.4.2. Let $(\Omega, \mathcal{F}, \mu)$ be a measure space. A finite collection $\mathcal{A}_1, \ldots, \mathcal{A}_n$ of subsets of $\mathcal{F}$ are independent if every selection of sets from these sets $A_1 \in \mathcal{A}_1, \ldots, A_n \in \mathcal{A}_n$ are independent. An infinite collection $\mathcal{A}_\theta, \theta \in \Theta$ of subsets of $\mathcal{F}$ are independent if every finite number of elements $\theta_1, \ldots \theta_k \in \Theta$, the sets $\mathcal{A}_{\theta_1}, \ldots, \mathcal{A}_{\theta_k}$ are independent.

Proposition E.4.1. *Independence of a collection of π-systems $\mathcal{A}_\theta, \theta \in \Theta$ implies independence of the generated σ-algebras $\sigma(\mathcal{A}_\theta), \theta \in \Theta$.*

Proof. We prove the result in the case that Θ is a finite set of size n, which we denote by $\Theta = \{1, 2, \ldots, n\}$. An extension for infinite Θ is available in [5].

We assume independence of the π-systems $\mathcal{A}_1, \ldots, \mathcal{A}_n$. First, note that $\mathcal{A}_i' = \mathcal{A}_i \cup \{\Omega\}$ are also π-systems. Fix $A_i' \in \mathcal{A}_i'$ for $i = 2, \ldots, n$ and consider the set $\mathcal{L}$ of sets L for which

$$\mathsf{P}(L \cap A_2' \cap \cdots \cap A_n') = \mathsf{P}(L)\,\mathsf{P}(A_2') \cdots \mathsf{P}(A_n').$$

We show next that $\mathcal{L}$ is a λ-system. The first condition holds since

$$P(\Omega \cap A_2' \cap \cdots \cap A_n') = 1 \cdot P(A_2') \cdots P(A_n') = P(\Omega) \cdot P(A_2') \cdots P(A_n').$$

The second condition holds since $U \in \mathcal{L}$ implies

$$\begin{aligned}
P(U^c \cap A_2' \cap \cdots \cap A_n') &= P((A_2' \cap \cdots \cap A_n') \setminus U) \\
&= P(A_2' \cap \cdots \cap A_n') - P(A_2' \cap \cdots \cap A_n' \cap U) \\
&= P(A_2') \cdots P(A_n') - P(A_2') \cdots P(A_n') P(U) \\
&= P(A_2') \cdots P(A_n') P(U^c).
\end{aligned}$$

The third property follows from the fact that for a disjoint sequence of sets $U_n \in \mathcal{L}, n \in \mathbb{N}$ implies

$$\begin{aligned}
P((\cup_n U_n) \cap A_2' \cap \cdots \cap A_n') &= P(\cup_n (U_n \cap A_2' \cap \cdots \cap A_n')) \\
&= \sum_n P(U_n \cap A_2' \cap \cdots \cap A_n') \\
&= \sum_n P(U_n) P(A_2') \cdots P(A_n') \\
&= P(A_2') \cdots P(A_n') P(\cup_n U_n).
\end{aligned}$$

Since $\mathcal{L}$ is a λ-system that also contains the π-system $\mathcal{A}_1'$, by Dynkin's theorem (Proposition E.3.4) we have that $\sigma(\mathcal{A}_1'), \{A_2'\}, \ldots, \{A_n'\}$ are independent. Since A_i' were selected arbitrarily from $\mathcal{A}_i'$, it follows that $\sigma(\mathcal{A}_1'), A_i, \ldots, A_i$ are independent as well. Repeating this argument multiple times we have that $\sigma(\mathcal{A}_1'), \sigma(\mathcal{A}_2'), \ldots, \sigma(\mathcal{A}_n)$ are independent. ∎

Corollary E.4.1. *Let $(\Omega, \mathcal{F}, P)$ be a probability measure space and A_{ij} be a potentially infinite array of independent events (here i, j vary over a finite or countably infinite set). Then the σ-algebras generated by the rows of the array $\mathcal{F}_i = \sigma(\{A_{ij} : \forall j\})$ are independent.*

Proof. We define the sets $\mathcal{C}_i$ of all finite intersections of sets in the i-row of the array and consider arbitrary sets $C_i \in \mathcal{C}_i$. Then for a finite set $I \subset \mathbb{N}$ of indices, we have

$$P\left(\bigcap_{i \in I} C_i\right) = P\left(\bigcap_{i \in I} \bigcup_j A_{ij}\right) = \prod_{i \in I} \prod_j P(A_{ij}) = \prod_{i \in I} P(C_i).$$

Note that above both i and j vary over a finite set, justifying the equality signs. This implies that the sets $\mathcal{C}_i$, for all i, are independent. Using the proposition above we have independence of $\sigma(\mathcal{C}_i) = \mathcal{F}_i$. ∎

E.5   Important Measure Functions*

We discuss three important measure functions: the discrete measure, the Lebesgue measure, and the Lebesgue-Stieltjes measure.

E.5.1   Discrete Measure Functions*

In this section, we consider a general method to define measures on a finite Ω or a countably infinite Ω. In this case, we use the σ-algebra 2^Ω — the power set of Ω.

We define the discrete measure on the measurable space $(\Omega, 2^\Omega)$ associated with a set of non-negative numbers $\{p_\omega : \omega \in \Omega\}$ as

$$\mu(A) = \sum_{\omega \in A} p_\omega.$$

The function $\mu : 2^\Omega \to \mathbb{R}$ satisfies $\mu(\emptyset) = 0$ and is countably additive, implying that it is a measure. If $\sum_{\omega \in \Omega} p_\omega = 1$, the discrete measure μ is also a probability measure P.

E.5.2   The Lebesgue Measure*

In the Section E.5.1 we saw our first example of a measure function, defined on a finite or countably infinite Ω. It is considerably harder to define a meaningful measure on $\mathbb{R}$ or on intervals. We describe in this section a measure function μ on the interval $\Omega = (0, 1]$, called the Lebesgue measure, that agrees with our intuitive notion of length. Specifically, if A is a disjoint union of intervals, $\mu(A)$ is the sum of the lengths of these intervals. The Lebesgue measure may also be defined for other intervals in $\mathbb{R}$, for the entire real line $\mathbb{R}$, and even for multidimensional Euclidean spaces $\mathbb{R}^d$. It is also the basis for the Lebesgue-Stieltjes measure that serves a fundamental role in probability theory.

We start by defining the σ-algebra that will be used to define the Lebesgue measure.

Definition E.5.1. The Borel σ-algebra associated with a metric space (Ω, d) is $\sigma(\mathcal{C})$ where $\mathcal{C}$ is the set of all open sets in the metric space (Ω, d). We denote this σ-algebra by $\mathcal{B}(\Omega)$ or simply $\mathcal{B}$ when no confusion arises.

Proposition E.5.1. *Let $\mathcal{A}$ be the set of all open balls in $\mathbb{R}^d$. Then $\sigma(\mathcal{A}) = \mathcal{B}(\mathbb{R}^d)$.*

Proof. Since all open balls are open sets we have $\sigma(\mathcal{A}) \subset \mathcal{B}(\mathbb{R}^d)$. Since $\mathbb{R}^d$ is second countable (Proposition B.4.5) every open set can be expressed as a countable union of open balls in $\mathcal{A}$, implying that all open sets are also in $\sigma(\mathcal{A})$ (since $\sigma(\mathcal{A})$ is closed under countable unions) and consequentially $\mathcal{B}(\mathbb{R}^d) \subset \sigma(\mathcal{A})$. ∎

Proposition E.5.2.

$$\mathcal{B}(\mathbb{R}) = \sigma\left(\{(-\infty, x] : x \in \mathbb{R}\}\right).$$

Proof. The previous proposition showed that the Borel sets in $\mathbb{R}$ are generated by the set of open intervals (a, b). It is thus sufficient to show (i) $(a, b) \in \sigma(\{(-\infty, x] :$

$x \in \mathbb{R}\}$), and (ii) $(-\infty, x] \in \mathcal{B}(\mathbb{R})$. These two assertions follow from the equations below.

$$(a, b) = \left(\bigcup_{n \in \mathbb{N}} (-\infty, b - 1/n] \right) \cap (-\infty, a]^c$$

$$(-\infty, x] = \bigcup_{n \in \mathbb{N}} (x, x + n)^c.$$

∎

Proposition E.5.3. *Let $\mathcal{A}$ be the set of all half-open intervals in $(\alpha, \beta]$ of the form $(a, b]$, $(\alpha \le a < b \le \beta)$ and $\mathcal{D}$ the set of all unions of a finite number of disjoint half open intervals from $\mathcal{A}$. Then $\mathcal{D}$ is an algebra in $\Omega = (\alpha, \beta]$, and $\sigma(\mathcal{D}) = \mathcal{B}((\alpha, \beta])$.*

Proof. With no loss of generality, we assume in the proof that $\alpha = 0$ and $\beta = 1$. We first show that $\mathcal{D}$ is an algebra over $\Omega = (0, 1]$. It is obvious that $(0, 1] \in \mathcal{A}$ and therefore $\Omega \in \mathcal{D}$. The complement of a disjoint union of half open intervals $(a, b]$ is a disjoint union of half open intervals. Similarly, a finite union of disjoint unions of half open intervals is a disjoint union of half open intervals. It thus follows that $\mathcal{D}$ is an algebra.

Proposition E.1.5 and Proposition E.5.1 imply that $\mathcal{B}((0, 1])$ is generated by the open balls in $(0, 1]$, or intervals (a, b) with $0 \le a < b \le 1$. To show that $\mathcal{B}((0, 1]) = \sigma(\mathcal{D})$ it suffices to show that (i) $(a, b) \in \sigma(\mathcal{D})$, and that (ii) $(a, b] \in \mathcal{B}((0, 1])$. The first claim holds since $(a, b) = \cup_{n \in \mathbb{N}}(a, b - 1/n]$ and the second holds since $(a, b] = \cap_{n \in \mathbb{N}}(a, b + 1/n) = (\cup_{n \in \mathbb{N}}(a, b + 1/n)^c)^c$. ∎

Proposition E.5.4. *Using the notations from the previous proposition, we define the Lebesgue measure μ on $\Omega = (\alpha, \beta]$ and the algebra $\mathcal{D}$ of half open intervals (see the proposition above) as follows:*

$$\mu(A) = \sum_{i=1}^{k} |b_i - a_i|$$

where A is represented by the disjoint union of intervals $A = \cup_{i=1}^{k}(a_i, b_i]$. The measure has a unique extension to the Borel σ-algebra $\mathcal{B}((\alpha, \beta])$.

Proof. With no loss of generality, we assume in the proof that $\alpha = 0$ and $\beta = 1$. The function μ is a measure since it assigns value 0 to the empty set, it is non-negative, and given a sequence of disjoint sets $D_n \in \mathcal{D}, n \in \mathbb{N}$ we have $\mu(\cup D_n) = \sum \mu(D_n)$.

Since $\mathcal{D}$ is an algebra and $\sigma(\mathcal{D}) = \mathcal{B}(0, 1])$ (by the proposition above), we can use Caratheodory's extension theorem (Proposition E.3.10) to assert that there is a unique extension of μ to the Borel σ-algebra $\mathcal{B}((\alpha, \beta])$. ∎

We make the following observations.

1. The Lebesgue measure may be defined on different types of intervals (α, β), $[\alpha, \beta]$, or $[\alpha, \beta)$ in accordance with the definition above on intervals of the form $[\alpha, \beta)$. Consistency of the Lebesgue measure over all of these intervals follows from the fact that the Lebesgue measure (length) of a single point is zero.

2. When $\beta - \alpha = 1$, the Lebesgue measure is also a probability measure. In other cases, we can define a probability measure associated with the Lebesgue measure on $(\alpha, \beta]$ as follows

$$P(A) = \frac{1}{\beta - \alpha} \mu(A).$$

   This probability function is the classical probability function on continuous spaces, defined in Section 1.4.

3. Letting $\alpha \to -\infty$ and $\beta \to +\infty$ generalizes the Lebesgue measure to $\Omega = \mathbb{R}$ and the σ-algebra $\mathcal{B}(\mathbb{R})$. Note that in this case $\mu(\Omega) = +\infty$, implying that the Lebesgue measure on $\mathbb{R}$ is not a probability measure, and cannot be converted to one using the normalization that is described above.

4. The Lebesgue measure can be generalized to subsets of higher dimensional Euclidean spaces $\mathbb{R}^d$. The resulting measure generalizes area and volume for the cases $d = 2$ and $d = 3$, respectively. These generalizations will be developed in Section F.5.

E.5.3   The Lebesgue-Stieltjes Measure*

The Lebesgue-Stieltjes measure is a useful generalization of the Lebesgue measure, where some regions of $\mathbb{R}$ have higher measure than others. The relation between measures across different regions of $\mathbb{R}$ is specified using a non-decreasing right continuous function.

Definition E.5.2. Let $F : (\alpha, \beta] \to [0, \infty)$ be a non-decreasing right continuous function. Using the definitions from Proposition E.5.3, we define the Lebesgue-Stieltjes measure on $\Omega = (\alpha, \beta]$ and the algebra $\mathcal{D}$ as

$$\mu_F(A) = \sum_{i=1}^{k} F(b_i) - F(a_i)$$

where $A \in \mathcal{D}$ is represented as a finite union of disjoint half open intervals $A = \cup_i (a_i, b_i]$.

Proposition E.5.5. *The Lebesgue-Stieltjes measure defined above on $\Omega = (\alpha, \beta]$ and the algebra $\mathcal{D}$ has a unique extension to $\mathcal{B}((\alpha, \beta])$.*

Proof. The proof is identical to the proof of Proposition E.5.4. ∎

As in the case of the Lebesgue measure, letting $\alpha \to -\infty$ and $\beta \to \infty$ generalizes the Lebesgue-Stieltjes measure to $\Omega = \mathbb{R}$ and the σ-algebra $\mathcal{B}(\mathbb{R})$.

E.6 Measurability of Functions*

Definition E.6.1. Let $(\Omega_1, \mathcal{F}_1)$ and $(\Omega_2, \mathcal{F}_2)$ be two measurable spaces. A function $f : \Omega_1 \to \Omega_2$ is $\mathcal{F}_1/\mathcal{F}_2$-measurable if

$$A \in \mathcal{F}_2 \qquad \text{implies} \qquad f^{-1}(A) \in \mathcal{F}_1.$$

If $\Omega_2 = \mathbb{R}^d$ we implicitly assume (unless stated otherwise) that $\mathcal{F}_2 = \mathcal{B}(\mathbb{R}^d)$. Similarly, if $\Omega_1 = \mathbb{R}^d$ we implicitly assume (unless stated otherwise) that $\mathcal{F}_1 = \mathcal{B}(\mathbb{R}^d)$.

Proposition E.6.1. *Let $(\Omega_1, \mathcal{F}_1)$ and $(\Omega_2, \sigma(\mathcal{C}))$ be two measurable spaces. A function $f : \Omega_1 \to \Omega_2$ is $\mathcal{F}_1/\mathcal{F}_2$-measurable if*

$$A \in \mathcal{C} \qquad \text{implies} \qquad f^{-1}(A) \in \mathcal{F}_1.$$

Proof. We have $f^{-1}(\Omega_2 \setminus A) = \Omega_2 \setminus f^{-1}(A)$ and $f^{-1}(\cup_n A_n) = \cup_n f^{-1}(A_n)$ (see Proposition A.2.1). This, together with the fact that $\mathcal{F}_1$ is a σ-algebra proves that $\{A : f^{-1}(A) \in \mathcal{F}_1\}$ is a σ-algebra that contains $\mathcal{C}$. Since $\sigma(\mathcal{C})$ is the smallest σ-algebra containing $\mathcal{C}$, the proposition follows. ∎

Proposition E.6.2. *The composition of two measurable functions is a measurable function.*

In the proposition above, there are three measurable spaces $(\Omega_i, \mathcal{F}_i)$, $i = 1, 2, 3$. If $f : \Omega_1 \to \Omega_2$ is $\mathcal{F}_1/\mathcal{F}_2$-measurable and $g : \Omega_2 \to \Omega_3$ is $\mathcal{F}_2/\mathcal{F}_3$-measurable, the proposition states that $f \circ g : \Omega_1 \to \Omega_3$ is $\mathcal{F}_1/\mathcal{F}_3$-measurable.

Proof. Let $A \in \mathcal{F}_3$. The measurability of g implies that $g^{-1}(A) \in \mathcal{F}_2$, and together with the measurability of f this implies $f^{-1}(g^{-1}(A)) \in \mathcal{F}_1$. ∎

Proposition E.6.3. *If a function $f : \mathbb{R}^m \to \mathbb{R}^n$ is continuous, then it is measurable.*

Proof. The σ-algebra $\mathcal{B}(\mathbb{R}^n)$ is generated by the set of all open sets. By Proposition E.6.1 it is then sufficient to show that for every open set A, $f^{-1}(A) \in \mathcal{B}(\mathbb{R}^m)$. Proposition B.3.4 states that $f^{-1}(A)$ is open, implying that $f^{-1}(A) \in \mathcal{B}(\mathbb{R}^m)$. ∎

The set of measurable functions is strictly larger than the set of continuous functions. For example, the indicator function $I_A : \mathbb{R} \to \mathbb{R}$, defined as $I_A(x) = 1$ if $x \in A$ and 0 otherwise, is not continuous (assuming $A \neq \Omega$ and $A \neq \emptyset$) but is measurable if $A \in \mathcal{B}(\mathbb{R})$.

Proposition E.6.4. *Let $f_n, n \in \mathbb{N}$ be measurable functions from $(\Omega, \mathcal{F})$ to $\mathbb{R}$. Then*

1. *The functions $\sup_n f_n$, $\inf_n f_n$, $\limsup_n f_n$, $\liminf_n f_n$ are measurable.*

2. *If $\lim_n f_n$ exists everywhere then $\lim_n f_n$ is measurable.*

Proof. Since f_n are measurable, $(-\infty, x] \in \mathcal{B}(\mathbb{R})$, and $\mathcal{F}$ is closed under countable intersections, we have

$$\left\{\omega : \sup_{n \in \mathbb{N}} f_n(\omega) \leq x\right\} = \bigcap_{n \in \mathbb{N}} \{\omega : f_n(\omega) \leq x\} = \bigcap_{n \in \mathbb{N}} f_n^{-1}((-\infty, x]) \in \mathcal{F}.$$

Note that this holds even for $x = +\infty$ or $x = -\infty$. The equation above, together with Proposition E.6.1 and Proposition E.5.2, imply that $\sup_n f_n$ is measurable. The proof for measurability of $\inf_n f_n$ is similar. The functions $\liminf$ and $\limsup$ are compositions of $\inf$ and $\sup$ functions (Definition B.2.6), which are measurable, and thus by Proposition E.6.2 $\liminf$ and $\limsup$ are measurable as well. If $\lim_n f_n$ exists, $\liminf_n f_n = \limsup_n f_n$ establishing that $\lim_n f_n$ is measurable. ∎

E.7 Notes

There are many textbooks that describe rigorously measure theory and its connection to probability theory. A few well known ones are [17, 10, 5, 1, 33, 25]. Our description is closest to [5], but is slightly simplified since we focus on probability measures rather than the general case.

Appendix F

Integration

Integration plays an important role in probability theory. We start by describing the Riemann integral, which is commonly taught in elementary calculus, and then describe the relationship between integration and differentiation. The next section covers the Lebesgue integral, which is technically harder than the Riemann integral and requires measure theory. The Lebesgue integral is needed for developing tools that are used to prove convergence results (Chapter 8). We conclude with several sections that extend integration and differentiation to multivariate functions.

F.1   Riemann Integral

Definition F.1.1. A partition of an interval $[a, b]$ is a finite set of points $Q = \{x_i : i = 0, \dots, n\}$ satisfying $a = x_0 \leq x_1 \leq \cdots \leq x_n = b$.

Definition F.1.2. Given a bounded function $f : [a, b] \to \mathbb{R}$ and a partition of $[a, b]$ we define

$$M_i = \sup_{x \in \Delta_i} f(x)$$

$$m_i = \inf_{x \in \Delta_i} f(x)$$

$$U(P, f) = \sum_{i=1}^{n} M_i |\Delta_i|$$

$$L(P, f) = \sum_{i=1}^{n} m_i |\Delta_i|.$$

where $\Delta_i = (x_{i-1}, x_i)$ and $|\Delta_i| = |x_{i-1} - x_i|$.

Figure F.1 shows a partition of four points and the corresponding m_i and M_i values. As the number of points in the partition increases the difference $|m_i - M_i|$ tend to decrease.

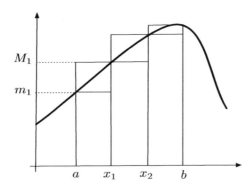

Figure F.1: A partition $\{a = x_0, x_1, x_2, b = a_3\}$ of $[a, b]$ and the corresponding m_1 and M_1 values.

Example F.1.1. *Consider the function $f(x) = x$ over $[0, 3]$ and the partition $Q = \{0, 1, 2, 3\}$. Then $|\Delta_i| = 1$, for all i, and $L(Q, f) = 1 \cdot 0 + 1 \cdot 1 + 1 \cdot 2$ and $U(Q, f) = 1 \cdot 1 + 1 \cdot 2 + 2 \cdot 3$. Clearly $M_i \geq m_i$, implying that $U(P, f) \geq L(P, f)$. As we add additional points to the partition the difference $U(P, f) - L(P, f)$ decays to zero.*

Definition F.1.3. Let $f : [a, b] \to \mathbb{R}$ be a bounded function. Then

$$\overline{\int_a^b} f(x)\, dx \stackrel{\text{def}}{=} \inf_Q U(Q, f)$$

$$\underline{\int_a^b} f(x)\, dx \stackrel{\text{def}}{=} \sup_Q L(Q, f),$$

where $\inf_Q$ and $\sup_Q$ above are taken over all possible partitions Q of $[a, b]$. If the two quantities above are finite and equal, we say that f is Riemann integrable over $[a, b]$ and denote the common value as

$$\int_a^b f(x)\, dx \stackrel{\text{def}}{=} \overline{\int_a^b} f(x)\, dx = \underline{\int_a^b} f(x)\, dx.$$

We sometimes consider integrals over the entire real line, achieved by letting $a \to \infty$, $b \to -\infty$ in the definition above.

Definition F.1.4. A partition Q is a refinement of a partition P if every point in P is also in Q. Given two partitions P_1, P_2, we define their common refinement as the partition composed of all the points that are either in P_1 or in P_2.

Proposition F.1.1. *If P' is a refinement of the partition P then*

$$L(P, f) \leq L(P', f)$$
$$U(P', f) \leq U(P, f).$$

Proof. We prove the first result. The proof of the second result is similar. We assume at first that P' contains only one more point x' ($x_i < x' < x_{i+1}$) than P. In this case $L(P', f) - L(P, f) = (m_i' \Delta_i' + m_{i+1}' \Delta_{i+1}') - (m_i \Delta_i)$, and since $m_i', m_{i+1}' \geq m$ we have $L(P', f) - L(P, f) \geq 0$. If P' contains several points in addition to P, we can repeat this argument several times to establish $L(P', f) - L(P, f) \geq 0$. ∎

Proposition F.1.2.

$$\overline{\int_a^b} f(x)\, dx \geq \underline{\int_a^b} f(x)\, dx.$$

Proof. Consider Q, a common refinement of two partitions P_1, P_2 of $[a, b]$. From Proposition F.1.1, it follows that

$$L(P_1, f) \leq L(Q, f) \leq U(Q, f) \leq U(P_2, f).$$

The above inequality holds for all P_1 and P_2. Taking supremum over all P_1 and infimum over all P_2 yields the desired result. ∎

Proposition F.1.3. *A bounded function $f : [a, b] \to R$ is Riemann integrable if and only if*

$$\forall \epsilon > 0, \exists Q \quad \text{such that} \quad U(Q, f) - L(Q, f) < \epsilon.$$

Proof. If f is Riemann integrable, then for all $\epsilon > 0$ there exists P_1, P_2 such that $U(P_2, f) - \int f\, dx < \epsilon/2$ and $\int f\, dx - L(P_1, f) < \epsilon/2$. If Q is the common refinement of P_1, P_2, then using Proposition F.1.2,

$$U(Q, f) \leq U(P_2, f) < \int f\, dx + \epsilon/2 < L(P_1, f) + \epsilon \leq L(Q, f) + \epsilon.$$

Conversely, we assume that for all $\epsilon > 0$ there exists Q, such that $U(Q, f) - L(Q, f) < \epsilon$. For all partitions P we have

$$L(P, f) \leq \underline{\int_a^b} f\, dx \leq \overline{\int_a^b} f\, dx \leq U(P, f).$$

and therefore for all $\epsilon > 0$

$$0 \leq \overline{\int_a^b} f\, dx - \underline{\int_a^b} f\, dx < \epsilon.$$

Since ϵ is arbitrary, the function f is Riemann integrable. ∎

Corollary F.1.1. *If $f : [a, b] \to \mathbb{R}$ is continuous, then it is Riemann integrable.*

Proof. Since $[a, b]$ is compact, Proposition B.3.5 implies that f is uniformly continuous on $[a, b]$. This implies that for all $\epsilon' < 0$ these exists $\delta > 0$ such that

$|x - y| < \delta$ implies $|f(x) - f(y)| < \epsilon'$. If P is a partition of $[a, b]$ such that $\delta_i < \delta$ for all i, then $M_i - m_i \leq \epsilon'$ and

$$U(P, f) - L(P, f) = \sum_{i=1}^{n}(M_i - m_i)\Delta_i \leq \epsilon' \sum_{i=1}^{n} \Delta_i = \epsilon'(b - a).$$

For all $\epsilon > 0$ we can set $\epsilon' = \epsilon/(b - a)$, implying that $U(P, f) - L(P, f) < \epsilon$, and by Proposition F.1.3, f is Riemann integrable.   ∎

Some standard properties of integrals are listed below.

Proposition F.1.4. *The following properties apply to the Riemann integral.*

1. *Linearity:*

$$\int_{a}^{b} af(x) + bg(x)\, dx = a \int_{a}^{b} f(x)\, dx + b \int_{a}^{b} g(x)\, dx.$$

2. *Interval Decomposition: for all $a \in (a, b)$ we have*

$$\int_{a}^{b} f(x)\, dx = \int_{a}^{q} f(x)\, dx + \int_{q}^{b} f(x)\, dx.$$

3.

$$\left| \int_{a}^{b} f(x)\, dx \right| \leq \int_{a}^{b} |f(x)|\, dx \leq \sup_{x \in [a,b]} |f(x)| \cdot |b - a|.$$

4. *If $f(x) \leq g(x)$ for all $x \in [a, b]$ then*

$$\int_{a}^{b} f\, dx \leq \int_{a}^{b} g\, dx.$$

Proof*. These properties follow from the corresponding properties of the Lebesgue integral (see Section F.3), and Proposition F.3.11, which implies that if f is Riemann integrable, then it is Lebesgue integrable with respect to the Lebesgue measure, and the two integrals agree in value.   ∎

Proposition F.1.5. *A function $f : [a, b] \to \mathbb{R}$ that is bounded and continuous except for a finite number of points at which it is perhaps discontinuous is Riemann integrable.*

Proof*. Let f be a bounded function that is continuous on $[a, b]$ except on k points. By the second property above (interval decomposition), we can decompose the integral to $k + 1$ integrals over continuous functions. Since each one of the integrals exists (see Corollary F.1.1), the integral of f exists as well.   ∎

The above proposition shows that functions that are continuous and bounded except at a finite number of points are Riemann integrable. It does not specify which functions are not Riemann integrable. Proposition F.3.5 strengthen this claim and shows that a bounded function f is Riemann integrable if and only if it is continuous almost everywhere (the phrase almost everywhere is made precise in Section F.3).

F.2 Relationship between Integration and Differentiation

As the following results indicate, integration and differentiation are in some sense opposite operations.

Proposition F.2.1. *Let $f : [a, b] \to \mathbb{R}$ be a bounded function and P a partition of $[a, b]$ such that $U(P, f) - L(P, f) < \epsilon$. Then*

$$\left| \sum_{i=1}^{n} f(t_i)|\Delta_i| - \int_{a}^{b} f \, dx \right| < \epsilon, \qquad \forall t_i \in \Delta_i, \ i = 1, \ldots, n.$$

Proof. The proposition follows from the fact that both $\sum_{i=1}^{n} f(t_i)|\Delta_i|$ and $\int_{a}^{b} f \, dx$ lie in the interval $[L(P, f), U(P, f)]$, whose length is ϵ. ∎

The following proposition formulates a very important connection between differentiation and integration. It leads to many useful integration techniques, and is important in probability theory in formulating a connection between the cdf and pdf of a continuous random variable.

Proposition F.2.2 (The Fundamental Theorem of Calculus). *Part 1: Let $f : [a, b] \to \mathbb{R}$ be a bounded integrable function. Then (i) $F(x) = \int_{a}^{x} f(t) \, dt$ is a continuous function on $[a, b]$, and (ii) if f is continuous then F is differentiable and*

$$F'(x) = \frac{d}{dx} \int_{a}^{x} f(t) \, dt = f(x).$$

Part 2: For any differentiable function F whose derivative $F'(t) = f(t)$ is Riemann integrable,

$$F(b) - F(a) = \int_{a}^{b} f(t) \, dt.$$

In the second part of the proposition above, the function F is called the anti-derivative of f. The second part of the proposition thus provides an easy way to compute the integral $\int_{a}^{b} f(t) \, dt$ of a function f with a known anti-derivative F.

Proof. We first prove part 1. For $x < y$ we have

$$|F(y) - F(x)| = \left| \int_{x}^{y} f(t) \, dt \right| \leq (y - x) \sup_{t \in [a,b]} |f(t)|.$$

Thus for all $\epsilon > 0$, if $|y - x| < \delta = \epsilon / \sup_{t \in [a,b]} |f(t)|$ then $|F(y) - F(x)| < \epsilon$ showing (i).

If f is continuous on $[a, b]$ then it is also uniformly continuous, and for all $\epsilon > 0$ we can select $\delta > 0$ such that $|x - x_0| < \delta$ implies $|f(x) - f(x_0)| < \epsilon$. It follows that

$$\left| \frac{F(x) - F(x_0)}{x - x_0} - f(x_0) \right| = \left| (x - x_0)^{-1} \int_{x_0}^{x} (f(r) - f(x_0)) \, dr \right| \leq \epsilon,$$

which implies (ii).

We now prove part 2. Given $\epsilon > 0$ we construct a partition $P = \{x_0, \ldots, x_n\}$ such that $U(P, f) - L(P, f) < \epsilon$. By the mean value theorem (Proposition D.1.5) there exists a point $r_i \in (x_{i-1}, x_i)$ such that

$$F(x_i) - F(x_{i-1}) = f(r_i)\Delta_i, \quad i = 1, \ldots, n.$$

Summing the above equations for all i we get

$$F(b) - F(a) = \sum_{i=1}^{n} f(r_i)|\Delta_i|$$

and using Proposition F.2.1 we have

$$\left| F(b) - F(a) - \int_a^b f(x)\,dx \right| < \epsilon.$$

Since $\epsilon > 0$ is arbitrary, the proposition follows.   ∎

Example F.2.1. *Since $cx^{n+1}/(n+1)$ is the anti-derivative of cx^n,*

$$\int_a^b cx^n\,dx = \frac{c}{n+1}x^{n+1}\Big|_a^b = \frac{c}{n+1}(b^{n+1} - a^{n+1}).$$

Example F.2.2. *Since $(1/c)e^{cx}$ is the anti-derivative of e^{cx},*

$$\int_a^b e^{cx}\,dx = (1/c)e^{cx}\Big|_a^b = (1/c)\left(e^{cb} - e^{ca}\right).$$

For example

$$\int_0^\infty e^{-r/2}\,dr = \left(-2e^{-r/2}\right)\Big|_0^\infty = -2(0 - 1) = 2. \tag{F.1}$$

Proposition F.2.3 (Change of Variables). *Let g be a strictly monotonic increasing and differentiable function whose derivative g' is continuous on $[a, b]$ and f be a continuous function. Then*

$$\int_{g(a)}^{g(b)} f(t)\,dt = \int_a^b f(g(t))g'(t)\,dt.$$

Proof. Since f, g, g' are continuous, $f(g(t))g'(t)$ is continuous as well and therefore Riemann integrable. Defining $F(t) = \int_{g(a)}^t f(t)\,dt$, we have

$$\int_a^b f(g(t))g'(t)\,dt = \int_a^b F'(g(t))g'(t)\,dt \tag{F.2}$$

$$= \int_a^b (F \circ g)'(t)\,dt \tag{F.3}$$

$$= (F \circ g)(b) - (F \circ g)(a) \tag{F.4}$$

$$= \int_{g(a)}^{g(b)} f(t)\,dt. \tag{F.5}$$

Above, we used the first part of the fundamental theorem of calculus in (F.2), the differentiation chain rule in (F.3), and the second part of the fundamental theorem of calculus in (F.4) and in (F.5). ∎

In many cases the integral on the right hand side above is difficult to compute but the integral on the left hand side is easier. In these cases the change of variables provides a useful tool for computing integrals.

Example F.2.3. *Using the variable transformation* $y = f(x) = (x - \mu)/\sigma$, $f'(x) = 1/\sigma$, *and*

$$\int_{(a-\mu)/\sigma}^{(b-\mu)/\sigma} \exp\left(-\frac{y^2}{2}\right) dy = \int_a^b \exp\left(-\frac{(x-\mu)^2}{2\sigma^2}\right) \frac{1}{\sigma} dx. \qquad \text{(F.6)}$$

We compute the integral on the left hand side in Example F.6.1 below (for $a \to -\infty, b \to \infty$*).*

Example F.2.4. *Using the variable transformation* $y = f(x) = -x^2/2$ *with* $f'(x) = -x$,

$$\int_a^b e^{-x^2/2}(-x)dx = \int_{-a^2/2}^{-b^2/2} e^y \, dy = e^{-b^2/2} - e^{-a^2/2}. \qquad \text{(F.7)}$$

where the left hand side above corresponds to the right hand side in Proposition F.2.3. For example,

$$\int_0^\infty e^{-x^2/2} x dx = -\int_0^{-\infty} e^y \, dy = -(e^y)\Big|_0^{-\infty} = -(0-1) = 1. \qquad \text{(F.8)}$$

Proposition F.2.4 (Integration by Parts). *For two differentiable functions* $f, g : \mathbb{R} \to \mathbb{R}$ *we have*

$$\int_a^b f(x)g'(x) \, dx = f(b)g(b) - f(a)g(a) - \int_a^b f'(x)g(x) \, dx.$$

Proof. The result follows from the second part of the Fundamental Theorem of Calculus (Proposition F.2.2), applied to $h(x) = f(x)g(x)$ and its derivative $h' = f'g + g'f$. ∎

Example F.2.5.

$$\int_a^b x^c e^{-x} \, dx = (-x^c e^{-x}) \Big|_a^b - c \int_a^b x^{c-1} e^{-x} \, dx.$$

For example, the integral below known as the Gamma function satisfies for $c > 1$

$$\Gamma(c) \overset{\text{def}}{=} \int_0^\infty x^{c-1} e^{-x}\, dx$$

$$= \left(-x^{c-1} e^{-x}\right)\Big|_0^\infty - \left(-(c-1) \int_0^\infty x^{c-2} e^{-x}\, dx\right)$$

$$= (0 - 0) + (c - 1) \int_0^\infty x^{c-2} e^{-x}\, dx$$

$$= (c - 1)\Gamma(c - 1).$$

where the last equality follows from L'Hospital's rule (Proposition D.1.7) that implies $\lim_{x \to \infty} x^{c-1}/e^x = 0$ for $c > 1$ (the limit $\lim_{x \to 0} x^{c-1}/e^x = 0$ follows from the facts that $x^{c-1} \to 0$ and $e^x \to 1$ and division is a continuous function that preserves limits). Applying the above equation recursively together with

$$\Gamma(1) = \int_0^\infty \exp(-t)\, dt = \left(-e^{-t}\right)\Big|_0^\infty = 1 - 0 = 1,$$

we get

$$\Gamma(n) = (n - 1)!, \qquad n \in \mathbb{N}.$$

Thus, for $x > 1$ the Gamma function extends the factorial function from the natural numbers to real values $x > 1$. The R code below graphs the Gamma function for the positive reals, as well as for negative values. The graph exhibits an unexpected behavior for $x < 1$, specifically: a non-monotonic behavior for $x < 1$ and discontinuities at non-positive integers.

```
plot.function(gamma, from = -2, to = 5, xlab = "$x$",
     ylab = "$\\Gamma(x)$", ylim = c(-10, 30))
```

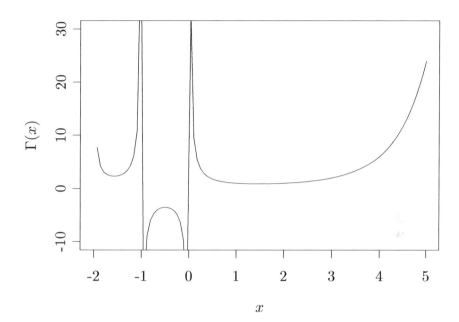

F.3   The Lebesgue Integral*

In the definitions and propositions below, we assume the presence of a measure space $(\Omega, \mathcal{F}, \mu)$ (Ω is an arbitrary set, $\mathcal{F}$ is a σ-algebra on Ω, and μ is a measure $\mu : \mathcal{F} \to \mathbb{R}$). We assume below and in the rest of this book that $0 \cdot \infty = 0$.

Definition F.3.1. A function $f : \Omega \to \mathbb{R}$ is simple if it has finite range, or in other words we can write f as

$$f(\omega) = \sum_{i=1}^{n} c_i I_{A_i}(\omega), \quad n \in \mathbb{N}, \quad c_i \in \mathbb{R}$$

for some sets $A_i \subset \Omega$, $i = 1, \ldots, n$ ($I_A(\omega) = 1$ if $\omega \in A$ and 0 otherwise).

Proposition F.3.1. *If $f : \Omega \to \mathbb{R}$ is a measurable function, then there exists a sequence of simple measurable functions $f_n : \Omega \to \mathbb{R}, n \in \mathbb{N}$ such that*

$$\begin{aligned} f(\omega) \geq 0 \qquad implies \qquad 0 \leq f_n(\omega) \nearrow f(\omega) \\ f(\omega) \leq 0 \qquad implies \qquad 0 \geq f_n(\omega) \searrow f(\omega). \end{aligned}$$

(Recall Definition B.2.12 regarding the notations $\nearrow, \searrow$.)

Proof. The sequence of functions

$$f_n(\omega) = \begin{cases} -n & -\infty \leq f(\omega) \leq -n \\ -(k-1)2^{-n} & -k2^{-n} < f(\omega) \leq -(k-1)2^{-n}, \quad 1 \leq k \leq n2^n \\ (k-1)2^{-n} & (k-1)2^{-n} \leq f(\omega) < k2^{-n}, \quad 1 \leq k \leq n2^n \\ n & n \leq f(\omega) \leq +\infty \end{cases}$$

has the necessary properties (see the following example for an illustration).   ∎

Example F.3.1. *In the case of $f(x) = x^2$, the functions f_1 and f_2 in the proof of the above proposition are*

$$f_1(\omega) = \begin{cases} 0 & 0 \leq f(\omega) < 1/2 \\ 1/2 & 1/2 \leq f(\omega) < 1 \\ 1 & 1 \leq f(\omega) \leq +\infty \end{cases},$$

$$f_2(\omega) = \begin{cases} 0 & 0 \leq f(\omega) < 1/4 \\ 1/4 & 1/4 \leq f(\omega) < 2/4 \\ 2/4 & 2/4 \leq f(\omega) < 3/4 \\ 3/4 & 3/4 \leq f(\omega) < 4/4 \\ 4/4 & 4/4 \leq f(\omega) < 5/4 \\ 5/4 & 5/4 \leq f(\omega) < 6/4 \\ 6/4 & 6/4 \leq f(\omega) < 7/4 \\ 7/4 & 7/4 \leq f(\omega) < 8/4 \\ 2 & 2 \leq f(\omega) \leq +\infty \end{cases}.$$

The figures below illustrate these functions in the case of $\Omega = \mathbb{R}$.

```
X = seq(-2, 2, length.out = 200)
f1 = rep(0, length.out = 200)
f1[X^2 > 1/2 & X^2 < 1] = 1/2
f1[X^2 > 1] = 1
qplot(X, X^2, xlab = "$x$", ylab = "$f(x)=x^2$",
    geom = "line", main = "Simple Function Approximation ($n=1$) ") +
    geom_area(aes(x = X, y = f1))
```

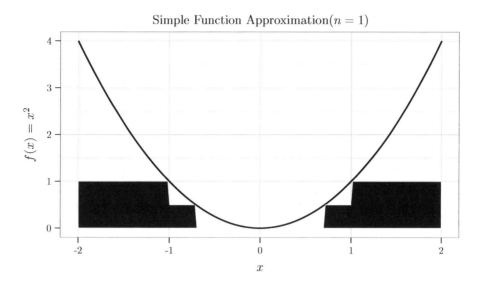

```
f2 = rep(0, length.out = 200)
f2[X^2 > 1/4 & X^2 < 2/4] = 1/4
f2[X^2 > 2/4 & X^2 < 3/4] = 2/4
f2[X^2 > 3/4 & X^2 < 4/4] = 3/4
f2[X^2 > 4/4 & X^2 < 5/4] = 4/4
f2[X^2 > 5/4 & X^2 < 6/4] = 5/4
f2[X^2 > 6/4 & X^2 < 7/4] = 6/4
f2[X^2 > 7/4 & X^2 < 8/4] = 7/4
f2[X^2 > 8/4] = 2
qplot(X, X^2, xlab = "$x$", ylab = "$f(x)=x^2$",
    geom = "line", main = "Simple Function Approximation ($n=2$)") +
    geom_area(aes(x = X, y = f2))
```

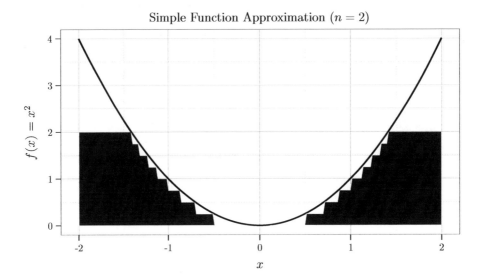

Simple Function Approximation ($n = 2$)

Definition F.3.2. The Lebesgue integral of a non-negative measurable simple function is

$$\int_E \sum_{i=1}^{n} \alpha_i I_{A_i} \, d\mu \overset{\text{def}}{=} \sum_{i=1}^{n} \alpha_i \, \mu(A_i \cap E), \qquad E \in \mathcal{F}.$$

Definition F.3.3. The Lebesgue integral of a non-negative measurable function is

$$\int_E f \, d\mu \overset{\text{def}}{=} \sup_g \int_E g \, d\mu, \qquad E \in \mathcal{F},$$

where the supremum ranges over all simple measurable functions g such that $0 \leq g \leq f$.

Note that if f is a non-negative measurable simple function, there are two possible definitions for $\int_E f \, d\mu$ (Definition F.3.2 and Definition F.3.3). Fortunately, they both agree as the supremum in the Definition F.3.3 is realized by $g = f$.

Proposition F.3.2. *Let f be a non-negative measurable function.*

1. *If $f \leq g$ then $\int_E f \, d\mu \leq \int_E g \, d\mu$.*

2. *If $A \subset B$ then $\int_A f \, d\mu \leq \int_B f \, d\mu$.*

3. *If $c \in [0, \infty)$ then $\int_E cf \, d\mu = c \int_E f \, d\mu$.*

4. *If $f(x) = 0$ for all $x \in E$ then $\int_E f \, d\mu = 0$.*

5. *If $\mu(E) = 0$ then $\int_E f \, d\mu = 0$.*

6. *$\int_E f \, d\mu = \int_\Omega (I_E f) \, d\mu$.*

Proof. These properties follow directly from Definitions F.3.2-F.3.3. The first property follows from the fact that the supremum in the definition of $\int_E f \, d\mu$ is less than or equal to the supremum in the definition of $\int_E g \, d\mu$. The second property holds since the supremum in the definition of $\int_A f \, d\mu$ is less than or equal than the supremum in the definition of $\int_B f \, d\mu$. The third property follows from the fact that the supremum in the definition of $\int_E cf \, d\mu$ is c times the supremum in the definition of $\int_E f \, d\mu$. The fourth property follows from the fact that the supremum in the definition of $\int_E f \, d\mu$ can only be the zero function. The fifth property follows since the integrals of all simple functions over E are zero. The sixth property follows directly from Definitions F.3.2-F.3.3.   ∎

Proposition F.3.3. *Let $(\Omega, \mathcal{F}, \mu)$ be a measure space. For two non-negative measurable simple functions $s, t : \Omega \to [0, \infty]$, we have*

$$\int_\Omega (s+t) \, d\mu = \int_\Omega s \, d\mu + \int_\Omega t \, d\mu$$

and $\phi(E) = \int_E s \, d\mu$ is a measure function on $(\Omega, \mathcal{F})$.

Proof. We assume that $s = \sum_{i=1}^k \alpha_i I_{A_i}$ and $t = \sum_{i=1}^l \beta_i I_{B_i}$. For all sets $E \in \mathcal{F}$, we have $\phi(E) \geq 0$ and $\phi(\emptyset) = 0$. If $E_n, n \in \mathbb{N}$ is a sequence disjoint sets in $\mathcal{F}$ with $\cup_n E_n = E$ then

$$\phi(E) = \int_E \sum_{i=1}^k \alpha_i I_{A_i} \, d\mu = \sum_{i=1}^k \alpha_i \mu(A_i \cap E) = \sum_{i=1}^k \sum_{n \in \mathbb{N}} \alpha_i \mu(A_i \cap E_n)$$
$$= \sum_{n \in \mathbb{N}} \phi(E_n),$$

proving that ϕ is a measure.

For all $E_{ij} = A_i \cap B_j$, $i = 1, \ldots, k$, $j = 1, \ldots, l$

$$\int_{E_{ij}} (s+t) \, d\mu = (\alpha_i + \beta_j) \mu(E_{ij}) = \int_{E_{ij}} s \, d\mu + \int_{E_{ij}} t \, d\mu.$$

Since ϕ is a measure, it is countable additive and so the above equation implies $\int_\Omega (s+t) \, d\mu = \int_\Omega s \, d\mu + \int_\Omega t \, d\mu$.   ∎

Proposition F.3.4 (Monotone Convergence Theorem).

$$0 \leq f_n \nearrow f \quad \text{implies} \quad \int_\Omega f_n \, d\mu \nearrow \int_\Omega f \, d\mu.$$

Proof. The function $f = \lim f_n$ is measurable by Proposition E.6.4, and since $\int f_n \, d\mu \nearrow$, we have $\int f_n \, d\mu \nearrow \alpha$ for some $\alpha \in [0, +\infty]$ (see Proposition B.2.1). Since $f_n \leq \lim f_n$, we have $\int f_n \, d\mu \leq \int_\Omega \lim_{n \to \infty} f_n \, d\mu$.

It remains to show the reverse inequality $\int f_n \, d\mu \geq \int_\Omega \lim_{n \to \infty} f_n \, d\mu$. For any non-negative simple measurable function s such that $s \leq \lim f_n$ and any

$c \in (0,1)$ we define $E_n = \{x : f_n(x) \geq cs(x)\}, n \in \mathbb{N}$. Note that $E_n \in \mathcal{F}, E_n \nearrow$, and $\Omega = \cup_n E_n$. By definition of E_n, we have for all $n \in \mathbb{N}$

$$\int_\Omega f_n \, d\mu \geq \int_{E_n} f_n \, d\mu \geq c \int_{E_n} s \, d\mu. \tag{F.9}$$

We let $n \to \infty$ and recall that $\int_{E_n} s \, d\mu$ is a measure (Proposition F.3.3) $\phi(E_n)$, which converges by Proposition E.2.1 to $\phi(\Omega)$. This implies that

$$\lim \int_\Omega f_n \, d\mu \geq \lim c \int_{E_n} s \, d\mu = c \int_\Omega s \, d\mu.$$

Since this holds for any $c \in (0,1)$ we also have

$$\lim \int_\Omega f_n \, d\mu \geq \int_\Omega s \, d\mu.$$

Finally, since this holds for all non-negative simple measurable function s such that $s \leq \lim f_n$, it also holds for their supremum, which defines the integral $\int_\Omega \lim f_n \, d\mu$.   ∎

Corollary F.3.1. *Let $f_n, n \in \mathbb{N}$ be a sequence of non-negative measurable functions whose sum $\sum_{n=1}^\infty f_n(x)$ converges for all x. Then*

$$\int_\Omega \sum_{n=1}^\infty f_n \, d\mu \ = \ \sum_{n=1}^\infty \int_\Omega f_n \, d\mu.$$

Proof. We first prove this result for a sum of two functions $f_1 + f_2$. Let s_k and t_k be the sequence of measurable functions that converge to f_1 and f_2 as in Proposition F.3.1. Then the sequence of simple functions $s_k + t_k$ converge to $f_1 + f_2$, and using the Monotone Convergence Theorem above together with Proposition F.3.3, we have

$$\int_\Omega f_1 + f_2 \, d\mu = \int_\Omega \lim_k (s_k + t_k) \, d\mu = \lim_k \int_\Omega s_k + t_k \, d\mu$$

$$= \lim_k \int_\Omega s_k \, d\mu + \lim_k \int_\Omega t_k \, d\mu = \int_\Omega f_1 \, d\mu + \int_\Omega f_2 \, d\mu.$$

By induction, we establish the result for a sum of N functions. Defining $g_N = \sum_{n=1}^N f_n$, we have that $g_N \nearrow \sum_{n=1}^\infty f_n$. Applying the Monotone Convergence

Theorem again yields

$$\int_\Omega \sum_{n=1}^\infty f_n \, d\mu = \int_\Omega \lim_{N\to\infty} g_N \, d\mu$$

$$= \lim_{N\to\infty} \int_\Omega g_N \, d\mu$$

$$= \lim_{N\to\infty} \sum_{n=1}^N \int_\Omega f_n \, d\mu$$

$$= \sum_{n=1}^\infty \int_\Omega f_n \, d\mu.$$

■

Proposition F.3.5. *Let $(\Omega, \mathcal{F}, \mu)$ be a measure space. For any two non-negative measurable functions $f, g : \Omega \to [0, \infty]$,*

1. *$\phi(E) = \int_E f \, d\mu$ is a measure function on $(\Omega, \mathcal{F})$, and*

2.
$$\int_\Omega g \, d\phi = \int_\Omega gf \, d\mu.$$

We say in this case that ϕ has a density f with respect to μ, and write $d\phi = f d\mu$.

Proof. It is easy to see that $\phi(E) \geq 0$ for all $E \in \mathcal{F}$ and $\phi(\emptyset) = 0$. We next show that ϕ is countable additive. Let $E_n, n \in \mathbb{N}$ be a sequence of disjoint sets whose union equals E. Corollary F.3.1 implies that

$$\phi(E) = \int_\Omega f I_E \, d\mu = \int_\Omega \sum_{n=1}^\infty f I_{E_n} \, d\mu = \sum_{n=1}^\infty \int f I_{E_n} \, d\mu = \sum_{n=1}^\infty \phi(E_n). \qquad \text{(F.10)}$$

The second statement above follows from the first statement whenever g is an indicator function $g(x) = c I_A(x)$. Similar arguments show that the second statement follows from the first statement whenever g is a simple function. The case of a general function g follows from approximating g by a sequence of simple functions $0 \leq s_1 \leq s_2 \leq \cdots$ (Proposition F.3.1) and using the Monotone Convergence Theorem twice

$$\int_\Omega g \, d\phi = \int_\Omega \lim_n s_n \, d\phi = \lim_n \int_\Omega s_n \, d\phi = \lim_n \int_\Omega s_n f \, d\mu = \int_\Omega gf \, d\mu.$$

■

Proposition F.3.6 (Fatou's Lemma). *For any sequence of non-negative measurable functions $f_n, n \in \mathbb{N}$,*

$$\int_\Omega \liminf_{n\to\infty} f_n \, d\mu \leq \liminf_{n\to\infty} \int_\Omega f_n \, d\mu.$$

Proof. Defining the function $g_k(x) = \inf\{f_i(x) : i \geq k\}$, we have $g_k \leq f_k$ and therefore for all $k \in \mathbb{N}$,

$$\int_\Omega g_k \, d\mu \leq \int_\Omega f_k \, d\mu.$$

Since $0 \leq g_1 \leq g_2 \leq \cdots$, and $g_k \to \liminf_n f_n$, we can apply the Monotone convergence theorem to the left hand side above to get

$$\int_\Omega \liminf_n f_n \, d\mu = \lim_k \int_\Omega g_k \, d\mu = \liminf_k \int_\Omega g_k \, d\mu \leq \liminf_k \int_\Omega f_k \, d\mu.$$

∎

The definition of the Lebesgue integral $\int_A f \, d\mu$ thus far has been restricted to non-negative functions f. We generalize this below to real valued functions $f : \Omega \to \mathbb{R}$.

Definition F.3.4. For a function $f : \Omega \to \mathbb{R}$ we define its positive and negative parts

$$f^+(\omega) = \begin{cases} f(\omega) & f(\omega) \geq 0 \\ 0 & f(\omega) < 0 \end{cases}$$

$$f^-(\omega) = \begin{cases} -f(\omega) & f(\omega) \leq 0 \\ 0 & f(\omega) > 0 \end{cases}.$$

This implies the following decomposition of an arbitrary function $f : \Omega \to \mathbb{R}$ to its positive and negative parts:

$$f(\omega) = f^+ + f^-.$$

Definition F.3.5. The Lebesgue integral of a measurable function $f : \Omega \to \mathbb{R}$ is

$$\int_\Omega f \, d\mu \overset{\text{def}}{=} \int_\Omega f^+ \, d\mu - \int_\Omega f^- \, d\mu$$

where $\int f^+ \, d\mu$ and $\int f^- \, d\mu$ are integrals of non-negative functions defined in Definition F.3.3. If the integral above is finite we say that f is μ-integrable or simply integrable.

If both $\int f^+ \, d\mu = \infty$ and $\int f^- \, d\mu = \infty$ in the definition above we say that the integral above does not exist.

Proposition F.3.7.

$$\left| \int_\Omega f \, d\mu \right| \leq \int_\Omega |f| \, d\mu.$$

Proof. The left hand side equals $\int f^+ \, d\mu - \int f^- \, d\mu$, a difference of two non-negative quantities. This is clearly smaller than the right hand side, which equals $\int |f^+| \, d\mu + \int |f^-| \, d\mu$.   ∎

Proposition F.3.8. *Let* f, g *be two measurable functions and* $\alpha, \beta \in \mathbb{R}$. *Then*

$$\int_\Omega \alpha f + \beta g \, d\mu = \alpha \int_\Omega f \, d\mu + \beta \int_\Omega g \, d\mu.$$

Proof. We decompose the proof into the case $\int_\Omega f + g \, d\mu = \int_\Omega f \, d\mu + \int_\Omega g \, d\mu$ and the case $\int_\Omega \alpha f \, d\mu = \alpha \int_\Omega f \, d\mu$. Both cases together imply the proposition above.

In the first case, if $\alpha > 0$ the result follows from applying Corollary F.3.1 to the positive-negative decomposition

$$\int \alpha f \, d\mu = \int (\alpha f)^+ \, d\mu - \int (\alpha f)^- \, d\mu = \alpha \int f^+ \, d\mu - \alpha \int f^- \, d\mu = \alpha \int f \, d\mu.$$

Similarly, if $\alpha < 0$

$$\int \alpha f \, d\mu = \int (\alpha f)^+ \, d\mu - \int (\alpha f)^- \, d\mu$$

$$= \int (-\alpha f^-) \, d\mu - \int (-\alpha f^+) \, d\mu$$

$$= -\alpha \int f^- \, d\mu - (-\alpha) \int f^+ \, d\mu = \alpha \int f \, d\mu.$$

In the second case, the result follows from applying Corollary F.3.1 to the positive-negative decomposition of $f + g$

$$\int f + g \, d\mu = \int f^+ - f^- + g^+ - g^- \, d\mu$$

$$= \int f^+ \, d\mu - \int f^- \, d\mu + \int g^+ \, d\mu - \int g^- \, d\mu$$

$$= \int f \, d\mu + \int g \, d\mu.$$

∎

Proposition F.3.9 (Dominated Convergence Theorem). *If* $f_n \to f$ *and* $|f_n(x)| \leq g(x)$ *for all* $n \in \mathbb{N}$ *for some integrable function* g, *then*

$$\int_\Omega \left| f_n - f \right| d\mu \to 0 \tag{F.11}$$

$$\int_\Omega f_n \, d\mu \to \int_\Omega f \, d\mu. \tag{F.12}$$

Proof. We denote $f = \lim f_n$ and observe that f_n is integrable since $|f_n| \leq g$ and g is integrable. Applying Fatou's lemma to the sequence of non-negative functions $2g - |f_n - f| \geq 0$ we get

$$\int 2g \, d\mu \leq \liminf_{n \to \infty} \int (2g - |f_n - f|) \, d\mu = \int 2g \, d\mu + \liminf_{n \to \infty} - \int |f_n - f| \, d\mu$$

$$= \int 2g \, d\mu - \limsup_{n \to \infty} \int |f_n - f| \, d\mu,$$

which implies

$$0 \geq \limsup_{n \to \infty} \int |f_n - f| \, d\mu.$$

Since $\int |f_n - f| \, d\mu$ is a sequence of non-negative values, the first result holds. The second result follows from applying Proposition F.3.7 to $f_n - f$:

$$0 = \lim_{n \to \infty} \int_{n \to \infty} |f_n - f| \, d\mu$$
$$\geq \lim_{n \to \infty} \left| \int f_n - f \, d\mu \right|$$
$$= \lim_{n \to \infty} \left| \int f_n \, d\mu - \int f \, d\mu \right|.$$

∎

Corollary F.3.2 (Bounded Convergence Theorem). *If $\mu(\Omega) < \infty$, $f_n \to f$, and for some finite M, $|f_n(x)| < M$ for all n and all x, then*

$$\int_{\Omega} f_n \, d\mu \to \int_{\Omega} f \, d\mu. \tag{F.13}$$

Proof. The result follows immediately from the second statement of the dominated convergence theorem above. ∎

Most of the propositions in this section are specified in terms of an integral of a function f over the measurable space Ω. These propositions remain true if the integrand f is replaced by $f I_A$, effectively creating new versions of these propositions where the integrals range over alternative sets A rather than Ω. As an example, consider a modified version of Proposition F.3.8, which states that

$$\int_A \alpha f + \beta g \, d\mu = \alpha \int_A f \, d\mu + \beta \int_A g \, d\mu.$$

Definition F.3.6. Let $(\Omega, \mathcal{F}, \mu)$ be a measure space. We say that a property holds almost everywhere, abbreviated a.e., if it holds on $\Omega \setminus S$ where $\mu(S) = 0$. In other words, the property holds everywhere except on a set of measure zero.

Proposition F.3.10. *Let f be a non-negative function. Then*

$$\int f \, d\mu = 0 \qquad implies \qquad f = 0 \text{ a.e.}.$$

Proof. Defining E_n to be the subset of Ω on which $f > 1/n$, we have $f \geq \sum_{n \in \mathbb{N}} (1/n) I_{E_n}$ and

$$0 = \int f \, d\mu \geq \int \sum_{n \in \mathbb{N}} (1/n) I_{E_n} \, d\mu = \sum_{n \in \mathbb{N}} (1/n) \mu(E_n).$$

It follows that $\mu(E_n) = 0$ for all $n \in \mathbb{N}$, implying that $\mu(\cup_{n \in \mathbb{N}} E_n) \leq \sum_{n \in \mathbb{N}} \mu(E_n) = 0$ and therefore $f = 0$ except on a set of measure zero. ∎

Corollary F.3.3. *If $\int (f - g)^2 \, d\mu = 0$ or $\int |f - g| \, d\mu = 0$ then $f = g$ a.e..*

F.3.1 Relation between the Riemann and the Lebesgue Integrals*

The following proposition states precisely when a function is Riemann integrable. It also states that if the Riemann integral exists then the Lebesgue integral exists as well and the two integrals agree in value.

Proposition F.3.11. *Let* $f : [a, b] \rightarrow \mathbb{R}$ *be a bounded function. Then it is Riemann integrable if and only if it is continuous a.e.. Furthermore, f then is also integrable with respect to the Lebesgue measure, and the two integrals agree in value.*

Proof. We use notations below from the section concerning Riemann integrals. To distinguish between the Riemann and the Lebesgue integrals we denote the former as $\int f \, dx$ and the latter as $\int f \, d\mu$ (μ here corresponds to the Lebesgue measure). All integrals in the proof below are over the interval $[a, b]$.

We construct a sequence of partitions $P_n, n \in \mathbb{N}$ such that P_k is a refinement of P_{k-1}, the distance between adjacent points in P_k is less than $1/k$, and

$$\lim_{k \to \infty} U(P_k, f) = \overline{\int f \, dx}$$

$$\lim_{k \to \infty} L(P_k, f) = \underline{\int f \, dx}.$$

(This is possible since the lower and upper Riemann integrals are defined using the limits above.) Defining the following two simple functions

$$L(x) = \begin{cases} m_i & x \in \Delta_i \\ f(a) & x = a \\ f(b) & x = b \end{cases} \tag{F.14}$$

$$U(x) = \begin{cases} M_i & x \in \Delta_i \\ f(a) & x = a \\ f(b) & x = b \end{cases}. \tag{F.15}$$

we have

$$L(P_k, f) = \int L_k \, d\mu$$

$$U(P_k, f) = \int U_k \, d\mu$$

$$L_1 \leq L_2 \leq \cdots \leq f \leq \cdots \leq U_2 \leq U_l \quad \text{(since } P_k \text{ refines } P_{k-1}\text{)}.$$

The last equation above implies that L_k and U_k converge to limit functions $L(x) = \lim_{n \to \infty} L_k(x)$ and $U(x) = \lim_{n \to \infty} U_k(x)$ and that $L \leq f \leq U$. Using the Monotone Convergence Theorem, we have

$$\int L \, d\mu = \underline{\int f \, dx}, \qquad \int U \, d\mu = \overline{\int f \, dx}.$$

Note that the above derivations assume only that f is bounded. The function f is Riemann integrable, if and only if the lower and upper integrals agree, which occurs if and only if $L = U$ a.e. (using the fact that $U - L \geq 0$ and Proposition F.3.10) or equivalently $L = f = U$. This proves that if f is a bounded Riemann integrable function, it is also Lebesgue integrable with respect to the Lebesgue measure and the two integrals agree.

If x is not equal to one of the points in the partition P_k, then $L(x) = U(x)$ if and only if $m_i = M_i$, which occurs if and only if f is continuous at x. Recalling that f is Riemann integrable if and only if $L = U$ a.e. (shown above), we have that f is Riemann integrable if and only if f is continuous a.e.. ∎

F.3.2   Transformed Measures*

Definition F.3.7. Consider a measure space $(\Omega, \mathcal{F}, \mu)$, and a measurable function $T : \Omega \to \Omega'$ where $(\Omega', \mathcal{F}')$ is a measurable space. The transformed measure μT^{-1} is a measure on $(\Omega', \mathcal{F}')$ defined as follows

$$\mu T^{-1}(A) = \mu(T^{-1}A), \qquad A \in \mathcal{F}'.$$

It is straightforward to verify that μT^{-1} satisfy the conditions of non-negativity and countable additivity, and is therefore a legitimate measure function. The following proposition uses this concept to construct the well-known change of variable formula for the Lebesgue integral.

Proposition F.3.12 (Change of Variable). *Let $f : \Omega' \to \mathbb{R}$ be a non-negative measurable function. Then*

$$\int_{T^{-1}(A')} f(Tx)\mu(dx) = \int_{A'} f(x')\mu T^{-1}(dx'), \qquad A \in \Omega'. \tag{F.16}$$

Equation (F.16) holds also for a real valued f (not necessarily non-negative) if $f \circ T$ is integrable with respect to μ.

Proof. If $f = I_A$ then $f \circ T = I_{T^{-1}A}$ and (F.16) holds by definition of the Lebesgue integral and the transformed measure. The linearity of the integral implies that (F.16) holds also for non-negative simple function. Equation (F.16) holds for non-negative function by constructing a sequence of simple functions $s_n, n \in \mathbb{N}$ such that $s_n \nearrow f$ and using the Monotone Convergence Theorem.

If f is a real valued function, we can apply (F.16) to $|f|$, which implies that f is integrable with respect to μT^{-1} if $f \circ T$ is integrable with respect to μ. The result then follows by decomposing f to its positive and negative parts and using the first part of the proposition on these two non-negative functions. ∎

Proposition F.2.3 derives a special case of the change of variables formula that is often useful in computing non-trivial integrals.

F.4 Product Measures*

Definition F.4.1. Let $(\Omega_t, \mathcal{F}_t)$, $t \in T$ be measurable spaces and consider the Cartesian product set (Definitions A.1.10, A.3.4) $\times_{t \in T} \Omega_t$. A set $A \subset \times_{t \in T} \Omega_t$ is a measurable rectangle if $A = \times_{t \in T} A_t$ for some $A_t \in \mathcal{F}_t$, for all $t \in T$. A set $A \subset \times_{t \in T} \Omega_t$ is a measurable cylinder if $A = A_s \times (\times_{t:t \neq s} \Omega_t)$, $A_s \in \mathcal{F}_s$, $s \in T$. If A_s above is open the set is called an open measurable cylinder and if $A_s = (a, b)$ the set is called a simple open measurable cylinder.

Measurable rectangles are products of sets in the corresponding σ-algebras $\mathcal{F}_t$, $t \in T$. Some examples of measurable rectangles in $\Omega = \mathbb{R}^3 = \mathbb{R} \times \mathbb{R} \times \mathbb{R}$ (assuming the the Borel σ-algebra on each copy of $\mathbb{R}$) are

$$(1, 2) \times \mathbb{R} \times [2, 3)$$
$$([1, 2] \cup [3, 4]) \times [2, 3] \times [4, 5]$$
$$\mathbb{R} \times \mathbb{R} \times \cup_{n \in \mathbb{N}} [n, n + 1/2).$$

For example, the set $\{\boldsymbol{x} : \sum_i x_i = 1\}$ is not a measurable rectangle. Figure 4.2 illustrates some sets in $\mathbb{R}^2$ that are measurable rectangles and some sets that are not.

Measurable cylinders $\{A_s \times (\times_{t \neq s} \Omega_t) : A_s \in \mathcal{F}_s, s \in T\}$ are special cases of measurable rectangles, in which only one of the products is allowed to differ from Ω_t. A general characterization of the measurable cylinders in $\mathbb{R}^3$ is

$$A \times \mathbb{R} \times \mathbb{R}, \qquad A \in \mathcal{B}(\mathbb{R})$$
$$\mathbb{R} \times A \times \mathbb{R}, \qquad A \in \mathcal{B}(\mathbb{R})$$
$$\mathbb{R} \times \mathbb{R} \times A, \qquad A \in \mathcal{B}(\mathbb{R}).$$

The following definitions constructs a new measurable space from several measurable spaces. The new measurable space, called the product measurable space, leads to the construction of the product measure, which is instrumental in studying probability over Euclidean spaces $\mathbb{R}^n$.

Definition F.4.2. Let $(\Omega_t, \mathcal{F}_t)$, $t \in T$ be measurable spaces. The product measurable space is the pair $(\Omega, \mathcal{F})$ of a set Ω and a σ-algebra defined by

$$\Omega = \times_{t \in T} \Omega_t \qquad \text{(see Definitions A.1.10, A.3.4)}$$
$$\mathcal{F} = \otimes_{t \in T} \mathcal{F}_t \overset{\text{def}}{=} \sigma \left(\{ A : A \text{ is a measurable cylinder in } \Omega \} \right).$$

Since σ(all open measurable cylinders) includes all measurable cylinders and therefore

$$\otimes_{t \in T} \mathcal{F}_t = \sigma \left(\{ A : A \text{ is an open measurable cylinder} \} \right).$$

Proposition F.4.1. *Recalling Definitions A.1.10, A.3.4, E.5.1, we have*

$$\mathcal{B}(\mathbb{R} \times \mathbb{R} \times \cdots) = \mathcal{B}(\mathbb{R}) \otimes \mathcal{B}(\mathbb{R}) \otimes \cdots$$
$$\mathcal{B}(\mathbb{R} \times \cdots \times \mathbb{R}) = \mathcal{B}(\mathbb{R}) \otimes \cdots \otimes \mathcal{B}(\mathbb{R}).$$

In other words, the σ-algebra $\mathcal{B}(\mathbb{R}) \otimes \cdots \otimes \mathcal{B}(\mathbb{R})$ is generated by open sets in $\mathbb{R}^d$ (and similarly for infinite products).

In the proposition above and elsewhere, whenever we refer to the set $\mathbb{R}^\infty$, we assume the metric structure defined in Example B.4.4.

Proof. The statement above may be written as $\sigma(\mathcal{L}_1) = \sigma(\mathcal{L}_2)$ where $\mathcal{L}_1$ is the set of open sets in $\mathbb{R}^d$ or R^∞ and $\mathcal{L}_2$ is the set of open measurable cylinders. It is sufficient to show that $\mathcal{L}_1 \subset \sigma(\mathcal{L}_2)$ and $\mathcal{L}_2 \subset \sigma(\mathcal{L}_1)$.

A set in $\mathcal{L}_2$ is an open measurable cylinder and is therefore an open set in $\mathbb{R}^d$ or in $\mathbb{R}^\infty$ (see Example B.4.4).

In the case of a finite number of products, let A be an open set in $\mathbb{R}^d$. Then for each $a \in A$ there exists an open ball $B_r(a')$ such that $a \in B_r(a') \subset A$. For each $a \in A$ and each open ball $B_r(a')$ containing a there exists a rectangle $\times_{i=1}^d (a_i, b_i)$ with rational endpoints such that $a \in \times_{i=1}^d (a_i, b_i) \subset B_r(a') \subset A$ and the union of all such open rectangles equals A. Since open rectangles are finite intersections of cylinder sets, and there are a countable number of rectangles with rational end-points, A is a countable union of finite intersections of cylinders, implying that $A \in \sigma(\mathcal{B}(\mathbb{R}) \otimes \cdots \otimes \mathcal{B}(\mathbb{R}))$.

The case of an infinite number of products proceeds similarly. By second countability of $\mathbb{R}^\infty$ (see Proposition B.4.5), an open set $A \subset R^\infty$ is a countable union of finite intersections of cylinders. It follows that $A = \sigma(\mathcal{L}_2)$.   ∎

Corollary F.4.1.

$$\mathcal{B}(\mathbb{R}^d) = \sigma(\{(a_1, b_1) \times \cdots \times (a_d, b_d) : a_i, b_i \in \mathbb{Q}, i = 1, \ldots, d\})$$

and similarly, $\mathcal{B}(R^\infty) = \sigma(\mathcal{A})$ where $\mathcal{A}$ is the set of all simple open measurable cylinders with rational endpoints.

Proof. This follows from the proof of the proposition above.   ∎

Proposition F.4.2. *A function $f : \Omega \to \mathbb{R}^d$, $f(x) = (f_1(x), \ldots, f_d(x))$ is measurable if $f_i : \Omega \to \mathbb{R}$ is measurable.*

Proof. Corollary F.4.1 implies that the σ-algebra $\mathcal{B}(\mathbb{R}^d)$ is generated by the set of all open rectangles. By Proposition E.6.1 it is then sufficient to show that for all open rectangles A, $f^{-1}(A) \in \mathcal{F}$. Because A is a rectangle, $f^{-1}(A) = \cap_{i=1}^d f_i^{-1}((a_i, b_i))$ which is a subset of $\mathcal{F}$ since f_i are measurable and since the σ-algebra of Ω is closed under finite intersections.   ∎

In the following, we concentrate on product of two spaces. This is done for the sake of simplicity, and the definitions and propositions carry over to a product of a finite number of spaces (for example, by induction).

Proposition F.4.3. *Let $(X, \mathcal{X})$ and $(Y, \mathcal{Y})$ be two measurable spaces. If $E \subset \mathcal{X} \otimes \mathcal{Y}$ then*

$$\{y : (x, y) \in E\} \in \mathcal{Y} \tag{F.17}$$

$$\{x : (x, y) \in E\} \in \mathcal{X}. \tag{F.18}$$

Proof. We prove the first statement. The proof of the second statement is similar. Defining the function $\phi : Y \to X \times Y$, $\phi(y) = (x_0, y)$, we have for each measurable rectangle $A \times B$,

$$\phi^{-1}(A \times B) = \begin{cases} B \in \mathcal{Y} & x_0 \in A \\ \emptyset \in \mathcal{Y} & x_0 \notin A \end{cases}.$$

Combined with Proposition E.6.1 and Definition F.4.2, this implies that ϕ is a measurable function. It follows that $\{y : (x, y) \in E\} = \phi^{-1}(E)$ is in $\mathcal{Y}$ for all $E \subset \mathcal{X} \otimes \mathcal{Y}$. ∎

Proposition F.4.4. *Let $(X, \mathcal{X})$ and $(Y, \mathcal{Y})$ be two measurable spaces and consider the measurable space $(X \times Y, \mathcal{X} \otimes \mathcal{Y})$. If $f : X \times Y \to \mathbb{R}$ is a measurable function on $X \times Y$, then the functions $g(x) = f(x, y_0)$ and $h(y) = f(x_0, y)$ are measurable on X and Y respectively, for each $x_0 \in X, y_0 \in Y$.*

Proof. Using the notations from the proof of the previous proposition, $h = f \circ \phi$, and since a composition of measurable functions is measurable (Proposition E.6.2), h is measurable. The proof of the measurability of g is similar. ∎

Proposition F.4.5 (Product Measure Theorem). *Let $(X, \mathcal{X}, \mathsf{P}_X)$ and $(Y, \mathcal{Y}, \mathsf{P}_Y)$ be two probability measure spaces. Then there is a unique probability measure $\mathsf{P}_{X \times Y}$ on the measurable space $(X \times Y, \mathcal{X} \otimes \mathcal{Y})$ such that for all measurable rectangles $A \times B$,*

$$\mathsf{P}_{X \times Y}(A \times B) = \mathsf{P}_X(A) \cdot \mathsf{P}_Y(B).$$

Proof. By Proposition F.4.3, for all measurable sets E and for all x_0, y_0, the sets $\{y : (x_0, y) \in E\}$ and $\{x : (x, y_0) \in E\}$ are measurable sets in Y and X, respectively. Since they are measurable sets we can apply the measures P_Y and P_X to them, obtaining the real valued functions $f_E(x) = \mathsf{P}_Y(\{y : (x, y) \in E\})$ and $g_E(y) = \mathsf{P}_X(\{x : (x, y) \in E\})$.

We denote the set of sets E for which f_E is measurable as $\mathcal{L}$, and show that it is a λ-system on $X \times Y$. We have $X \times Y \in \mathcal{L}$ since $f_E(x) = \mathsf{P}_Y(\{y : (x, y) \in X \times Y\}) = \mathsf{P}_Y(Y)$. If $A \in \mathcal{L}$, then $f_{E^c}(x)$ is measurable and therefore $f_E(x) = \mathsf{P}_Y(\{y : (x, y) \in A^c\}) = \mathsf{P}_Y(Y) - f_{E^c}(x)$ is measurable as well. If $A_n, n \in \mathbb{N}$ is a sequence of disjoint sets in $\mathcal{L}$ then $f_{\cup A_n}(x) = \mathsf{P}_Y(\{y : (x, y) \in \cup_n A_n\}) = \sum_{n \in \mathbb{N}} f_{A_n}(x)$ implying that $\cup A_n$ is in $\mathcal{L}$ as well. We have thus shows that $\mathcal{L}$ is a λ system. This derivation also shows that $f_E(x)$ and $g_E(x)$ are measure functions for all x.

The function f_E is also measurable for all measurable rectangles $E = A \times B$, since $f_{A \times B}(x) = I_A(x) \mathsf{P}_X(B)$. $\mathcal{L}$ is a λ-system containing the π-system of measurable rectangles, which generates the σ algebra $\mathcal{X} \otimes \mathcal{Y}$. It follows from Dynkin's Theorem (Proposition E.3.4) that $\mathcal{X} \otimes \mathcal{Y} \subset \mathcal{L}$ implying that f_E is measurable. The proof of measurability of g_E is similar.

Since for all $E \in \mathcal{X} \otimes \mathcal{Y}$, the functions f_E and g_E are measurable, we can define the following functions that assign real values to sets in $\mathcal{X} \otimes \mathcal{Y}$

$$\mathsf{P}_{X \times Y}^{(1)}(E) = \int_X \mathsf{P}_X(\{y : (x,y) \in E\}) \, d\mathsf{P}_X, \qquad \text{(F.19)}$$

$$\mathsf{P}_{X \times Y}^{(2)}(E) = \int_Y \mathsf{P}_Y(\{y : (x,y) \in E\}) \, d\mathsf{P}_Y . \qquad \text{(F.20)}$$

For measurable rectangles, $f_{A \times B}(x) = I_A(x) \mathsf{P}_X(B)$ and $g_{A \times B}(x) = I_B(x) \mathsf{P}_Y(A)$, implying that $\mathsf{P}_{X \times Y}^{(1)}(A \times B) = \mathsf{P}_{X \times Y}^{(2)}(A \times B) = \mathsf{P}_X(A) \cdot \mathsf{P}_Y(B)$. Since measurable rectangles generate $\mathcal{X} \otimes \mathcal{Y}$, it follows from Corollary E.3.1 that $\mathsf{P}_{X \times Y}^{(1)}$ agrees with $\mathsf{P}_{X \times Y}^{(2)}$ on $\mathcal{X} \otimes \mathcal{Y}$. It also follows that if there is any other measure $\mathsf{P}_{X \times Y}^{(3)}$ which assigns the value $\mathsf{P}_{X \times Y}^{(3)} = (A \times B) = \mathsf{P}_X(A) \cdot \mathsf{P}_Y(B)$ on measurable rectangles, then $\mathsf{P}_{X \times Y}^{(3)}$ agrees with $\mathsf{P}_{X \times Y}^{(1)}$ and $\mathsf{P}_{X \times Y}^{(2)}$ on $\mathcal{X} \otimes \mathcal{Y}$. ∎

We conclude with a few comments.

1. We usually denote the product measure of P_X and P_Y as $\mathsf{P}_X \times \mathsf{P}_Y$.

2. The product measure corresponds to independence between the random variables X and Y. Other probability measures are also available on the product space $(X \times Y, \mathcal{X} \otimes \mathcal{Y})$ (see Chapters 4 and 5).

3. The product measure theorem can be generalized to measures that are not probability functions. See [5] for a more general version.

4. The product measure theorem can be generalized for products of more than two spaces. The generalization is straightforward since the definitions above are associative:

$$(X \times Y) \times Z = X \times (Y \times Z)$$
$$(\mathcal{X} \otimes \mathcal{Y}) \otimes \mathcal{Z} = \mathcal{X} \otimes (\mathcal{Y} \otimes \mathcal{Z})$$
$$(\mathsf{P}_X \times \mathsf{P}_Y) \times \mathsf{P}_Z = \mathsf{P}_X \times (\mathsf{P}_Y \times \mathsf{P}_Z).$$

Further extensions exist for a product of an infinite number of measure spaces (both countably infinite and uncountably infinite).

F.5 Integration over Product Spaces*

We consider in this section a measure space that is a product space of two measure spaces $(X, \mathcal{X}, \mathsf{P}_X)$ and $(Y, \mathcal{Y}, \mathsf{P}_Y)$. As stated in the previous section, the product space is the set $X \times Y$ (Cartesian product), endowed with the product σ-algebra $\mathcal{X} \otimes \mathcal{Y}$, and the product measure $\mathsf{P}_{X \times Y} = \mathsf{P}_X \times \mathsf{P}_Y$. In this section we consider integration over product spaces and relate it to integrals over the component spaces. Although the section emphasizes products of two spaces, the results generalize to products of three or more spaces.

For notational convenience, we denote integration with respect to the measure P_X on X as $\int f\, P_X(dx)$ and integration with respect to the measure P_Y on Y as $\int f\, P_Y(dy)$.

Proposition F.5.1 (Fubini's Theorem). *Let $(X, \mathcal{X}, P_X)$ and $(Y, \mathcal{Y}, P_Y)$ be two probability measure spaces. Then for all integrable functions $f : X \times Y \to \mathbb{R}$*

$$\int_Y \left(\int_X f\, dP_X \right) dP_Y = \int_{X \times Y} f\, d(P_X \times P_Y) = \int_X \left(\int_Y f\, dP_Y \right) dP_X. \quad \text{(F.21)}$$

We make the following comments.

1. In the expression

$$\int_X \left(\int_Y f\, dP_Y \right) dP_X,$$

   the inner integral is with respect to P_Y over y, keeping x fixed. It is thus a function of x, which is then integrated in the outer integral with respect to P_X. The expression

$$\int_Y \left(\int_X f\, dP_X \right) dP_Y$$

   has a similar interpretation.

2. Fubini's theorem may be extended for measures that are not probability functions.

3. In the proof below, we omit a verification that the iterated integrals are finite. A more careful proof is available for example in [5].

Proof. We prove the first equality. The proof of the second equality is similar. By Proposition F.4.4, the function $f(\cdot, x_0)$ is measurable, and therefore the integral $\int_X f\, dP_X$ is a well defined function of y.

We assume first that f is a non-negative function. If $f = I_E$, then (F.22) follows from Proposition F.4.5 (in this case the inner integral $\int_X f\, dP_X$ is $P_X(A)$, which is constant in y, implying that the outer integral is $P_X(A)\, P_Y(B)$. If f is a simple function, $\int_X f\, dP_X$ is a linear function of measurable and integrable functions, which is a measurable and integrable function. Also, due to the linearity of the Lebesgue integral, the fact that (F.22) holds for each of the indicator functions in the simple function linear combination implies that (F.22) holds for the simple function f itself. For a general non-negative function f, the integrals in (F.22) are monotone limits of integrals of simple functions. Applying the monotone convergence theorem to the left hand side (twice) and to the central term (once) establishes that (F.22) holds for a non-negative integrable f.

$$\int_Y \int_X f \, d\mathsf{P}_X \, d\mathsf{P}_Y = \int_Y \sup_{s:s \le f} \int_X s \, d\mathsf{P}_X \, d\mathsf{P}_Y$$

$$= \sup_{s:s \le f} \int_Y \int_X s \, d\mathsf{P}_X \, d\mathsf{P}_Y$$

$$= \sup_{s:s \le f} \int_{X \times Y} s \, d(\mathsf{P}_X \times \mathsf{P}_Y)$$

$$= \int_{X \times Y} f \, d(\mathsf{P}_X \times \mathsf{P}_Y).$$

To prove (F.22) for a general function f, we decompose f into its positive and negative components $f = f^+ - f^-$. The result follows from the case of non-negative f and due to the linearity of the Lebesgue integral. ∎

Fubini's theorem is particularly useful for decomposing product integrals of decomposable functions $f(x, y) = g(x)h(y)$ into a product of an integral of $\int_X g \, d\mathsf{P}_X$ and integral of $\int_Y h \, d\mathsf{P}_Y$.

Corollary F.5.1. *Let $(X, \mathcal{X}, \mathsf{P}_X)$ and $(Y, \mathcal{Y}, \mathsf{P}_Y)$ be two probability measure spaces. Then for all integrable functions $f : X \to \mathbb{R}$ and $g : Y \to \mathbb{R}$*

$$\int_{X \times Y} (fg) \, d(\mathsf{P}_X \times \mathsf{P}_Y) = \int_X f \, d\mathsf{P}_X \cdot \int_Y g \, d\mathsf{P}_Y. \qquad (F.22)$$

F.5.1   The Lebesgue Measure over $\mathbb{R}^d$*

Whenever we encounter in this book integrals over $\mathbb{R}^d$, we assume the integration is with respect to the product Lebesgue measure on $\mathbb{R}^d$. This measure is simply the product measure (see previous two sections) of d copies of the Lebesgue measure spaces $(\mathbb{R}, \mathcal{B}, \mu)$. Fubini's theorem implies that such integrals may be expressed using a sequence of iterated one dimensional integrals with respect to the Lebesgue measure. For example, in two dimensions we have

$$\int_{[a,b] \times [c,d]} f(x_1, x_2) \, d\boldsymbol{x} = \int_a^b \left(\int_c^d f(x_1, x_2) \, dx_2 \right) dx_1.$$

Fubini's theorem indicates that the product Lebesgue measure assigns to rectangles the integral value of their area (in two dimensions) or their volume (in three or higher dimensions). The product Lebesgue measure corresponds to the multivariate Riemann integral.

We denote integrals with respect to the product Lebesgue measure as $\int f(\boldsymbol{x}) \, d\boldsymbol{x}$ or as $\int \cdots \int f(\boldsymbol{x}) \, d\boldsymbol{x}$, where the bold face emphasizes the vector nature of $\boldsymbol{x}$.

F.6 Multivariate Differentiation and Integration

Many of the definitions above generalize to functions $f : \mathbb{R}^n \to \mathbb{R}$ and even functions $\boldsymbol{f} : \mathbb{R}^n \to \mathbb{R}^m$. We denote the latter in bold-face to emphasize that $\boldsymbol{f}(\boldsymbol{x})$ is a vector. Its components are denoted by subscript functions: $\boldsymbol{f}(\boldsymbol{x}) = (f_1(\boldsymbol{x}), \ldots, f_m(\boldsymbol{x}))$, where $f_i : \mathbb{R}^n \to \mathbb{R}$, $i = 1, \ldots, m$.

Definition F.6.1. The partial derivative $\partial f(\boldsymbol{x})/\partial x_j$ of a function $f : \mathbb{R}^n \to \mathbb{R}$ is

$$\frac{\partial f(\boldsymbol{x})}{\partial x_j} = \lim_{t \to x_j} \frac{f(x_1, \ldots, x_{j-1}, t, x_{j+1}, \ldots, x_n) - f(\boldsymbol{x})}{t - x}.$$

In other words, the partial derivative $\partial f(\boldsymbol{x})/\partial x_j$ is the regular derivative $dg(x)/dx$ of the function $g(x)$ obtained by setting $x = x_j$ and fixing the remaining components of $\boldsymbol{x}$ as constants.

Definition F.6.2. We define the following generalization of first and second order derivatives.

1. For a function $\boldsymbol{f} : \mathbb{R}^n \to \mathbb{R}^m$, we define its derivative matrix $\nabla \boldsymbol{f} \in \mathbb{R}^{m \times n}$ given by $[\nabla \boldsymbol{f}]_{ij} = \partial f_i(\boldsymbol{x})/\partial x_j$. This matrix is sometimes called the Jacobian matrix and is often denoted by the letter J. If $m = 1$, we call the resulting $1 \times d$ matrix the gradient vector.

2. For a function $f : \mathbb{R}^n \to \mathbb{R}$, we define its directional derivative along a vector $\boldsymbol{v} \in \mathbb{R}^n$ as $D_{\boldsymbol{v}}(\boldsymbol{x}) = \boldsymbol{v}^\top \nabla f(\boldsymbol{x})$.

3. For a function $f : \mathbb{R}^n \to \mathbb{R}$, we define the second derivative or Hessian matrix $\nabla^2 f \in \mathbb{R}^{n \times n}$ as the matrix of second order partial derivatives

$$[\nabla^2 f]_{ij} = \frac{\partial^2 f(\boldsymbol{x})}{\partial x_i \partial x_j} \stackrel{\text{def}}{=} \frac{\partial}{\partial x_j} \frac{\partial f(\boldsymbol{x})}{\partial x_i}.$$

Proposition F.6.1. *For $\boldsymbol{f} : \mathbb{R}^n \to \mathbb{R}^m$, $\boldsymbol{g} : \mathbb{R}^m \to \mathbb{R}^s$, we have*

1. $\nabla(\boldsymbol{g}(\boldsymbol{f}(\boldsymbol{x}))) = (\nabla \boldsymbol{g})(\boldsymbol{f}(\boldsymbol{x}))\nabla \boldsymbol{f}(\boldsymbol{x})$,

2. $\nabla(\boldsymbol{f}(\boldsymbol{x})^\top \boldsymbol{g}(\boldsymbol{x})) = \boldsymbol{g}(\boldsymbol{x})^\top \nabla \boldsymbol{f}(\boldsymbol{x}) + \boldsymbol{f}(\boldsymbol{x})^\top \nabla \boldsymbol{g}(\boldsymbol{x})$, *and*

3. *if $\nabla \boldsymbol{f}(\boldsymbol{x})$ is continuous over $\boldsymbol{x} \in B_r(\boldsymbol{y})$, then for some $t \in B_r(\boldsymbol{0})$,*

$$\boldsymbol{f}(\boldsymbol{y} + \boldsymbol{t}) = \boldsymbol{f}(\boldsymbol{y}) + \int_0^1 \nabla \boldsymbol{f}(\boldsymbol{y} + z\boldsymbol{t}) \, dz \, \boldsymbol{t}.$$

The third part of the proposition above is called the multivariate mean value theorem. The integral in it represents a matrix whose entries are the integrals of the corresponding argument, and thus the second term on the right hand side of the third statement is a product of a matrix and a (column) vector. The entire equation is therefore a vector equation with vectors on both sides of the equality symbol.

Proof. Statements 1 and 2 are restatements of the chain rule and and the product rule for partial derivative in matrix form. To prove statement 3, consider the following vector equation (integrals over vectors are interpreted as vectors containing the corresponding component-wise integrals)

$$\int_0^1 (\nabla f(y + zt))t\, dz = \int_0^1 \frac{df(y + zt)}{dz}\, dz = f(y + t) - f(y),$$

where the first equality follows from the chain rule (statement 1 above) and the second equality follows from the vector form of the fundamental theorem of calculus (Proposition F.2.2). $\blacksquare$

The multivariate mean value theorem (part 3 of the proposition above) forms the most basic multivariate Taylor series approximation. Specifically, it implies that if the derivatives are bounded, we have the following approximation of f using a constant function.

$$f(y + t) = f(y) + o(1), \qquad \text{as} \qquad t \to 0$$

Applying the multivariate mean value theorem twice, we get the proposition below, which implies the following linear approximation of f

$$f(y + t) = f(y) + \nabla f(y)t + o(\|t\|^{-1}) \approx f(y) + \nabla f(y)t, \qquad \text{as} \qquad t \to 0.$$

Proposition F.6.2 (Multivariate Taylor Series Theorem (second order))**.** *Let $f : \mathbb{R}^d \to \mathbb{R}$ be a function whose higher order derivatives exist in $B_r(y)$. Then for all $t \in B_r(0)$,*

$$f(y + t) = f(y) + \nabla f(y)t + t^\top \int_0^1 \int_0^1 z\nabla^2 f(y + zwt)\, dzdw\, t. \qquad \text{(F.23)}$$

Proof. Applying twice the multivariate mean value theorem, we have

$$\nabla f(y + zt) = \nabla f(y) + \int_0^1 \nabla^2 zf(y + zt)\, dz\, t$$

$$f(y + t) = f(y) + \int_0^1 \nabla f(y + zt)\, dz\, t$$

$$= f(y) + \int_0^1 \nabla f(y)\, dz\, t + t^\top \int_0^1 \int_0^1 \nabla^2 f(y + zwt)z\, dzdw\, t$$

$$= f(y) + \nabla f(y)t + t^\top \int_0^1 \int_0^1 \nabla^2 f(y + zwt)z\, dzdw\, t.$$

$\blacksquare$

The Taylor series above is given with an integral remainder. The remainder may also be expressed in a differential way:

$$f(x) = f(\alpha) + (\nabla f(\alpha))^\top (x - \alpha) + \frac{1}{2}(x - \alpha)^\top \nabla^2 f(\beta)(x - \alpha)$$

for some β on the line segment connecting x and α.

Proposition F.6.3 (Multivariate Change of Variables). *Let $\boldsymbol{f}$ be a differentiable function $\boldsymbol{f} : \mathbb{R}^k \to \mathbb{R}^k$ whose Jacobian matrix is non-singular in the domain of $\boldsymbol{f}$, and g be a continuous function. Then*

$$\int_{\boldsymbol{f}(A)} g(\boldsymbol{x}) \, d\boldsymbol{x} = \int_A g(\boldsymbol{f}(\boldsymbol{y})) \, |\det J(\boldsymbol{f}(\boldsymbol{y}))| \, d\boldsymbol{y}.$$

A proof of Proposition F.6.3 appears for example in [37, Chapter 10]. We describe below two examples that use the multivariate change of variables method to solve a multivariate integral. Both examples have application in probability theory and are used in the book elsewhere. A third example is available in the proof of Proposition 5.3.1.

Example F.6.1. *The transformation from the Cartesian coordinates (x, y) to polar coordinates (r, θ) is*

$$r = \sqrt{x^2 + y^2}$$
$$\theta = \tan^{-1}(y/x)$$

and the inverse transformation $(r, \theta) \mapsto (x, y)$ is given by

$$x = r \cos \theta$$
$$y = r \sin \theta.$$

The space $\mathbb{R}^2$ is realized by either $(x, y) \in \mathbb{R}^2$ or by $(r, \theta) \in [0, \infty) \times [0, 2\pi)$. The Jacobian of the mapping $(r, \theta) \mapsto (x, y)$ is

$$J = \begin{pmatrix} \cos \theta & -r \sin \theta \\ \sin \theta & r \cos \theta \end{pmatrix} \qquad \det J = r(\cos^2 \theta + \sin^2 \theta) = r \cdot 1 = r.$$

An important application of the polar coordinates change of variable is the following calculation of the Gaussian integral

$$\left(\int_{\mathbb{R}} e^{-x^2/2} dx \right)^2 = \left(\int_{\mathbb{R}} e^{-x^2/2} dx \right) \left(\int_{\mathbb{R}} e^{-y^2/2} dy \right) = \iint_{\mathbb{R}^2} e^{-(x^2+y^2)/2} dx dy$$

$$= \int_0^{2\pi} \int_0^\infty e^{-r^2/2} r \, dr d\theta = 2\pi \int_0^\infty e^{-r^2/2} r \, dr = 2\pi \cdot 1 \quad \text{(F.24)}$$

where the last equality follows from Example F.2.4. We therefore have

$$\int_{\mathbb{R}} e^{-x^2/2} dx = \sqrt{2\pi}, \tag{F.25}$$

showing that the Gaussian pdf from Section 3.9 integrates to one.

Example F.6.2. *The Beta function is defined as follows*

$$B(\alpha, \beta) \overset{\text{def}}{=} \int_0^1 x^{\alpha-1}(1-x)^{\beta-1} \, dx, \qquad \alpha, \beta > 0.$$

Recalling the definition of the Gamma function $\Gamma(x)$ in Definition 3.10.1, we have

$$\Gamma(\alpha)\Gamma(\beta) = \int_0^\infty \int_0^\infty u^{\alpha-1} v^{\beta-1} e^{-u-v} \, du dv$$

$$= \int_0^\infty \int_0^1 (zt)^{\alpha-1} (z(1-t))^{\beta-1} e^{-z} \, z dt dz$$

$$= \int_0^\infty e^{-z} (z)^{\alpha+\beta-1} \, dz \int_0^1 t^{\alpha-1} (1-t)^{\beta-1} \, dt$$

$$= \Gamma(\alpha+\beta) B(\alpha,\beta),$$

where we used the variable transformation $(u,v) \mapsto (zt)$ expressed by $u = zt$, $v = z(1-t)$ and $z = u+v$, $t = u/(u+v)$, whose jacobian is

$$\left| \det \begin{pmatrix} t & z \\ 1-t & -z \end{pmatrix} \right| = |-tz - z + tz| = z.$$

The derivation above implies the following relationship between the Beta and Gamma functions

$$B(\alpha,\beta) = \frac{\Gamma(\alpha)\Gamma(\beta)}{\Gamma(\alpha+\beta)}.$$

F.7   Notes

Our description of the Riemann integral follows [37] and [46] where more details may be found. Chapter 8 follows up with more advanced material that is directly relevant to probability theory.

Bibliography

[1] R. A. Ash and C. A. Doleans-Dade. *Probability and Measure Theory*. Academic Press, second edition, 1999.

[2] J. O. Berger. *Statistical Decision Theory and Bayesian Analysis*. Springer, 1985.

[3] K. N. Berk. A central limit theorem for m-dependent random variables with unbounded m. *The Annals of Probability*, 1(2):352–354, 1973.

[4] A. C. Berry. The accuracy of the gaussian approximation to the sum of independent variates. *Transactions of the American Mathematical Society*, 49(1):122–136, 1941.

[5] P. Billingsley. *Probability and Measure*. Wiley, third edition, 1995.

[6] Y. Bishop, S. Fienberg, and P. Holland. *Discrete multivariate analysis: theory and practice*. MIT press, 1975.

[7] D. Blackwell and J. MacQueen. Ferguson distributions via polya urn schemes. *The Annals of Statistics*, 1, 1973.

[8] D. Blei, A. Ng, and M. Jordan. Latent dirichlet allocation. *Journal of Machine Learning Research*, 3:993–1022, 2003.

[9] D. D. Boos. A converse to scheffé's theorem. *The Annals of Statistics*, 13(1):423–427, 1985.

[10] L. Breiman. *Probability*. SIAM, 1992.

[11] L. D. Brown. *Fundamentals of Statistical Exponential Families*, volume 9 of *Lecture Notes-Monograph Series*. IMS Press, 1986.

[12] T. M. Cover and J. A. Thomas. *Elements of Information Theory*. John Wiley & Sons, second edition, 2005.

[13] A. DasGupta. *Asymptotic Theory of Statistics and Probability*. Springer, 2008.

[14] A. DasGupta. *Fundamentals of Probability: A First Course*. Springer, 2010.

[15] J. Davidson. *Stochastic limit theory: An introduction for econometricians.* Oxford University Press, USA, 1994.

[16] W. Feller. *An Introduction to Probability Theory and its Application,* volume 1. John Wiley and Sons, third edition, 1968.

[17] W. Feller. *An Introduction to Probability Theory and its Application,* volume 2. John Wiley and Sons, second edition, 1971.

[18] T. S. Ferguson. A bayesian analysis of some nonparametric problems. *The Annals of Statistics,* 1(4):209–230, 1973.

[19] T. S. Ferguson. *A Course in Large Sample Theory.* Chapman & Hall, 1996.

[20] P. Halmos. *Naive Set Theory.* Springer, 1998.

[21] C. Forbes M. Evans N. Hastings and B. Peacock. *Statistical Distributions.* Wiley, fourth edition, 2010.

[22] W. Hoeffding and H. Robbins. The central limit theorem for dependent random variables. *Duke Mathematical Journal,* 15:773–780, 1948.

[23] R. Horn and C. R. Johnson. *Matrix Analysis.* Cambridge University Press, 1990.

[24] R. Horn and C. R. Johnson. *Topics in Matrix Analysis.* Cambridge University Press, 1994.

[25] O. Kallenberg. *Foundations of Modern Probability.* Springer, second edition, 2002.

[26] S. Karlin and H. M. Taylor. *A First Course in Stochastic Processes.* Academic Press, second edition, 1975.

[27] A. N. Kolmogorov and S. V. Fomin. *Elements of the Theory of Functions and Functional Analysis.* Dover, 1999.

[28] G. J. McLachlan and D. Peel. *Finite mixture models.* Wiley-Interscience, 2000.

[29] D. McLeish. Dependent central limit theorems and invariance principles. *The Annals of Probability,* 2:620–628, 1974.

[30] J. Munkres. *General Topology.* Prentice Hall, second edition, 2000.

[31] A. V. Oppenheim, A. Willsky, and S. Hamid. *Signals and Systems.* Prentice Hall, second edition, 1996.

[32] C. E. Rasmussen and C. K. I. Williams. *Gaussian Processes for Machine Learning.* MIT Press, 2005.

[33] S. I. Resnick. *A Probability Path.* Birkhauser, 1999.

[34] J. P. Romano and M. Wolf. A more general central limit theorem for m-dependent random variables with unbounded m. *Statistics and Probability Letters*, 47:115–124, 2000.

[35] K. Rosen. *Discrete Mathematics and Its Applications*. McGraw Hill, sixth edition, 2006.

[36] Sheldon M. Ross. *Introduction to Probability Models*. Academic Press, tenth edition, 2009.

[37] W. Rudin. *Principles of Mathematical Analysis*. McGraw-Hill, third edition, 1976.

[38] G. A. Seber and A. J. Lee. *Linear Regression Analysis*. Wiley Interscience, 2003.

[39] G. A. F. Seber. *A Matrix Handbook for Statisticians*. Wiley, 2007.

[40] R. J. Serfling. *Approximation Theorems of Mathematical Statistics*. John Wiley, 1980.

[41] C. E. Shannon. A mathematical theory of communication. *Bell System Technical Journal*, 27, 1948.

[42] R. P. Stanley. *Enumerative Combinatorics*, volume 1. Cambridge University Press, 2000.

[43] G. Strang. *Introduction to Linear Algebra*. Wellesley Cambridge Press, fourth edition, 2009.

[44] T. J. Sweeting. On a converse to scheffé's theorem. *The Annals of Statistics*, 14(3):1252–1256, 1986.

[45] G. Thomas, M. D. Weir, and J. Hass. *Thomas' Calculus*. Addison Wesley, twelfth edition, 2009.

[46] W. F. Trench. *Introduction to Real Analysis*. Pearson, 2003.

[47] A. W. van der Vaart. *Asymptotic Statistics*. Cambridge University Press, 1998.

[48] N. A. Weiss. *A Course in Probability*. Addison Wesley, 2005.

[49] H. Wickham. *ggplot2: Elegant Graphics for Data Analysis*. Springer, 2009.

Index

Made in the USA
San Bernardino, CA
07 May 2016